Sound and Recording:
An Introduction

Further Focal Press titles by Francis Rumsey

Stereo Sound for Television (1989)
Tapeless Sound Recording (1990)
Digital Audio Operations (1991)
MIDI Systems and Control, 2nd Ed. (1994)
The Digital Interface Handbook, 2nd Ed. (1995)
The Audio Workstation Handbook (1996)

Other titles in the Focal Press Music Technology Series, edited by Francis Rumsey

Acoustics and Psychoacoustics (1996), David Howard and
    James Angus

Sound Synthesis and Sampling (1996), Martin Russ

Reviews of the previous editions of
Sound & Recording: An Introduction

'... *surely one of the best written books on the subject
        available*'   Studio Sound

'... *I can recommend that students should stay up past
        bedtime reading it*'   Ken Pohlmann, AES Journal

Sound and Recording:
An Introduction

Third edition

Francis Rumsey BMus (Tonmeister), PhD
Tim McCormick BA, LTCL

)41
A division of Reed Educational and Professional Publishing Ltd

ℛ A member of the Reed Elsevier plc group

OXFORD   BOSTON   JOHANNESBURG
MELBOURNE   NEW DELHI   SINGAPORE

First published 1992
Reprinted 1994
Second edition 1994
Reprinted 1995, 1996
Third edition 1997
Reprinted 1998

British Library Cataloguing in Publication Data
A catalogue record for this book is available from the British Library

Library of Congress Cataloguing in Publication Data
A catalogue record for this book is available from the Library of Congress

ISBN 0 240 51487 4

Printed and bound in Great Britain by
Redwood Books, Trowbridge, Wiltshire

Contents

Fact File Directory

Preface to the Second Edition

One of the greatest dangers in writing a book at an introductory level is to sacrifice technical accuracy for the sake of simplicity. In writing *Sound and Recording: An Introduction* we have gone to great lengths not to fall into this trap, and have produced a comprehensive introduction to the field of audio, intended principally for the newcomer to the subject, which is both easy to understand and technically precise. We have written the book that we would have valued when we first entered the industry, and as such it represents a readable reference, packed with information. Many books stop after a vague overview, just when the reader wants some clear facts about a subject, or perhaps assume too much knowledge on the reader's behalf. Books by contributed authors often suffer from a lack of consistency in style, coverage and technical level. Furthermore, there is a tendency for books on audio to be either too technical for the beginner or, alternatively, subjectively biased towards specific products or operations. There are also quite a number of American books on sound recording which, although good, tend to ignore European trends and practices. We hope that we have steered a balanced course between these extremes, and have deliberately avoided any attempt to dictate operational practice.

Sound and Recording: An Introduction is definitely biased towards an understanding of 'how it works', as opposed to 'how to work it', although technology is never discussed in an abstract manner but related to operational reality. Although we have included a basic introduction to acoustics and the nature of sound perception, this is not a book on acoustics or musical acoustics (there are plenty of those around). It is concerned with the principles of audio recording and reproduction, and has a distinct bias towards the professional rather than the consumer end of the market. The coverage of subject matter is broad, including chapters on digital audio, timecode synchronisation and MIDI, amongst other more conventional subjects, and there is comprehensive coverage of commonly misunderstood subjects such as the decibel, balanced lines, reference levels and metering systems.

This second edition of the book has been published only two years after the first, and the subject matter has not changed significantly enough in the interim to warrant major modifications to the existing chapters. The key difference between the second and first editions is the addition of a long chapter on stereo recording and reproduction. This important topic is covered in considerable detail, including historical developments, principles of stereo reproduction, surround sound and stereo microphone techniques. Virtually every recording or broadcast happening today is made in stereo, and although surround sound has had a number of notable 'flops' in the past it is likely to become considerably more important in the next ten years. Stereo and surround sound are used extensively in film, video and television production, and any new audio engineer should be familiar with the principles.

Since this is an introductory book, it will be of greatest value to the student of sound recording or music technology, and to the person starting out on a career in sound engineering or broadcasting. The technical level has deliberately been kept reasonably low for this reason, and those who find this frustrating probably do not need the book! Nonetheless, it is often valuable for the seasoned audio engineer to go back to basics. Further reading suggestions have been made in order that the reader may go on to a more in-depth coverage of the fields introduced here, and some of the references are considerably more technical than this book. Students will find these suggestions valuable when planning a course of study.

Francis Rumsey
Tim McCormick

July 1994

Preface to the Third Edition

Since the first edition of *Sound and Recording* some of the topics have advanced quite considerably, particularly the areas dependent on digital and computer technology. Consequently I have rewritten the chapters on digital recording and MIDI (Chapters 10 and 15), and have added a larger section on mixer automation in Chapter 7. Whereas the first edition of the book was quite 'analogue', I think that there is now a more appropriate balance between analogue and digital topics. Although analogue audio is by no means dead (sound will remain analogue for ever!), most technological developments are now digital.

I make no apologies for leaving in the chapter on record players, although some readers have commented that they think it is a waste of space. People still use record players, and there is a vast store of valuable material on LP record. I see no problem with keeping a bit of history in the book – you never know, it might come in useful one day when everyone has forgotten (and some may never have known) what to do with vinyl disks. It might even appease the faction of our industry that continues to insist that vinyl records are the highest fidelity storage medium ever invented.

Francis Rumsey
Guildford

Chapter 1

What is sound?

1.1 A vibrating source

Sound is produced when an object (the source) vibrates and causes the air around
it to move. Consider the sphere shown in Figure 1.1. It is a pulsating sphere which
could be imagined as something like a squash ball, and it is pulsating regularly so
that its size oscillates between being slightly larger than normal and then slightly
smaller than normal. As it pulsates it will alternately compress and then rarefy the

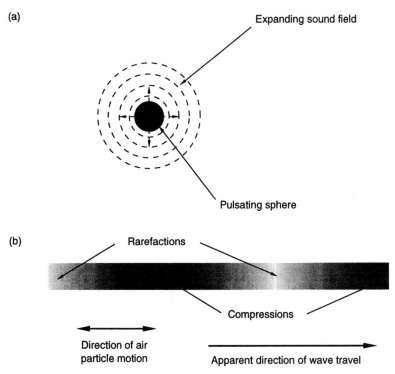

Figure 1.1 (a) A simple sound source can be imagined as like a pulsating sphere radiating spherical
waves. (b) The longitudinal wave thus created is a succession of compressions and rarefactions of
the air

1

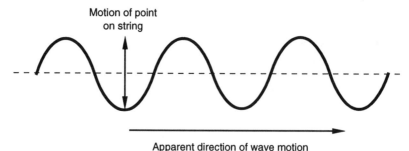

Figure 1.2 In a transverse wave the motion of any point on the wave is at right angles to the apparent direction of motion of the wave

surrounding air, resulting in a series of compressions and rarefactions travelling away from the sphere, rather like a three-dimensional version of the ripples which travel away from a stone dropped into a pond. These are known as *longitudinal waves* since the air particles move in the same dimension as the direction of wave travel. The alternative to longitudinal wave motion is *transverse* wave motion (see Figure 1.2), such as is found in vibrating strings, where the motion of the string is at right angles to the direction of apparent wave travel.

1.2 Characteristics of a sound wave

The rate at which the source oscillates is the *frequency* of the sound wave it produces, and is quoted in *hertz* (Hz) or cycles per second (cps). 1000 hertz is termed 1 *kilohertz* (1 kHz). The amount of compression and rarefaction of the air which results from the sphere's motion is the *amplitude* of the sound wave, and is related to the loudness of the sound when it is finally perceived by the ear (see Chapter 2). The distance between two adjacent peaks of compression or rarefaction as the wave travels through the air is the *wavelength* of the sound wave, and is often represented by the Greek letter *lambda* (λ). The wavelength depends on how fast the sound wave travels, since a fast-travelling wave would result in a greater distance between peaks than a slow-travelling wave, given a fixed time between compression peaks (i.e.: a fixed frequency of oscillation of the source).

As shown in Figure 1.3, the sound wave's characteristics can be represented on a graph, with amplitude plotted on the vertical axis and time plotted on the horizontal axis. It will be seen that both positive and negative ranges are shown on the vertical axis: these represent compressions (+) and rarefactions (−) of the air. This graph represents the *waveform* of the sound. For a moment, a source vibrating in a very simple and regular manner is assumed, in so-called *simple harmonic motion*, the result of which is a simple sound wave known as a *sine wave*. The most simple vibrating systems oscillate in this way, such as a mass suspended from a spring, or a pendulum swinging on a fulcrum (see also section 1.7 on phase). It will be seen that the frequency (f) is the inverse of the time between peaks or troughs of the wave ($f = 1/t$). So the shorter the time between oscillations of the source, the higher the frequency. The human ear is capable of perceiving sounds with frequencies between approximately 20 Hz and 20 kHz (see section 2.2); this is known as the *audio frequency range* or *audio spectrum*.

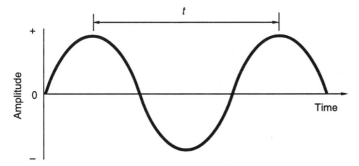

Figure 1.3 A graphical representation of a sinusoidal sound waveform. The period of the wave is represented by t, and its frequency by $1/t$

1.3 How sound travels in air

Air is made up of gas molecules and has an elastic property (imagine putting a thumb over the end of a bicycle pump and compressing the air inside – the air is springy). Longitudinal sound waves travel in air in somewhat the same fashion as a wave travels down a row of up-ended dominoes after the first one is pushed over. The half-cycle of compression created by the vibrating source causes successive air particles to be moved in a knock-on effect, and this is normally followed by a balancing rarefaction which causes a similar motion of particles in the opposite direction.

It may be appreciated that the net effect of this is that individual air particles do not actually travel – they oscillate about a fixed point – but the result is that a wave is formed which appears to move away from the source. The speed at which it moves away from the source depends on the density and elasticity of the substance through which it passes, and in air the speed is relatively slow compared with the speed at which sound travels through most solids. In air the speed of sound is approximately 340 metres per second (m s^{-1}), although this depends on the temperature of the air. At freezing point the speed is reduced to nearer 330 m s^{-1}. In steel, to give an example of a solid, the speed of sound is approximately 5100 m s^{-1}.

The frequency and wavelength of a sound wave are related very simply if the speed of the wave (usually denoted by the letter c) is known:

$$c = f\lambda \quad \text{or} \quad \lambda = c/f$$

To show some examples, the wavelength of sound in air at 20 Hz (the low-frequency or LF end of the audio spectrum), assuming normal room temperature, would be:

$$\lambda = 340/20 = 17 \text{ metres}$$

whereas the wavelength of 20 kHz (at the high-frequency or HF end of the audio spectrum) would be 1.7 cm. Thus it is apparent that the wavelength of sound ranges from being very long in relation to most natural objects at low frequencies, to quite short at high frequencies. This is important when considering how sound behaves when it encounters objects – whether the object acts as a barrier or whether the sound bends around it (see Fact File 1.5).

1.4 Simple and complex sounds

In the foregoing example, the sound had a simple waveform – it was a sine wave or *sinusoidal* waveform – the type which might result from a very simple vibrating system such as a weight suspended on a spring. Sine waves have a very pure sound because they consist of energy at only one frequency, and are often called *pure tones*. They are not heard very commonly in real life (although they can be generated electrically) since most sound sources do not vibrate in such a simple manner. A person whistling or a recorder (a simple wind instrument) produces a sound which approaches a sinusoidal waveform. Most real sounds are made up of a combination of vibration patterns which result in a more complex waveform. The more complex the waveform, the more like noise the sound becomes, and when the waveform has a highly random pattern the sound is said to be *noise* (see section 1.6).

The important characteristic of sounds which have a definite pitch is that they are

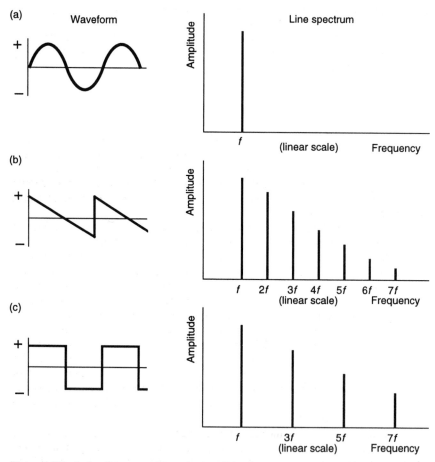

Figure 1.4 Equivalent line spectra for a selection of simple waveforms. (a) The sine wave consists of only one component at the fundamental frequency f. (b) The sawtooth wave consists of components at the fundamental and its integer multiples, with amplitudes steadily decreasing. (c) The square wave consists of components at odd multiples of the fundamental frequency

repetitive: that is, the waveform, no matter how complex, repeats its pattern in the same way at regular intervals. All such waveforms can be broken down into a series of components known as *harmonics*, using a mathematical process called Fourier analysis (after the mathematician Joseph Fourier). Some examples of equivalent line spectra for different waveforms are given in Figure 1.4. This figure shows another way of depicting the characteristics of the sound graphically – that is, by drawing a so-called *line spectrum* which shows frequency along the horizontal axis and amplitude up the vertical axis. The line spectrum shows the relative strengths of different frequency components which make up a sound. Where there is a line there is a frequency component. It will be noticed that the more complex the waveform the more complex the corresponding line spectrum.

For every waveform, such as that shown in Figure 1.3, there is a corresponding line spectrum: waveforms and line spectra are simply two different ways of showing the characteristics of the sound. Figure 1.3 is called a *time-domain* plot, whilst the line spectrum is called a *frequency-domain* plot. Unless otherwise stated, such frequency-domain graphs in this book will cover the audio-frequency range, from 20 Hz at the lower end to 20 kHz at the upper end.

In a reversal of the above breaking-down of waveforms into their component frequencies it is also possible to *construct* or *synthesise* waveforms by adding together the relevant components.

1.5 Frequency spectra of repetitive sounds

As will be seen in Figure 1.4, the simple sine wave has a line spectrum consisting of only one component at the frequency of the sine wave. This is known as the *fundamental* frequency of oscillation. The other repetitive waveforms, such as the square wave, have a fundamental frequency as well as a number of additional components above the fundamental. These are known as *harmonics*, but may also be referred to as *overtones* or *partials*.

Harmonics are frequency components of a sound which occur at integer multiples of the fundamental frequency, that is at twice, three times, four times and so on. Thus a sound with a fundamental of 100 Hz might also contain harmonics at 200 Hz, 400 Hz and 600 Hz. The reason for the existence of these harmonics is that most simple vibrating sources are capable of vibrating in a number of harmonic *modes* at the same time. Consider a stretched string, as shown in Figure 1.5. It may be made to vibrate in any of a number of modes, corresponding to integer multiples of the fundamental frequency of vibration of the string (the concept of a 'standing wave' is introduced in section 1.13, below). The fundamental corresponds to the mode in which the string moves up and down as a whole, whereas the harmonics correspond to modes in which the vibration pattern is divided into points of maximum and minimum motion along the string (these are called *antinodes* and *nodes*). It will be seen that the second mode involves two peaks of vibration, the third mode three peaks, and so on.

In accepted terminology, the fundamental is also the *first harmonic*, and thus the next component is the *second harmonic*, and so on. Confusingly, the second harmonic is also known as the *first overtone*. For the waveforms shown in Figure 1.4, the fundamental has the highest amplitude, and the amplitudes of the harmonics decrease with increasing frequency, but this will not always be the case with real sounds since many waveforms have line spectra which show the harmonics to be

(a)

(b)

(c)

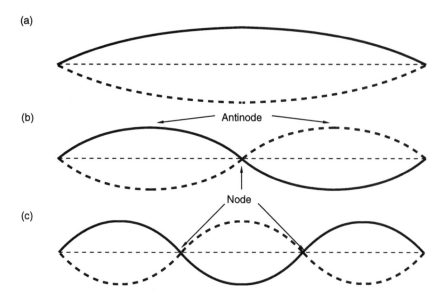

Figure 1.5 Modes of vibration of a stretched string. (a) Fundamental. (b) Second harmonic. (c) Third harmonic

higher in amplitude than the fundamental. It is also quite feasible for there to be harmonics missing in the line spectrum, and this depends entirely on the waveform in question.

It is also possible for there to be overtones in the frequency spectrum of a sound which are *not* related in a simple integer-multiple fashion to the fundamental. These cannot correctly be termed harmonics, and they are more correctly referred to as *overtones* or *inharmonic partials.* They tend to arise in vibrating sources which have a complicated shape, and which do not vibrate in simple harmonic motion but have a number of repetitive modes of vibration. Their patterns of oscillation are often unusual, such as might be observed in a bell or a percussion instrument. It is still possible for such sounds to have a recognisable pitch, but this depends on the strength of the fundamental. In bells and other such sources, one often hears the presence of several strong inharmonic overtones.

1.6 Frequency spectra of non-repetitive sounds

Non-repetitive waveforms do not have a recognisable pitch and sound noise-like. Their frequency spectra are likely to consist of a collection of components at unrelated frequencies, although some frequencies may be more dominant than others. The analysis of such waves to show their frequency spectra is more complicated than with repetitive waves, but is still possible using a mathematical technique called *Fourier transformation*, the result of which is a frequency-domain plot of a time-domain waveform.

Single, short pulses can be shown to have continuous frequency spectra which extend over quite a wide frequency range, and the shorter the pulse the wider its frequency spectrum but usually the lower its total energy (see Figure 1.6). Random

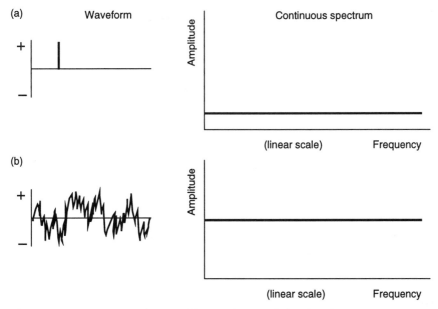

Figure 1.6 Frequency spectra of non-repetitive waveforms. (a) Pulse. (b) Noise

waveforms will tend to sound like hiss, and a completely random waveform in which the frequency, amplitude and phase of components are equally probable and constantly varying is called *white noise*. A white noise signal's spectrum is flat, when averaged over a period of time, right across the audio-frequency range (and theoretically above it). White noise has equal energy for a given bandwidth, whereas another type of noise, known as *pink noise*, has equal energy per *octave*. For this reason white noise sounds subjectively to have more high-frequency energy than pink noise.

1.7 Phase

Two waves of the same frequency are said to be 'in phase' when their compression (positive) and rarefaction (negative) half-cycles coincide exactly in time and space (see Figure 1.7). If two in-phase signals of equal amplitude are added together, or superimposed, they will sum to produce another signal of the same frequency but twice the amplitude. Signals are said to be out of phase when the positive half-cycle of one coincides with the negative half-cycle of the other. If these two signals are added together they will cancel each other out, and the result will be no signal.

Clearly these are two extreme cases, and it is entirely possible to superimpose two sounds of the same frequency which are only partially in phase with each other. The resultant wave in this case will be a partial addition or partial cancellation, and the phase of the resulting wave will lie somewhere between that of the two components (see Figure 1.7(c)).

Phase differences between signals can be the result of time delays between them. If two signals start out at sources equidistant from a listener at the same time as each other then they will be in phase by the time they arrive at the listener (although see

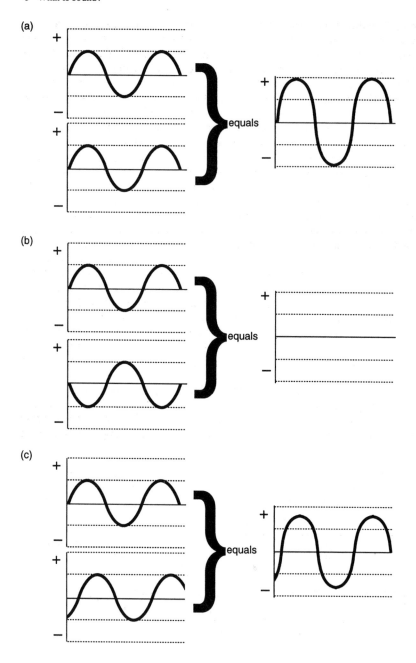

Figure 1.7 (a) When two identical in-phase waves are added together, the result is a wave of the same frequency and phase but twice the amplitude. (b) Two identical out-of-phase waves add to give nothing. (c) Two identical waves partially out of phase add to give a resultant wave with a phase and amplitude which is the point-by-point sum of the two

section 2.5 on directional perception). If one source is more distant than the other then it will be delayed, and the phase relationship between the two will depend upon the amount of delay (see Figure 1.8). A useful rule-of-thumb is that sound travels about 30 cm (1 foot) per millisecond, so if the second source in the above example were 1 metre (just over 3 ft) more distant than the first it would be delayed by just over 3 ms. The resulting phase relationship between the two signals, it may be appreciated, would depend on the frequency of the sound, since at a frequency of around 330 Hz the 3 ms delay would correspond to one wavelength and thus the delayed signal would be *in* phase with the undelayed signal. If the delay had been half this (1.5 ms) then the two signals would have been out of phase at 330 Hz.

Phase is often quoted as a number of degrees relative to some reference, and this must be related back to the nature of a sine wave. A diagram is the best way to illustrate this point, and looking at Figure 1.9 it will be seen that a sine wave may be considered as a graph of the vertical position of a rotating spot on the outer rim of a disk (the amplitude of the wave), plotted against time. The height of the spot rises and falls regularly as the circle rotates at a constant speed. The sine wave is so called because the spot's height is directly proportional to the mathemetical *sine* of the angle of rotation of the disk, with zero degrees occurring at the origin of the graph and at the point shown on the disk's rotation in the diagram. The vertical amplitude scale on the graph goes from minus one (maximum negative amplitude) to plus one (maximum positive amplitude), passing through zero at the half-way point. At 90° of rotation the amplitude of the sine wave is maximum positive (the sine of 90° is +1), and at 180° it is zero (sin 180° = 0). At 270° it is maximum negative (sin 270° = −1), and at 360° it is zero again. Thus in one cycle of the sine wave the circle has passed through 360° of rotation.

It is now possible to go back to the phase relationship between two waves of the same frequency. If each cycle is considered as corresponding to 360°, then one can say just how many degrees one wave is ahead of or behind another by comparing the 0° point on one wave with the 0° point on the other (see Figure 1.10). In the

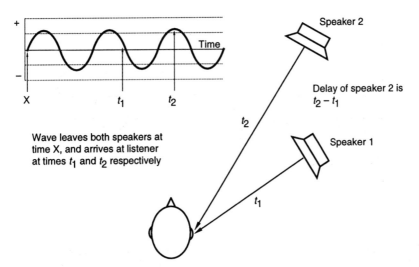

Figure 1.8 If the two loudspeakers in the drawing emit the same wave at the same time, the phase difference between the waves at the listener's ear will be directly related to the delay $t_2 - t_1$

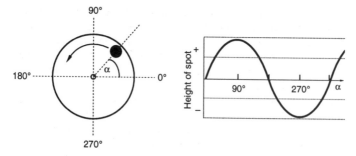

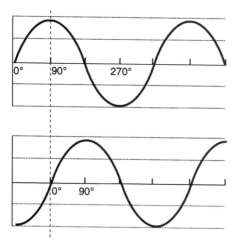

Figure 1.9 The height of the spot varies sinusoidally with the angle of rotation of the wheel. The phase angle of a sine wave can be understood in terms of the number of degrees of rotation of the wheel

Figure 1.10 The lower wave is 90° out of phase with the upper wave

example wave 1 is 90° out of phase with wave 2. It is important to realise that phase is only a relevant concept in the case of continuous repetitive waveforms, and has little meaning in the case of impulsive or transient sounds where time difference is the more relevant quantity. It can be deduced from the foregoing discussion that (a) the higher the frequency, the greater the phase difference which would result from a given time delay between two signals, and (b) it is possible for there to be more than 360° of phase difference between two signals if the delay is great enough to delay the second signal by more than one cycle. In the latter case it becomes difficult to tell how many cycles of delay have elapsed unless a discontinuity arises in the signal, since a phase difference of 360° is indistinguishable from a phase difference of 0°.

1.8 Sound in electrical form

Although the sound that one hears is due to compression and rarefaction of the air, it is often necessary to convert sound into an electrical form in order to perform operations on it such as amplification, recording and mixing. As detailed in Fact File 3.1 and Chapter 4, it is the job of the microphone to convert sound from an *acoustical* form into an *electrical* form. The process of conversion will not be described here, but the result is important because if it can be assumed for a moment that the microphone is perfect then the resulting electrical waveform will be exactly the same shape as the acoustical waveform which caused it.

The equivalent of the amplitude of the acoustical signal in electrical terms is the *voltage* of the electrical signal. If the voltage at the output of a microphone were to be measured whilst the microphone was picking up an acoustical sine wave, one would measure a voltage which changed sinusoidally as well. Figure 1.11 shows this situation, and it may be seen that an acoustical compression of the air corresponds to a positive-going voltage, whilst an acoustical rarefaction of the air corresponds to a negative-going voltage. (This is the norm, although some sound reproduction systems introduce an *absolute phase reversal* in the relationship between acoustical phase and electrical phase, such that an acoustical compression becomes equivalent to a negative voltage. Some people claim to be able to hear the difference.)

The other important quantity in electrical terms is the *current* flowing down the wire from the microphone. Current is the electrical equivalent of the air particle motion discussed in section 1.3. Just as the acoustical sound wave was carried in the motion of the air particles, so the electrical sound wave is carried in the motion of tiny charge carriers which reside in the metal of a wire (these are called *electrons*). When the voltage is positive the current moves in one direction, and when it is negative the current moves in the other direction. Since the voltage generated by a microphone is repeatedly alternating between positive and negative, in sympathy with the sound wave's compression and rarefaction cycles, the current similarly changes direction each half cycle. Just as the air particles in section 1.3 did not actually go anywhere in the long term, so the electrons carrying the current do not go anywhere either – they simply oscillate about a fixed point. This is known as *alternating current* or AC.

A useful analogy to the above (both electrical and acoustical) exists in plumbing. If one considers water in a pipe fed from a header tank, as shown in Figure 1.12, the voltage is equivalent to the pressure of water which results from the header tank, and the current is equivalent to the rate of flow of water through the pipe. The only

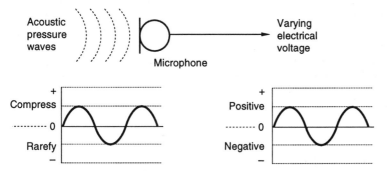

Figure 1.11 A microphone converts variations in acoustical sound pressure into variations in electrical voltage. Normally a compression of the air results in a positive voltage and a rarefaction results in a negative voltage

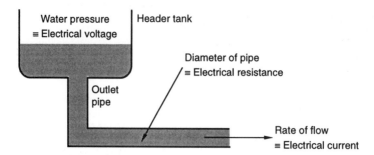

Figure 1.12 There are parallels between the flow of water in a pipe and the flow of electricity in a wire, as shown in this drawing

difference is that the diagram is concerned with a *direct current* situation in which the direction of flow is not repeatedly changing. The quantity of *resistance* should be introduced here, and is analogous to the diameter of the pipe. Resistance impedes the flow of water through the pipe, as it does the flow of electrons through a wire and the flow of acoustical sound energy through a substance. For a fixed voltage (or water pressure in this analogy), a high resistance (narrow pipe) will result in a small current (a trickle of water), whilst a low resistance (wide pipe) will result in a large current. The relationship between voltage, current and resistance was established by Ohm, in the form of Ohm's law, as described in Fact File 1.1. There is also a relationship between power and voltage, current and resistance.

In AC systems, resistance is replaced by *impedance*, a complex term which contains both resistance and *reactance* components. The reactance part varies with the frequency of the signal; thus the impedance of an electrical device also varies with the frequency of a signal. Capacitors (basically two conductive plates separated by an insulator) are electrical devices which present a high impedance to low-frequency signals and a low impedance to high-frequency signals. They will not pass direct current. Inductors (basically coils of wire) are electrical devices which present a high impedance to high-frequency signals and a low impedance to low-frequency signals. Capacitance is measured in farads, inductance in henrys.

1.9 Displaying the characteristics of a sound wave

Two devices can be introduced at this point which illustrate graphically the various characteristics of sound signals so far described. It would be useful to (a) display the waveform of the sound, and (b) display the frequency spectrum of the sound. In other words (a) the time-domain signal and (b) the frequency-domain signal.

An *oscilloscope* is used for displaying the waveform of a sound, and a *spectrum analyser* is used for showing which frequencies are contained in the signal and their amplitudes. Examples of such devices are pictured in Figure 1.13. Both devices accept sound signals in electrical form and display their analyses of the sound on a screen. The oscilloscope displays a moving spot which scans horizontally at one of a number of fixed speeds from left to right and whose vertical deflection is controlled by the voltage of the sound signal (up for positive, down for negative). In this way it plots the waveform of the sound as it varies with time. Many oscilloscopes have two inputs and can plot two waveforms at the same time, and this

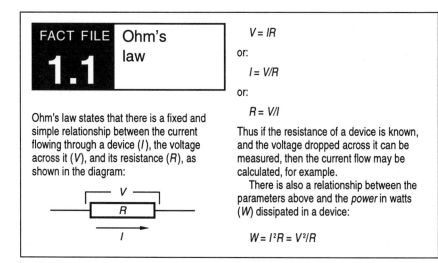

FACT FILE 1.1 Ohm's law

Ohm's law states that there is a fixed and simple relationship between the current flowing through a device (I), the voltage across it (V), and its resistance (R), as shown in the diagram:

$$V = IR$$

or:

$$I = V/R$$

or:

$$R = V/I$$

Thus if the resistance of a device is known, and the voltage dropped across it can be measured, then the current flow may be calculated, for example.

There is also a relationship between the parameters above and the *power* in watts (W) dissipated in a device:

$$W = I^2R = V^2/R$$

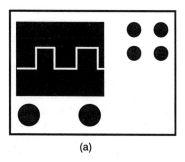

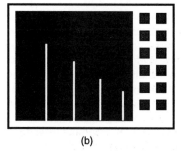

(a) (b)

Figure 1.13 (a) An oscilloscope displays the *waveform* of an electrical signal by means of a moving spot which is deflected up by a positive signal and down by a negative signal. (b) A spectrum analyser displays the *frequency spectrum* of an electrical waveform in the form of lines representing the amplitudes of different spectral components of the signal

can be useful for comparing the relative phases of two signals (see section 1.7).

The spectrum analyser works in different ways depending on the method of spectrum analysis. A real-time analyser displays a constantly updating line spectrum, similar to those depicted earlier in this chapter, and shows the frequency components of the input signal on the horizontal scale together with their amplitudes on the vertical scale.

1.10 The decibel

The unit of the decibel is used widely in sound engineering, often in preference to other units such as volts, watts, or other such absolute units, since it is a convenient way of representing the *ratio* of one signal's amplitude to another's. It also results in numbers of a convenient size which approximate more closely to one's subjective impression of changes in the amplitude of a signal, and it helps to compress the range of values between the maximum and minimum sound levels encountered in real signals. For example, the range of sound intensities (see section 1.11) which can be handled by the human ear covers about fourteen powers of ten, from 0.000 000 000 001 W m^{-2} to around 100 W m^{-2}, but the equivalent range in decibels is only from 0 to 140 dB.

Some examples of the use of the decibel are given in Fact File 1.2. The relationship between the decibel and human sound perception is discussed in more detail in Chapter 2. Operating levels in recording equipment are discussed further in sections 3.5, 7.5 and 8.5.

Decibels are not only used to describe the ratio between two signals, or the level of a signal above a reference, but they are also used to describe the *voltage gain* of a device. For example, a microphone amplifier may have a gain of 60 dB, which is the equivalent of multiplying the input voltage by a factor of 1000, as shown in the example below:

20 log 1000/1 = 60 dB

1.11 Sound power and sound pressure

A simple sound source, such as the pulsating sphere used at the start of this chapter, radiates sound power *omnidirectionally* – that is, equally in all directions, rather like a three-dimensional version of the ripples moving away from a stone dropped in a pond. The sound source generates a certain amount of power, measured in *watts*, which is gradually distributed over an increasingly large area as the wavefront travels further from the source; thus the amount of power per square metre passing through the surface of the imaginary sphere surrounding the source gets smaller with increasing distance (see Fact File 1.3). For practical purposes the intensity of the direct sound from a source drops by 6 dB for every doubling in distance from the source (see Figure 1.14).

The amount of acoustical power generated by real sound sources is surprisingly small, compared with the number of watts of electrical power involved in lighting a light bulb, for example. An acoustical source radiating 20 watts would produce a sound pressure level close to the threshold of pain if a listener was close to the source. Most everyday sources generate fractions of a watt of sound power, and this energy is eventually dissipated into heat by absorption (see below). The amount of heat produced by the dissipation of acoustic energy is relatively insignificant – the chances of increasing the temperature of a room by shouting are slight, at least in the physical sense.

Acoustical power is sometimes confused with the power output of an amplifier used to drive a loudspeaker, and audio engineers will be familiar with power outputs from amplifiers of many hundreds of watts. It is important to realise that loudspeakers are very inefficient devices – that is, they only convert a small proportion of their electrical input power into acoustical power. Thus, even if the input to a loudspeaker was to be, say, 100 watts electrically, the acoustical output power might only be perhaps 1 watt, suggesting a loudspeaker that is only 1% efficient. The remaining power would be dissipated as heat in the voice coil.

Sound *pressure* is the effect of sound power on its surroundings. To use a central heating analogy, sound power is analogous to the heat energy generated by a radiator into a room, whilst sound pressure is analogous to the temperature of the air in the room. The temperature is what a person entering the room would feel, but the heat-

FACT FILE

1.2 The decibel

Basic decibels

The decibel is based on the logarithm of the ratio between two numbers. It describes how much larger or smaller one value is than the other. It can also be used as an absolute unit of measurement if the reference value is fixed and known. Some standardised references have been established for decibel scales in different fields of sound engineering (see below).

The decibel is strictly ten times the logarithm to the base ten of the ratio between the *powers* of two signals:

$$dB = 10 \log_{10} (P_1/P_2)$$

For example, the difference in decibels between a signal with a power of 1 watt and one of 2 watts is 10 log (2/1) = 3 dB.

If the decibel is used to compare values other than signal powers, the relationship to signal power must be taken into account. Voltage has a square relationship to power (from Ohm's law: $W = V^2/R$); thus to compare two voltages:

$$dB = 10 \log (V_1^2/V_2^2), \text{ or } 10 \log (V_1/V_2)^2,$$
$$\text{or } 20 \log (V_1/V_2)$$

For example, the difference in decibels between a signal with a voltage of 1 volt and one of 2 volts is 20 log (2/1) = 6 dB. So a doubling in voltage gives rise to an increase of 6 dB, and a doubling in power gives rise to an increase of 3 dB. A similar relationship applies to acoustical sound pressure (analogous to electrical voltage) and sound power (analogous to electrical power).

Decibels with a reference

If a signal level is quoted in decibels, then a reference must normally be given, otherwise the figure means nothing; e.g.: 'Signal level = 47 dB' cannot have a meaning unless one knows that the signal is 47 dB above a known point. '+8 dB ref. 1 volt' has a meaning since one now knows that the level is 8 dB higher than 1 volt, and thus one could calculate the voltage of the signal.

There are exceptions in practice, since in some fields a reference level is accepted as implicit. Sound pressure levels (SPLs) are an example, since the reference level is defined worldwide as 2×10^{-5} N m^{-2} (20 µPa). Thus to state 'SPL = 77 dB' is probably acceptable, although confusion can still arise due to misunderstandings over such things as weighting curves (see Fact File 1.4). In sound recording, 0 dB or 'zero level' is a nominal reference level used for aligning equipment and setting recording levels, often corresponding to 0.775 volts (0 dBu) although this is subject to variations in studio centres in different locations. (Some studios use +4 dBu as their reference level, for example.) '0 dB' does not mean 'no signal', it means that the signal concerned is at the same level as the reference.

Often a letter is placed after 'dB' to denote the reference standard in use (e.g.: 'dBm'), and a number of standard abbreviations are in use, some examples of which are given below. Sometimes the suffix denotes a particular frequency weighting characteristic used in the measurement of noise (e.g.: 'dBA').

Abbrev.	Ref. level
dBV	1 volt
dBu	0.775 volt (Europe)
dBv	0.775 volt (USA)
dBm	1 milliwatt (see section 13.9)
dBA	dB SPL, A-weighted response

A full listing of suffixes is given in CCIR Recommendation 574-1, 1982.

Useful decibel ratios to remember (voltages or SPLs)

It is more common to deal in terms of voltage or SPL ratios than power ratios in audio systems. Here are some useful dB equivalents of different voltage or SPL relationships and multiplication factors:

dB	Multiplication factor
0	1
+3	$\sqrt{2}$
+6	2
+20	10
+60	1000

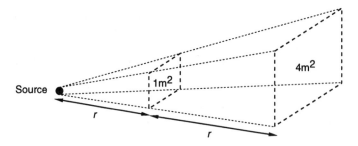

Figure 1.14 The sound power which had passed through 1m² of space at distance *r* from the source will pass through 4m² at distance 2*r*, and thus will have one quarter of the intensity

FACT FILE 1.3	The inverse-square law

The law of decreasing power per unit area (intensity) of a wavefront with increasing distance from the source is known as the *inverse square law*, because intensity drops in proportion to the inverse square of the distance from the source. Why is this? It is because the sound power from a point source is spread over the surface area of a sphere (*S*), which from elementary maths is given by:

$$S = 4\pi r^2$$

where *r* is the distance from the source or the radius of the sphere, as shown in the diagram.

If the original power of the source is *W* watts, then the intensity, or power per unit area (*I*) at distance *r* is:

$$I = W/(4\pi r^2)$$

For example, if the power of a source was 0.1 watt, the intensity at 4 m distance would be:

$$I = 0.1 \div (4 \times 3.14 \times 16) \quad 0.0005 \text{ W m}^{-2}$$

The sound intensity level (SIL) of this signal in decibels can be calculated by comparing it with the accepted reference level of 10^{-12} W m⁻²:

$$SIL\,(\text{dB}) = 10 \log ((5 \times 10^{-4}) \div (10^{-12}))$$
$$= 87 \text{ dB}$$

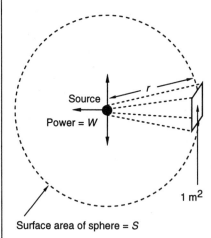

Source
Power = W

Surface area of sphere = *S*

1 m²

generating radiator is the source of power. Sound pressure level (SPL) is measured in *newtons per square metre* (N m⁻²). A convenient reference level is set for sound pressure and intensity measurements, this being referred to as 0 dB. This level of 0 dB is approximately equivalent to the threshold of hearing (the quietest sound perceivable by an average person) at a frequency of 1 kHz, and corresponds to an SPL of 2×10^{-5} N m^{-2}, which in turn is equivalent to an intensity of approximately 10^{-12} W m^{-2} in the free field (see below).

Sound pressure levels are often quoted in dB (e.g.: SPL = 63 dB means that the SPL is 63 dB above 2×10^{-5} N m^{-2}). The SPL in dB may not accurately represent the *loudness* of a sound, and thus a subjective unit of loudness has been derived from research data, called the *phon*. This is discussed further in Chapter 2. Some methods of measuring sound pressure levels are discussed in Fact File 1.4.

FACT FILE 1.4 Measuring SPLs

Typically a sound pressure level (SPL) meter is used to measure the level of sound at a particular point. It is a device which houses a high-quality omnidirectional (pressure) microphone (see section 4.4.1) connected to amplifiers, filters and a meter (see diagram).

Weighting filters
The microphone's output voltage is proportional to the SPL incident upon it, and the weighting filters may be used to attenuate low and high frequencies according to a standard curve such as the 'A'-weighting curve, which corresponds closely to the sensitivity of human hearing at low levels (see Chapter 2). SPLs quoted simply in 'dB' are usually unweighted – in other words all frequencies are treated equally – but SPLs quoted in 'dBA' will have been A-weighted and will correspond more closely to the *perceived* loudness of the signal. A-weighting was originally designed to be valid up to a loudness of 55 phons, since the ear's frequency response becomes flatter at higher levels; between 55 and 85 phons the 'B' curve was intended to be used; and above 85 phons the 'C' curve was used. Nowadays, standards suggest that the 'A' curve may be

used at any SPL, principally for ease of comparability of measurements, although the 'C' curve is still used in vehicle acoustics measurements.

Noise criterion or rating (NC or NR)
Noise levels are often measured in rooms by comparing the level of the noise across the audible range with a standard set of curves called the *noise criteria (NC)* or *noise rating (NR)* curves. These curves set out how much noise is acceptable in each of a number of narrow frequency bands for the noise to meet a certain criterion. The noise criterion is then that of the nearest curve above which none of the measured results rise. NC curves are used principally in the USA, whereas NR curves are used principally in Europe. They allow a considerably higher noise level in low-frequency bands than in high-frequency bands, since the ear is less sensitive at low frequencies.

In order to measure the NC or NR of a location it is necessary to connect the measuring microphone to a set of filters or a spectrum analyser which is capable of displaying the SPL in one octave or one-third octave bands.

Further reading
British Standard 5969. *Specification for sound level meters.*
British Standard 6402. *Sound exposure meters.*

Mic   Amplifier   Amplifier   Meter

Weighting filters

1.12 Free and reverberant fields

The free field in acoustic terms is an acoustical area in which there are no reflections. Truly free fields are rarely encountered in reality, because there are nearly always reflections of some kind, even if at a very low level. If the reader can imagine the sensation of being suspended out-of-doors, way above the ground, away from any buildings or other surfaces, then he or she will have an idea of the experience of a free-field condition. The result is an acoustically 'dead' environment. Acoustic experiments are sometimes performed in *anechoic chambers*, which are rooms specially treated so as to produce almost no reflections at any frequency – the surfaces are totally absorptive – and these attempt to create near free-field conditions.

In the free field all the sound energy from a source is radiated away from the source and none is reflected; thus the inverse square law (Fact File 1.3) entirely dictates the level of sound at any distance from the source. Of course the source may be directional, in which case its *directivity factor* must be taken into account. A source with a directivity factor of 2 on its axis of maximum radiation radiates twice as much power in this direction as it would have if it had been radiating omnidirectionally. The directivity *index* of a source is measured in dB, giving the above example a directivity index of 3 dB. If calculating the intensity at a given distance from a directional source (as shown in Fact File 1.3), one must take into account its directivity factor on the axis concerned by multiplying the power of the source by the directivity factor before dividing by $4\pi r^2$.

In a room there is both direct and reflected sound. At a certain distance from a source contained within a room the acoustic field is said to be *diffuse* or *reverberant*, since reflected sound energy predominates over direct sound. A short time after the source has begun to generate sound a diffuse pattern of reflections will have built up throughout the room, and the reflected sound energy will become roughly constant at any point in the room. Close to the source the direct sound energy is still at quite a high level, and thus the reflected sound makes a smaller contribution to the total. This region is called the *near field*. (It is popular in sound recording to make use of so-called 'near-field monitors', which are loudspeakers mounted quite close to the listener, such that the direct sound predominates over the effects of the room.)

The exact distance from a source at which a sound field becomes dominated by reverberant energy depends on the reverberation time of the room, and this in turn depends on the amount of absorption in the room, and the room's volume (see Fact File 1.5). Figure 1.15 shows how the SPL changes as distance increases from a source in three different rooms. Clearly, in the acoustically 'dead' room, the conditions approach that of the free field (with sound intensity dropping at close to the expected 6 dB per doubling in distance), since the amount of reverberant energy is very small. The *critical distance* at which the contribution from direct sound equals that from reflected sound is further from the source than when the room is very reverberant. In the reverberant room the sound pressure level does not change much with distance from the source because reflected sound energy predominates after only a short distance. This is important in room design, since although a short reverberation time may be desirable in a recording control room, for example, it has the disadvantage that the change in SPL with distance from the speakers will be quite severe, requiring very highly powered amplifiers and heavy-duty speakers to provide the necessary level. A slightly longer reverberation time makes the room less disconcerting to work in, and relieves the requirement on loudspeaker power.

FACT FILE 1.5 Absorption, reflection and RT

Absorption

When a sound wave encounters a surface some of its energy is absorbed and some reflected. The *absorption coefficient* of a substance describes, on a scale from 0 to 1, how much energy is absorbed. An absorption coefficient of 1 indicates total absorption, whereas 0 represents total reflection. The absorption coefficient of substances varies with frequency.

The total amount of absorption present in a room can be calculated by multiplying the absorption coefficient of each surface by its area and then adding the products together. All of the room's surfaces must be taken into account, as must people, chairs and other furnishings. Tables of the performance of different substances are available in acoustics references (see *Recommended further reading*). Porous materials tend to absorb high frequencies more effectively than low frequencies, whereas resonant membrane- or panel-type absorbers tend to be better at low frequencies. Highly tuned artificial absorbers (Helmholtz absorbers) can be used to remove energy in a room at specific frequencies. The trends in absorption coefficient are shown in the diagram below.

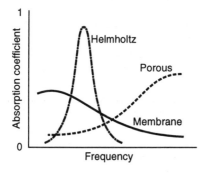

Reflection

The size of an object in relation to the wavelength of a sound is important in determining whether the sound wave will bend round it or be reflected by it. When an object is large in relation to the wavelength the object will act as a partial barrier to the sound, whereas when it is small the sound will bend or diffract around it. Since sound wavelengths in air range from approximately 18 metres at low frequencies to just over 1 cm at high frequencies, most commonly encountered objects will tend to act as barriers to sound at high frequencies but will have little effect at low frequencies.

Reverberation time

W. C. Sabine developed a simple and fairly reliable formula for calculating the reverberation time (RT_{60}) of a room, assuming that absorptive material is distributed evenly around the surfaces. It relates the volume of the room (V) and its total absorption (A) to the time taken for the sound pressure level to decay by 60 dB after a sound source is turned off.

$$RT_{60} = (0.16V)/A \text{ seconds}$$

In a large room where a considerable volume of air is present, and where the distance between surfaces is large, the absorbtion of the air becomes more important, in which case an additional component must be added to the above formula:

$$RT_{60} = (0.16V)/(A+xV) \text{ seconds}$$

where x is the absorption factor of air, given at various temperatures and humidities in acoustics references.

The Sabine formula has been subject to modifications by such people as Eyring, in an attempt to make it more reliable in extreme cases of high absorption, and it should be realised that it can only be a guide.

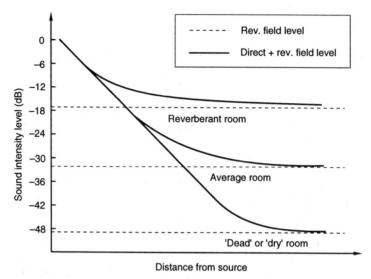

Figure 1.15 As the distance from a source increases direct sound level drops but reverberant sound level remains roughly constant. The resultant sound level experienced at different distances from the source depends on the reverberation time of the room, since in a reverberant room the level of reflected sound is higher than in a 'dead' room

1.13 Standing waves

The wavelength of sound varies considerably over the audible frequency range, as indicated in Fact File 1.5. At high frequencies, where the wavelength is small, it is appropriate to consider a sound wavefront rather like light – as a ray. Similar rules apply, such as the angle of incidence of a sound wave to a wall is the same as the angle of reflection. At low frequencies where the wavelength is comparable with the dimensions of the room it is necessary to consider other factors, since the room behaves more as a complex resonator, having certain frequencies at which strong pressure peaks and dips are set up in various locations.

Standing waves or *eigentones* (sometimes also called *room modes*) may be set up when half the wavelength of the sound or a multiple is equal to one of the dimensions of the room (length, width or height). In such a case (see Figure 1.16) the reflected wave from the two surfaces involved is in phase with the incident wave and a pattern of summations and cancellations is set up, giving rise to points in the room at which the sound pressure is very high, and other points where it is very low. For the first mode (pictured), there is a peak at the two walls and a trough in the centre of the room. It is easy to experience such modes by generating a low-frequency sine tone into a room from an oscillator connected to an amplifier and loudspeaker placed in a corner. At selected low frequencies the room will resonate strongly and the pressure peaks may be experienced by walking around the room. There are always peaks towards the boundaries of the room, with troughs distributed at regular intervals between them. The positions of these depend on whether the mode has been created between the walls or between the floor and ceiling. The frequencies (*f*) at which the strongest modes will occur is given by:

$$f = (c/2) \times (n/d)$$

where c is the speed of sound, d is the dimension involved (distance between walls or floor and ceiling), and n is the number of the mode.

A more complex formula can be used to predict the frequencies of all the modes in a room, including those secondary modes formed by reflections between four and six surfaces (*oblique* and *tangential* modes). The secondary modes typically have lower amplitudes than the primary modes (the *axial* modes) since they experience greater absorption. The formula is:

$$f = (c/2) \sqrt{((p/L)^2 + (q/W)^2 + (r/H)^2)}$$

where p, q and r are the mode numbers for each dimension (1, 2, 3 ...) and L, W and H are the length, width and height of the room. For example, to calculate the first axial mode involving only the length, make $p = 1$, $q = 0$ and $r = 0$. To calculate the first oblique mode involving all four walls, make $p = 1$, $q = 1$, $r = 0$, and so on.

Some quick sums will show, for a given room, that the modes are widely spaced at low frequencies and become more closely spaced at high frequencies. Above a certain frequency, there arise so many modes per octave that it is hard to identify them separately. As a rule-of-thumb, modes tend only to be particularly problematical up to about 200 Hz. The larger the room the more closely spaced the modes. Rooms with more than one dimension equal will experience so-called *degenerate modes* in which modes between two dimensions occur at the same frequency, resulting in an even stronger resonance at a particular frequency than otherwise. This is to be avoided.

Since low-frequency room modes cannot be avoided, except by introducing total absorption, the aim in room design is to reduce their effect by adjusting the ratios between dimensions to achieve an even spacing. A number of 'ideal' mode-spacing

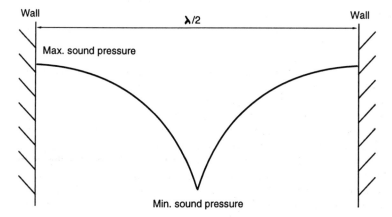

Figure 1.16 When a standing wave is set up between two walls of a room there arise points of maximum and minimum pressure. The first simple mode or *eigentone* occurs when half the wavelength of the sound equals the distance between the boundaries, as illustrated, with pressure maxima at the boundaries and a minimum in the centre

FACT FILE 1.6   Echoes and reflections

Early reflections
Early reflections are those echoes from nearby surfaces in a room which arise within the first few milliseconds (up to about 50 ms) of the direct sound arriving at a listener from a source (see the diagram). It is these reflections which give the listener the greatest clue as to the size of a room, since the delay between the direct sound and the first few reflections is related to the distance of the major surfaces in the room from the listener. Artificial reverberation devices allow for the simulation of a number of early reflections before the main body of reverberant sound decay, and this gives different rever-beration programs the characteristic of different room sizes.

Echoes
Echoes may be considered as discrete reflections of sound arriving at the listener after about 50 ms from the direct sound. These are perceived as separate arrivals, whereas those up to around 50 ms are normally integrated by the brain with the first arrival, not being perceived consciously as echoes. Such echoes are normally caused by more distant surfaces which are strongly reflective, such as a high ceiling or distant rear wall. Strong echoes are usually annoying in critical listening situations and should be suppressed by dispersion and absorption.

Flutter echoes
A flutter echo is sometimes set up when two parallel reflective surfaces face each other in a room, whilst the other surfaces are absorbent. It is possible for a wavefront to become 'trapped' into bouncing back and forth between these two surfaces until it decays, and this can result in a 'buzzing' or 'ringing' effect on transients (at the starts and ends of impulsive sounds such as hand claps).

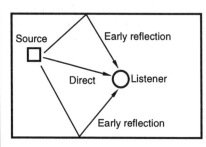

criteria have been developed by acousticians, but there is not the space to go into these in detail here. Larger rooms are generally more pleasing than small rooms, since the mode spacing is closer at low frequencies, and individual modes tend not to stick out so prominently, but room size has to be traded off against the target reverberation time. Making walls non-parallel does not prevent modes from forming (since oblique and tangential modes are still possible); it simply makes their frequencies more difficult to predict.

The practical difficulty with room modes results from the unevenness in sound pressure throughout the room at mode frequencies. Thus a person sitting in one position might experience a very high level at a particular frequency whilst other listeners might hear very little. A room with prominent LF modes will 'boom' at certain frequencies, and this is unpleasant and undesirable for critical listening. The response of the room modifies the perceived frequency response of a loudspeaker, for example, such that even if the loudspeaker's own frequency response may be acceptable it may become unacceptable when modified by the resonant character-istics of the room.

Room modes are not the only results of reflections in enclosed spaces, and some other examples are given in Fact File 1.6.

Recommended further reading

General

Eargle, J. (1990) *Music, Sound, Technology*. Van Nostrand Rheinhold
Howard, D. and Angus, J. (1996) *Acoustics and Psychoacoustics*. Focal Press
Rossing, T. D. (1989) *The Science of Sound*, 2nd Edition. Addison-Wesley

Architectural acoustics

Egan, M. D. (1988) *Architectural Acoustics*. McGraw–Hill
Rettinger, M. (1988) *Handbook of Architectural Acoustics and Noise Control*. TAB Books
Templeton, D. and Saunders, D. (1987) *Acoustic Design*. Butterworth Architecture

Musical acoustics

Benade, A. H. (1976) *Fundamentals of Musical Acoustics*. Oxford University Press
Campbell, M. and Greated, C. (1987) *The Musician's Guide to Acoustics*. Dent
Hall, D. E. (1991) *Musical Acoustics*, 2nd Edition. Brooks/Cole Publishing Co.

Chapter 2

Auditory perception

In this chapter the mechanisms by which sound is perceived will be introduced. The human ear often modifies the sounds presented to it before they are presented to the brain, and the brain's interpretation of what it receives from the ears will vary depending on the information contained in the nervous signals. An understanding of loudness perception is important when considering such factors as the perceived frequency balance of a reproduced signal, and an understanding of directional perception is relevant to the study of stereo recording techniques. Below, a number of aspects of the hearing process will be related to the practical world of sound recording and reproduction.

2.1 The hearing mechanism

Although this is not intended to be a lesson in physiology, it is necessary to investigate the basic components of the ear, and to look at how information about sound signals is communicated to the brain. Figure 2.1 shows a diagram of the ear mechanism, not anatomically accurate but showing the key mechanical components. The outer ear consists of the pinna (the visible skin and bone structure) and the auditory canal, and is terminated by the *tympanic membrane* or 'ear drum'. The middle ear consists of a three-bone lever structure which connects the tympanic membrane to the inner ear via the oval window (another membrane). The inner ear is a fluid-filled bony spiral device known as the *cochlea*, down the centre of which runs a flexible membrane known as the basilar membrane. The cochlea is shown here as if 'unwound' into a straight chamber for the purposes of description. At the end of the basilar membrane, furthest from the middle ear, there is a small gap called the *helicotrema* which allows fluid to pass from the upper to the lower chamber. There are other components in the inner ear, but those noted above are the most significant.

The ear drum is caused to vibrate in sympathy with the air in the auditory canal when excited by a sound wave, and these vibrations are transferred via the bones of the middle ear to the inner ear, being subject to a multiplication of force of the order of 15:1 by the lever arrangement of the bones. The lever arrangement, coupled with the difference in area between the tympanic membrane and the oval window, helps to match the impedances of the outer and inner ears so as to ensure optimum transfer of energy. Vibrations are thus transferred to the fluid in the inner ear in which pressure waves are set up. The basilar membrane is not uniformly stiff along its

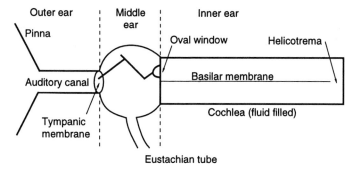

Figure 2.1 A simplified mechanical diagram of the ear

length (it is narrow and stiff at the oval window end and wider and more flexible at the far end), and the fluid is relatively incompressible; thus a high-speed pressure wave travels through the fluid and a pressure difference is created across the basilar membrane.

2.2 Frequency perception

The motion of the basilar membrane depends considerably on the frequency of the sound wave, there being a peak of motion which moves closer towards the oval window the higher the frequency (see Figure 2.2).

At low frequencies the membrane has been observed to move as a whole, with the maximum amplitude of motion at the far end, whilst at higher frequencies there arises a more well-defined peak. It is interesting to note that for every octave (i.e.: for every doubling in the frequency) the position of this peak of maximum vibration moves a similar length up the membrane, and this may explain the human preference for displaying frequency-related information on a logarithmic frequency scale, which represents increase in frequency by showing octaves as equal increments along a frequency axis.

Frequency information is transmitted to the brain in two principal ways. At low frequencies hair cells in the inner ear are stimulated by the vibrations of the basilar membrane, causing them to discharge small electrical impulses along the auditory nerve fibres to the brain. These impulses are found to be synchronous with the sound waveform, and thus the period of the signal can be measured by the brain. Not all

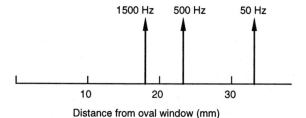

Figure 2.2 The position of maximum vibration on the basilar membrane moves towards the oval window as frequency increases

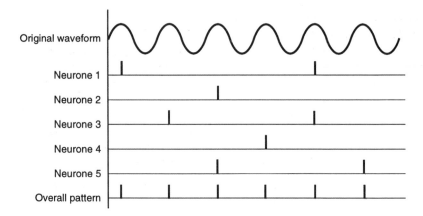

Figure 2.3 Although each neurone does not normally fire on every cycle of the causatory sound wave, the outputs of a combination of neurones firing on different cycles represent the period of the wave

nerve fibres are capable of discharging once per cycle of the sound waveform (in fact most have spontaneous firing rates of a maximum of 150 Hz with many being much lower than this). Thus at all but the lowest frequencies the period information is carried in a *combination* of nerve fibre outputs, with at least a few firing on every cycle (see Figure 2.3). There is evidence to suggest that nerve fibres may re-trigger faster if they are 'kicked' harder – that is, the louder the sound the more regularly they may be made to fire. Also, whilst some fibres will trigger with only a low level of stimulation, others will only fire at high sound levels.

The upper frequency limit at which nerve fibres appear to cease firing synchronously with the signal is around 4 kHz, and above this frequency the brain relies increasingly on an assessment of the position of maximum excitation of the membrane to decide on the pitch of the signal. There is clearly an overlap region in the middle-frequency range, from about 200 Hz upwards, over which the brain has both synchronous discharge information and 'position' information on which to base its measurement of frequency. It is interesting to note that one is much less able to determine the precise musical pitch of a note when its frequency is above the synchronous discharge limit of 4 kHz.

The frequency selectivity of the ear has been likened to a set of filters, and this concept is described in more detail in Fact File 2.1. It should be noted that there is an unusual effect whereby the perceived pitch of a note is related to the loudness of the sound, such that the pitch shifts slightly with increasing sound level. This is sometimes noticed as loud sounds decay, or when removing headphones, for example. The effect of 'beats' may also be noticed when two pure tones of very similar frequency are sounded together, resulting in a pattern of addition and cancellation as they come in and out of phase with each other. The so-called 'beat frequency' is the difference frequency between the two signals, such that signals at 200 Hz and 201 Hz would result in a cyclic modulation of the overall level, or beat, at 1 Hz. Combined signals slightly further apart in frequency result in a 'roughness' which disappears once the frequencies of the two signals are further than a critical band apart.

FACT FILE
2.1
Critical bandwidth

The basilar membrane appears to act as a rough mechanical spectrum analyser, providing a spectral analysis of the incoming sound to an accuracy of between one-fifth and one-third of an octave in the middle frequency range (depending on which research data is accepted). It acts rather like a bank of overlapping filters of a fixed bandwidth. This analysis accuracy is known as the *critical bandwidth*, that is the range of frequencies passed by each notional filter.

The critical band concept is important in understanding hearing because it helps to explain why some signals are 'masked' in the presence of others (see Fact File 2.3). Fletcher, working in the 1940s, suggested that only signals lying within the same critical band as the wanted signal would be capable of masking it, although other work on masking patterns seems to suggest that a signal may have a masking effect on frequencies well above its own.

With complex signals, such as noise or speech for example, the total loudness of the signal depends to some extent on the number of critical bands covered by a signal. It can be demonstrated by a simple experiment that the loudness of a constant power signal does not begin to increase until its bandwidth extends over more than the relevant critical bandwidth, which appears to support the previous claim. (A useful demonstration of this phenomenon is to be found on the Compact Disc entitled *Auditory Demonstrations* described at the end of this chapter.)

Although the critical band concept helps to explain the first level of frequency analysis in the hearing mechanism, it does not account for the fine frequency selectivity of the ear which is much more precise than one-third of an octave. One can detect *changes* in pitch of only a few hertz, and in order to understand this it is necessary to look at the ways in which the brain 'sharp-ens' the aural tuning curves. For this the reader is referred to Moore (1989), as detailed at the end of this chapter.

2.3 Loudness perception

The subjective quantity of 'loudness' is not directly related to the SPL of a sound signal (see section 1.11). The ear is not uniformly sensitive at all frequencies, and a set of curves has been devised which represents the so-called equal-loudness contours of hearing (see Fact File 2.2). This is partially due to the resonances of the outer ear which have a peak in the middle-frequency region, thus increasing the effective SPL at the ear drum over this range.

The unit of loudness is the *phon*. If a sound is at the threshold of hearing (just perceivable) it is said to have a loudness of 0 phons, whereas if a sound is at the threshold of pain it will probably have a loudness of around 140 phons. Thus the ear has a dynamic range of approximately 140 phons, representing a range of sound pressures with a ratio of around 10 million to one between the loudest and quietest sounds perceivable. As indicated in Fact File 1.4, the 'A'-weighting curve is often used when measuring sound levels because it shapes the signal spectrum to represent more closely the subjective loudness of low-level signals. A noise level quoted in dBA is very similar to a loudness level in phons.

To give an idea of the loudnesses of some common sounds, the background noise of a recording studio might be expected to measure at around 20 phons, a low-level conversation perhaps at around 50 phons, a busy office at around 70 phons, shouted

FACT FILE	Equal-
2.2	loudness contours

Fletcher and Munson devised a set of curves to show the sensitivity of the ear at different frequencies across the audible range. They derived their results from tests on a large number of subjects who were asked to adjust the level of test tones until they appeared equally as loud as a reference tone with a frequency of 1 kHz. The test tones were spread across the audible spectrum. From these results could be drawn curves of average 'equal loudness', indicating the SPL required at each frequency for a sound to be perceived at a particular loudness level (see diagram).

Loudness is measured in *phons*, the zero phon curve being that curve which passes through 0 dB SPL at 1 kHz – in other words, the threshold of hearing curve. All points along the 0 phon curve will sound equally loud, although clearly a higher SPL is required at extremes of the spectrum than in the middle. The so-called Fletcher–Munson curves are not the only equal-loudness curves in existence – Robinson and Dadson, amongst others, have published revised curves based upon different test data. The shape of the curves depends considerably on the type of sound used in the test, since filtered noise produces slightly different results to sine tones.

It will be seen that the higher-level curves are flatter than the low-level curves, indicating that the ear's frequency response changes with signal level. This is important when considering monitoring levels in sound recording (see text).

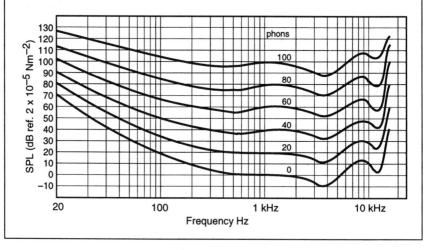

speech at around 90 phons, and a full symphony orchestra playing loudly at around 120 phons. These figures of course depend on the distance from the sound source, but are given as a guide.

The loudness of a sound depends to a great extent on its nature. Broad-band sounds tend to appear louder than narrow-band sounds, because they cover more critical bands (see Fact File 2.1), and distorted sounds appear psychologically to be louder than undistorted sounds, perhaps because one associates distortion with system overload. If two music signals are played at identical levels to a listener, one with severe distortion and the other without, the listener will judge the distorted signal to be louder.

A further factor of importance is that the threshold of hearing is raised at a particular frequency in the presence of another sound at a similar frequency. In other words, one sound may 'mask' another – a principle described in more detail in Fact File 2.3.

In order to give the impression of a doubling in perceived loudness, an increase of some 9–10 dB is required. Although 6 dB represents a doubling of the actual sound pressure, the hearing mechanism appears to require a greater increase than this for the signal to appear to be twice as loud. Another subjective unit, rarely used in practice, is that of the *sone*: 1 sone is arbitrarily aligned with 40 phons, and 2 sones is twice as loud as 1 sone, representing approximately 49 phons; 3 sones is three times as loud, and so on. Thus the sone is a true indication of the relative loudness of signals on a linear scale, and sone values may be added together to arrive at the total loudness of a signal in sones.

FACT FILE 2.3 Masking

Most people have experienced the phenomenon of masking, although it is often considered to be so obvious that it does not need to be stated. As an example: it is necessary to raise your voice in order for someone to hear you if you are in noisy surroundings. The background noise has effectively raised the perception threshold so that a sound must be louder before it can be heard. If one looks at the masking effect of a pure tone, it will be seen that it raises the hearing threshold considerably for frequencies which are the same as or higher than its own (see diagram). Frequencies below the masking tone are less affected. The range of

frequencies masked by a tone depends mostly on the area of the basilar membrane set into motion by the tone, and the pattern of motion of this membrane is more extended towards the HF end than towards the LF end. If the required signal produces more motion on the membrane than the masking tone produces at that point then it will be perceived.

The phenomenon of masking has many practical uses in audio engineering. It is used widely in noise reduction systems, since it allows the designer to assume that low-level noise which exists in the same frequency band as a high-level music signal will be effectively masked by the music signal. It is also used in digital audio data compression systems, since it allows the designer to use lower resolution in some frequency bands where the increased noise will be effectively masked by the wanted signal.

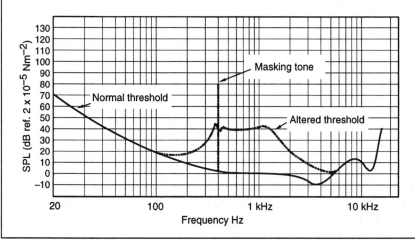

The ear is by no means a perfect transducer; in fact it introduces considerable distortions into sound signals due to its non-linearity. At high signal levels, especially for low-frequency sounds, the amount of harmonic and intermodulation distortion (see Appendix 1) produced by the ear can be high.

2.4 Practical implications of equal-loudness contours

The non-linear frequency response of the ear presents the sound engineer with a number of problems. Firstly, the perceived frequency balance of a recording will depend on how loudly it is replayed, and thus a balance made in the studio at one level may sound different when replayed in the home at another. In practice, if a recording is replayed at a much lower level than that at which it was balanced it will sound lacking in bass and extreme treble – it will sound thin and lacking warmth. Conversely, if a signal is replayed at a higher level than that at which it was balanced it will have an increased bass and treble response, sounding boomy and overbright.

A 'loudness' control is often provided on hi-fi amplifiers to boost low and high frequencies for low-level listening, but this should be switched out at higher levels. Rock-and-roll and heavy-metal music often sounds lacking in bass when replayed at moderate sound levels beacause it is usually balanced at extremely high levels in the studio.

Some types of noise will sound louder than others, and hiss is usually found to be most prominent due to its considerable energy content at middle–high frequencies. Rumble and hum may be less noticeable because the ear is less sensitive at low frequencies, and a low-frequency noise which causes large deviations of the meters in a recording may not sound particularly loud in reality. This does not mean, of course, that rumble and hum are acceptable.

Recordings equalised to give a strong mid-frequency content often sound rather 'harsh', and listeners may complain of listening fatigue, since the ear is particularly sensitive in the range between about 1 and 5 kHz.

2.5 Directional perception

Directional hearing principles become important when considering such matters as stereo sound reproduction and when designing PA rigs for large auditoria, since an objective in both these cases is to give the illusion of a sound coming from a particular direction.

Directional hearing may coarsely be divided into three planes: the lateral plane from left to right, the front–back or 'median' plane, and the vertical plane. Front–back distinction becomes more important when considering microphone techniques which aim to offer some degree of 'surround-sound' information, and vertical distinction completes the picture if one is ever to attempt a complete reconstruction of the original sound field, as is the case in full 'periphonic' reproduction, such as can be achieved with the system known as *Ambisonics*.

In the discussion of directional hearing the existence of a phase difference between the two ears is often quoted, when what is really meant is a difference in time of arrival (TOA) of the sound. When considering the mechanisms of hearing it is often convenient to use pure tones as a signal source example, but it should be remembered that one rarely listens to pure tones in real life, since real sounds are usually of a more complex, noise-like nature.

One's ability to perceive directionality in sound depends almost entirely on the fact that two ears are involved, although there are more minor effects which may be

---

FACT FILE 2.4   Lateral position judgement

The brain relies on differences between the signals from the two ears to detect the position of a source in the lateral plane. These can be summarised as follows with respect to the diagram:

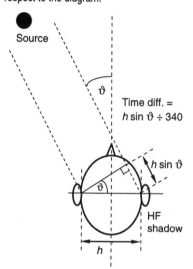

Source

Time diff. = $h \sin \vartheta \div 340$

$h \sin \vartheta$

HF shadow

h

Level
If a source is close to the head, the extra distance travelled by the sound wave to the more distant ear will result in a small drop in level due to inverse-square-law decay (see Fact File 1.3). For distant sources the additional distance between the ears is tiny compared with the distance already travelled, and thus the level drop can be considered as insignificant.

Frequency response
The more distant ear will be 'shadowed' by the head, this effect being most noticeable at middle to high frequencies where the wavelength of the sound is similar to or smaller than the head dimensions. The result of this shadowing is that the distant ear will suffer a modification in its frequency response, mainly involving attenuation of the high-frequency content of the signal. The HF attenuation becomes more severe as the angle of the sound source away from the front increases, reaching a maximum at 90° to the front.

The outer ear also modifies the frequency response of signals arriving from different angles, due to the unusual shape of the pinna and its resonant properties.

Time
Owing to the spacing between the ears, off-centre sources will be delayed at the more distant ear. The maximum delay occurs at 90° to the front, where the time difference between the ears is approximately 600 microseconds (0.6 ms). The delay is related to the sine of the angle of offset from the front. The brain is able to measure the difference in arrival time of neural discharge from the two ears and use this in its determination of the source's position.

Phase
If the source is emitting a continuous repetitive waveform, then it is possible for the time difference between the ears to be considered as a phase difference. Above the frequency at which the distance between the ears amounts to half a wavelength of the sound the phase cue becomes ambiguous, since it becomes difficult for the brain to tell which ear is lagging and which leading. For example, the question arises – is the left ear 330° in advance of the right ear, or is it 30° behind? Also, to talk in terms of phase is confusing since the phase difference is dependent on the frequency of the signal. For a fixed position of the source the phase angle between the two ear signals will be different for every frequency. Similarly, if the frequency remains constant but the source moves, the phase angle will change also.

perceived even if only one ear is involved. Principally, lateral distinction of direction is based on sound amplitude and TOA differences between the two ears. These factors are detailed in Fact File 2.4.

Initially it is hard to see what distinguishes a point source at a given number of degrees off-centre in front of the head from the same source at the same angle behind the head, since the TOA difference will be the same and so, to a large extent, will the shadowing effect of the head. Even so, it is clear that the freedom to move the head plays an important part in localisation in this plane, since, in subjective tests, people whose heads have been held perfectly still have been found to display greater difficulty in front–back distinction. The reason for this is that even a small change in the rotation angle of the head will alter the TOA difference between the ears, and for a given direction of rotation this difference will either get smaller or greater depending on whether the source is in front of or behind the head.

A further factor is the rôle that sight plays in this distinction, since the eyes are used much more to determine the location of a source in front, whereas only the ears may be used to determine the location of a source behind. If a source cannot be seen then it must be behind (ignoring the vertical for the moment)! This factor must not be overplayed since it is still possible to determine front–back location with the eyes closed, although there is some evidence to show that blind people have better aural localisation to the front than sighted people.

A final factor is the effect of the pinna and head on sounds from the rear, since the pinna's size is such as to act as a partial barrier to very high-frequency sounds from the rear, changing the spectral emphasis of a sound from the rear when compared with the same sound from the front. Further, the head-related transfer function (HRTF) is slightly different for rear sounds than for frontal sounds (this being the distortion of phase and amplitude evident in the sounds reaching the two ears due to the interference effects of the head).

Localisation in the vertical plane is concerned partially with the effect of reflections from the ground and from the shoulders, in addition to pinna and HRTF effects, since sounds at different angles of elevation will reach the ears directly and will also be reflected, reaching the ear slightly later, via indirect paths. For sounds from above, the difference in path length between the sound reflected from the shoulders and the floor compared with the direct path will result in cancellation and addition at different frequencies (in the same way as discussed above with reflections from the pinna, although reflections from the shoulders and floor will result in notches and peaks which are more widely spaced, and which extend to a lower frequency, due to the distances involved). It is partially in the comparison of these frequency spectra with stored templates that the brain will locate sounds in the vertical plane.

The memory of learned situations, and also the expectation that particular sounds will emanate from certain directions, is also important for vertical localisation, since there are not many sounds that come from below, as a general rule.

2.6 The precedence effect

So far only the single point source has been considered, resulting in delays of up to 0.6 ms between the ears (the so-called 'binaural delay'), but it is also necessary to consider the situation in which more than one source exists, since this is one of the keys to understanding stereo sound reproduction. Haas and others, studying the

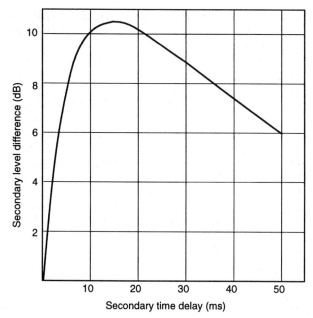

Figure 2.4 The Haas or precedence effect describes how a delayed secondary source must be louder than the primary source if it is to appear equally loud

effects of echoes on the perception of direction of a source, showed that if two sources were emitting similar sounds the perceived direction of the sound tended towards the advanced (in time) source, and that the delays over which the phenomenon was noticed extended up to around 50 ms, a much greater delay than that over which the binaural effect applies. For delays up to about 50 ms the sounds from the two sources are 'fused' together by the brain, appearing as one source with the perceived location towards that of the first arrival. Beyond 50 ms the brain begins to perceive the sounds as distinct, and the second appears as an 'echo' of the first. With single clicks, the effect breaks down more quickly than with complex sounds, allowing delays of only up to around 5 ms before the 'fusing' effect disappears.

The so-called 'Haas effect curve' (see Figure 2.4) shows that for the delayed source to appear equally as loud as the undelayed source it must be a certain number of decibels higher in amplitude to compensate for the precedence advantage of the first arrival, the peak of the effect being at a delay of around 15 ms, where the delayed source must be 11 dB louder than the undelayed source in order to be perceived as equally loud. It is therefore possible to see the beginnings of a possibility for trading amplitude difference against time difference to achieve the same directional effect.

2.7 Implications for stereo sound reproduction

The aim of stereo sound reproduction is to give the impression of directionality and space in sound emitted from two or more loudspeakers, or over headphones. Stereo microphones are discussed further in section 4.7, but some principles will be introduced here.

Firstly, the impression of directionality in reproduced sound may be given using a combination of time and level differences between two channels. Owing to the precedence effect described above, if the right loudspeaker signal is delayed slightly in relation to the left loudspeaker then the sound will appear to come from somewhere towards the left, depending on the amount of delay. Depending on the type of signal, the signal will appear to be fully left when there is a delay of between about 2 and 4 ms. Spaced microphone techniques work on this principle.

If a level difference is introduced between two loudspeakers the signal will appear to be towards the louder speaker, with the signal appearing to be fully left or right with approximately 18 dB of level difference. Coincident microphone techniques and 'pan-potted' stereo work on this principle. Stereo impressions can be given using a combination of both time and level difference, and the diagram in Figure 2.5 illustrates the approximate trade-off between the two for a sound to appear in a certain position.

Binaural microphone techniques attempt to capture the full collection of inter-aural differences described in Fact File 2.4, by using small pressure microphones

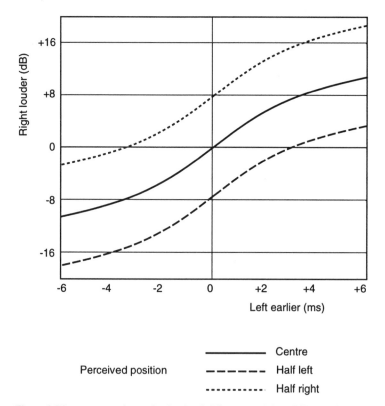

Figure 2.5 In stereo sound reproduction level difference and time difference between left and right channels may to some extent be traded off against each other in order for a source to be perceived in a particular position on the 'sound stage'

placed at the entrances of the ear canals of either a real or dummy head. When replayed on headphones such techniques can be capable of reproducing very life-like images, together with an impression of placing the listener directly in the original sound field. Such techniques are most successful when the head and ears used to make the recording are as similar as possible to the listener's own characteristics, although headphone inadequacies can modify the situation consid-erably, often resulting in poor reproduction.

Recommended further reading

Blauert, J. (1983) *Spatial Hearing*. Translated by J.S. Allen. MIT Press
Eargle, J. (1986) ed. *Stereophonic Techniques – An Anthology*. Audio Engineering Society
Howard, D. and Angus, J. (1996) *Acoustics and Psychoacoustics*. Focal Press
Moore, B. C. J. (1989) *An Introduction to the Psychology of Hearing*. Academic Press
Tobias, J. (1970) ed. *Foundations of Modern Auditory Theory*. Academic Press

Recommended listening

Auditory Demonstrations (Compact Disc). Philips Cat. No. 1126-061. Available from the Acoustical Society of America.

Chapter 3

A guide to the audio signal chain

Sound signals start life as vibrations carried through the air, and this is also how they end up after they have passed through the recording and reproduction process with all its stages. In broadcasting the signal is not only recorded and reproduced, it is also transmitted by means of radio waves. In the following chapter an overview will be given of the various stages involved in the recording and broadcasting chain, and in subsequent chapters the technical and operational aspects involved in audio systems will be examined in greater detail. An introduction to the differences in level which may be encountered at stages in the signal chain is also included here. More detailed coverage will be given to the matter of line-up, metering and signal levels in appropriate chapters.

3.1 A short history

3.1.1 Early recording machines

When Edison and Berliner first developed recording machines in the last years of the nineteenth century they involved little or no electrical apparatus. Certainly the recording and reproduction process itself was completely mechanical or 'acoustic', the system making use of a small horn terminated in a stretched, flexible diaphragm attached to a stylus which cut a groove of varying depth into the malleable tin foil on Edison's 'phonograph' cylinder or of varying lateral deviation in the wax on Berliner's 'gramophone' disk (see Figure 3.1). On replay, the undulations of the groove caused the stylus and diaphragm to vibrate, thus causing the air in the horn to move in sympathy, thus reproducing the sound – albeit with a very limited frequency range and very distorted.

Cylinders for the phonograph could be recorded by the user, but they were difficult to duplicate for mass production, whereas disks for the gramophone were normally replay only, but they could be duplicated readily for mass production. For this reason disks fairly quickly won the day as the mass-market pre-recorded music medium. There was no such thing as magnetic recording tape at the time, and thus recordings were made directly on to a master disk, lasting for the duration of the side of the disk – a maximum of around 4 minutes – with no possibility for editing. Recordings containing errors were either remade or they were passed with mistakes intact. A long item of music would be recorded in short sections with gaps to change the disk, and possibilities arose for discontinuities between the sections, as well as

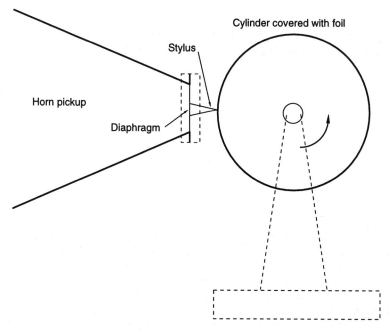

Figure 3.1 The earliest phonograph used a rotating foil-covered cylinder and a stylus attached to a flexible diaphragm. The recordist spoke or sang into the horn causing the stylus to vibrate, thus inscribing a modulated groove into the surface of the soft foil. On replay the modulated groove would cause the stylus and diaphragm to vibrate, resulting in a sound wave being emitted from the horn

variations in pitch and tempo.

Owing to the deficiencies of the acoustic recording process, instruments had to be grouped quite tightly around the pickup horn in order for them to be heard on the recording, and often louder instruments were substituted for quieter ones (the double bass was replaced by the tuba, for example) in order to correct for the poor frequency balance. It is perhaps partly because of this that much of the recorded music of the time consisted of vocal soloists and small ensembles, since these were easier to record than large orchestras.

3.1.2 Electrical recording

During the 1920s, when broadcasting was in its infancy, electrical recording became more widely used, based on the principles of electromagnetic transduction (see Fact File 3.1). The possibility for a microphone to be connected remotely to a recording machine meant that microphones could be positioned in more suitable places, connected by wires to a complementary transducer at the other end of the wire which drove the stylus to cut the disk. Even more usefully, the outputs of microphones could be mixed together before being fed to the disk cutter, allowing greater flexibility in the balance. Basic variable resistors could be inserted into the signal chain in order to control the levels from each microphone, and valve amplifiers would be used to increase the electrical level so that it would be suitable to drive the cutting stylus (see Figure 3.2).

FACT FILE

3.1

Electro-magnetic transducers

Electromagnetic transducers facilitate the conversion of acoustic signals into electrical signals, and they also act to convert electrical signals back into acoustic sound waves. The principle is very simple: if a wire can be made to move in a magnetic field, perpendicular to the lines of flux linking the poles of the magnet, then an electric current is induced in the wire (see diagram). The direction of motion governs the direction of current flow in the wire, and thus if the wire can be made to move back and forth then an alternating current can be induced in the wire, related in frequency and amplitude to the motion of the wire. Conversely, if a current is made to flow through a wire which cuts the lines of a magnetic field then the wire will move.

It is a short step from here to see how acoustic sound signals may be converted into electrical signals and vice versa. A

simple moving-coil microphone, as illustrated in Fact File 4.1, involves a wire moving in a magnetic field, by means of a coil attached to a flexible diaphragm which vibrates in sympathy with the sound wave. The output of the microphone is an alternating electrical current, whose frequency is the same as that of the sound wave which caused the diaphragm to vibrate. The amplitude of the electrical signal generated depends on the mechanical characteristics of the transducer, but is proportional to the *velocity* of the coil.

Vibrating systems, such as transducer diaphragms, with springiness (compliance) and mass, have a *resonant frequency* (a natural frequency of free vibration). If the driving force's frequency is below this resonant frequency then the motion of the system depends principally on its stiffness; at resonance the motion is dependent principally on its damping (resistance); and above resonance it is mass controlled. Damping is used in transducer diaphragms to control the amplitude of the resonant response peak, and to ensure a more even response around resonance. Stiffness and mass control are used to ensure as flat a frequency response as possible in the relevant frequency ranges.

Exactly the reverse process occurs in a loudspeaker, where an alternating current is fed *into* a coil attached to a diaphragm, there being a similar magnet around the coil. This time the diaphragm moves in sympathy with the frequency and magnitude of the incoming electrical audio signal, causing compression and rarefaction of the air.

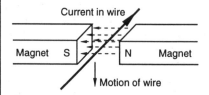

The sound quality of electrical recordings shows a marked improvement over acoustic recordings, with a wider frequency range and a greater dynamic range (see Appendix 1). Experimental work took place both in Europe and the USA on stereo recording and reproduction, but it was not to be until much later that stereo took its place as a common consumer format, nearly all records and broadcasts being in mono at that time.

3.1.3 Later developments

During the 1930s work progressed on the development of magnetic recording equipment, and examples of experimental wire recorders and tape recorders began

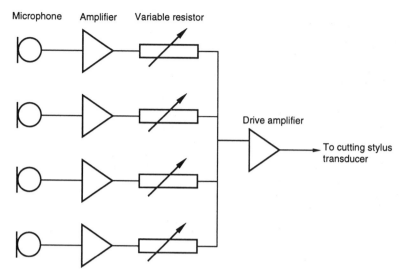

Figure 3.2 Early electrical recording made it possible for the electrical signals from microphones to be fed through variable resistors to alter their gains, and then combined to feed an electrical cutting stylus

to appear, based on the principle of using a current flowing through a coil to create a magnetic field which would in turn magnetise a moving metal wire or tape coated with magnetic material. The 1940s, during wartime, saw the introduction of the first AC-biased tape recorders (see section 8.2) which brought with them good sound quality and the possibility for editing. Tape itself, though, was first made of paper coated with metal oxide which tended to deteriorate rather quickly, and only later of plastics which proved longer lasting and easier to handle.

In the 1950s the microgroove LP record appeared, with markedly lower surface noise and improved frequency response, having a playing time of around 25 minutes per side. This was an ideal medium for the distribution of commercial stereo recordings, which began to appear in the late 1950s, although it was not until the 1960s that stereo really took hold. In the early 1960s the first multitrack tape recorders appeared, the Beatles making use of an early four-track recorder for their *Sergeant Pepper's Lonely Hearts Club Band* album. The machine offered the unprecedented flexibility of allowing sources to be recorded separately, and the results in the stereo mix are panned very crudely to left and right in somewhat 'gimmicky' stereo.

Mixing equipment in the 1950s and 1960s was often quite basic, compared with today's sophisticated consoles, and rotary faders were the norm. There simply was not the quantity of tracks involved as exists today.

Recent recording history has seen the birth of high-quality digital recording (see Chapter 10), with Compact Disc and digital tape systems for consumer use, allowing full frequency-range recordings with minimal distortions of all kinds to be distributed commercially. Broadcast sound is now capable of high quality, either using FM or digital transmissions with radio and television. Studio recording has the benefit of an almost unlimited number of tracks and outboard effects if required.

3.2 The modern recording chain

In a modern music recording, sound signals begin life either as real acoustic sources (e.g.: vocals, a piano, acoustic guitars) or as sources of electrical sound signals (e.g.: synthesisers, electric guitars, drum machines). The latter may be amplified and reproduced via a loudspeaker, but can also be fed directly into a mixer (see Chapter 6). Acoustic sources will be picked up by microphones and fed into the mic inputs of a mixer (which incorporates amplifiers to raise the low-voltage output from microphones), whilst other sources usually produce so-called 'line level' outputs (see section 3.4) which can be connected to the mixer without extra amplification.

In the mixer, sources are combined in proportions controlled by the engineer and recorded. It is at this point that different techniques diverge, since much 'classical' music recording makes use of very straightforward recording techniques, whereas 'pop' sessions usually involve multitrack recording, separate mixing down of the multitrack tape, the addition of effects and the compilation of the album or single master.

3.2.1 'Straight-to-stereo'

Figure 3.3 shows the typical signal chain through the stages of a production which is being recorded 'straight-to-stereo', such as a classical music recording for CD or broadcast. Microphone sources are mixed 'live' without recording to multitrack tape, creating a stereo *session master*, either analogue or digital, which is the collection of original recordings, often consisting of a number of *takes* of the musical material. The balance between the sources must be correct at this stage, and often only a small number of carefully positioned microphones is used. The session master tapes will then proceed to the editing stage where takes are assembled in an artistically satisfactory manner, under the control of the producer, to create a final master which will be transmitted or made into a commercial release. This final master could be made into a number of production masters which will be used to make LPs, cassettes and CDs. An LP master requires special equalisation to prepare

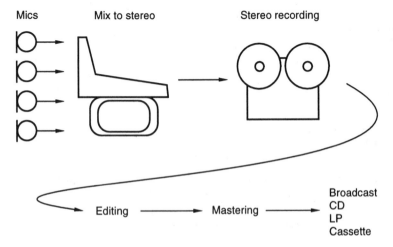

Figure 3.3 A diagrammatic representation of 'straight-to-stereo' music production

it for cutting on to disk, whereas a cassette master is usually made on to a half-inch 'loop-bin master' for high-speed duplication. Digital masters require the addition of 'PQ subcodes' to mark the starts and ends of tracks, as well as other cueing information. This stage is known, not surprisingly, as 'mastering'.

In the case of 'straight-to-stereo' productions the mixing console used may be a simpler affair than that used for multitrack recording, since the mixer's job is to take multiple inputs and combine them to a single stereo output, with intervening processing such as equalisation. This method of production is clearly cheaper and less time consuming than multitrack recording, but requires skill to achieve a usable balance quickly. It also limits flexibility in post-production. Occasionally, classical music is recorded on to multitrack tape, especially in the case of complex operas or large-force orchestral music with a choir and soloists, where to get a correct balance at the time of the session could be costly and time-consuming. In such a case, the production process becomes more similar to the pop recording situation described below.

3.2.2 Multitrack recording

'Pop' music is rarely recorded live, except at live events such as concerts, but is created in the recording studio. Acoustic and electrical sources are fed into a mixer and recorded on to multitrack tape, often a few tracks at a time, gradually building up a *montage* of sounds (see Figure 3.4). The resulting tape thus contains a collection of individual sources on multiple tracks which must subsequently be mixed into the final format (usually stereo). Individual songs or titles are recorded in separate places on the tape, to be compiled later.

It is not so common these days to record multitrack pop title in 'takes' for later editing, as with classical music, since mixer automation allows the engineer to work on a song in sections for automatic execution in sequence by a computer (see

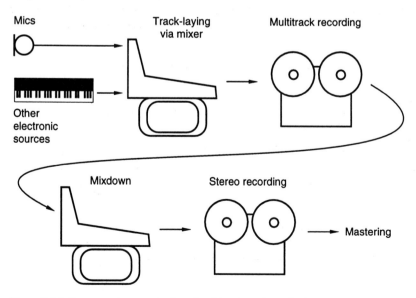

Figure 3.4 A diagrammatic representation of multitrack music production

Chapter 15). In any case, multitrack tape machines have comprehensive 'drop-in' facilities for recording short inserted sections on individual tracks without introducing clicks, and a pop-music master is usually built up by laying down backing tracks for a complete song (drums, keyboards, rhythm guitars, etc.) after which lead lines are overdubbed using drop-in facilities. Occasionally multitrack tapes are edited early on during a recording session to compile an acceptable backing track from a number of takes, after which further layers are added, but this is considered by some to be a potentially risky business.

Considerable use is made these days of computer-sequenced electronic instruments, under MIDI control (see Chapter 15), often in conjunction with multitrack tape or disk recording. The computer controlling the electronic instruments is synchronised to the tape machine using timecode (see Chapter 16) and the outputs of the instruments are fed to the mixer to be combined with the non-sequenced sources. Increasingly, MIDI-controlled sources are not recorded on to multitrack tape at all, the information about 'which-notes-occurred-when' being stored in the computer to be synchronised with the multitrack recorder only during the mixdown. This has the obvious disadvantage that all the same electronic instruments are required in the control room on mixdown as were used in the session, but may have the advantage of reducing the number of tracks required on tape and allows first-generation sound from sequenced sources to be fed into the mix. This technique is more common in 'home studio' and budget sessions than in 'no-expense-spared' music production, since in the latter the producer may wish to take a tape to the other side of the world for mixdown and will tend to feel happier if he or she knows that everything he or she needs is stored on one tape.

Once the session is completed, the multitrack master tape is mixed down. This is often done somewhere different from the original session, and involves feeding the outputs of each track into individual line inputs of the mixer, treating each track as if it were an original source. The balance between the tracks, and the positioning of the tracks in the stereo image, can then be carried out at leisure (within the budget constraints of the project!), often without all the musicians present, under control of the producer. During the mixdown, further post-production takes place such as the addition of effects from outboard equipment (see Chapter 14) to enhance the mix. An automation system is often used to memorise fader and mute movements on the console, since the large number of channels involved in modern recording makes it difficult if not impossible for the engineer to mix a whole song correctly in one go.

Following mixdown, the stereo master which results will be edited very basically, in order to compile titles in the correct order for the production master. The compiled tape will then be mastered for the various distribution media, as described in section 3.2.1.

3.2.3 Film sound

In commercial film production, sound is often recorded separately to picture, using a quarter-inch portable tape recorder such as the Nagra IV-S. A synchronising signal is also recorded on the sound tape, related to the speed of the film camera, to allow for the subsequent locking of sound to picture in post-production. Sound tracks from field recordings are copied on to sprocket-holed magnetic tape in the studio in order that they may be edited in the same way as pictures (see Figure 3.5). Individual sprocketed sound 'tracks' can be locked together very simply using a device known

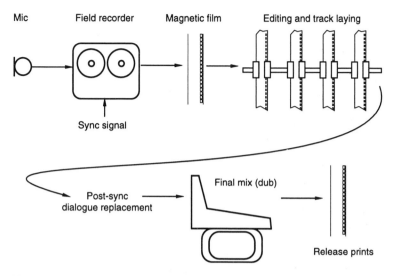

Figure 3.5 A diagrammatic representation of film sound production

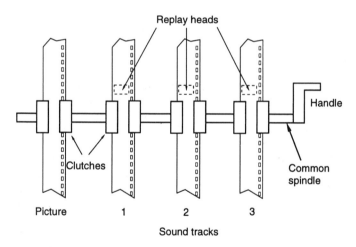

Figure 3.6 Film sound tracks are often synchronised during editing using a mechanical device such as that shown here

as a 'Steenbeck' which involves a mechanical linkage between multiple sprockets which can be declutched and slipped physically in relation to each other – thus the sound tracks can be moved relatively in time (see Figure 3.6).

In film sound editing the original sound and various added tracks are cut to match the edited picture, with blank stock added to make all the tracks the same length, and to make the timing of entries correct. Overdubbed dialogue may be added on a separate track to improve the clarity of scenes where the original sound was poor, and music and sound effects (FX) are added on further tracks. A dubbing sheet is built up showing which tracks are active at which times, and what is contained in each section. During the 'dub', the tracks are all synchronised on a number of

Time	Picture	Dialogue 1	Dialogue 2	FX1	Music
33'15"	Long shot restuarant	Margaret: 'Over here wtr.!'		Bkgnd. restrnt.	Quiet Palm Court Orch. 1920s dance music
33'24"	Close-up – Nigel		Nigel: 'Mine's a G&T, etc...'		
33'30"			Waiter: 'Just a moment sir.'		
33'33"	Nigel + Margaret		M. 'Now, about our little problem...'		

Figure 3.7 A fictitious example of a simple dubbing chart, showing picture action, two dialogue tracks, an effects track and a music track

sprocketed film transports, electrically locked together, and the resulting sound is mixed against the dubbing sheet so as to be suitable for the anticipated release format (see Figure 3.7). Often sound is produced in a number of formats depending on the destination, there being a number of different cinema sound systems, such as Academy Mono, 35 mm Dolby Stereo, and 70 mm Dolby Stereo.

More recently, digital hard-disk editing systems (see section 10.9) have been adopted for film sound dubbing, and the number of tracks available on such systems is increasing to a point where fairly complex productions are possible. The independence of tracks in such systems has close parallels with traditional film sound techniques, and thus the technology is fairly easily assimilated by film sound editors.

3.2.4 Sound for video

Professional video recorders (VTRs) have a limited number of sound tracks, and until recently the sound quality has been only average. The recent development of digital VTRs has brought with it an improvement in the number and quality of sound tracks, but it is still normal for additional tracks to be required for all but the simplest programme productions.

Original sound in a video production may be recorded directly on to the audio tracks of the VTR concerned, since there is not the tradition of separate sound recording in this field, as there is with film sound. Original sound is then edited alongside pictures, using an electronic editing system, often using a separate timecode-locked quarter-inch tape recorder or solid-state store to 'lay off' sound around edit points so that sound overlaps may be made between outgoing and incoming takes (see Figure 3.8). The edited master may then proceed to post-production for 'sweetening' or 'dubbing', during which ambience tracks, overdub-bed dialogue, effects and music may be added. Traditionally a multitrack tape machine has been used for this purpose, but increasingly hard-disk editors are used (see section 10.9) since they offer greater flexibility in track positioning and allow a library of effects to be accessed quickly and easily.

Once the dub is completed, the result is mixed and 'layed back' on to the edited video tape in synchronism with the picture, the original edited sound now being replaced by the dubbed sound which includes parts of the original.

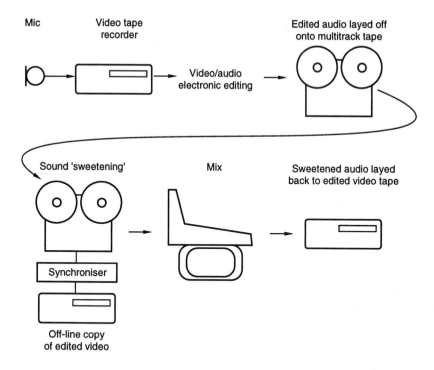

Figure 3.8 A diagrammatic representation of sound production for video

3.3 Broadcast distribution

Radio and television sound may go through any of the production processes introduced above, but will experience the additional problems of distribution around a potentially large network of studio centres and transmitter sites, requiring that it is carried over a variety of links, both analogue and digital.

Over its life, a typical television sound signal from an outside broadcast location may have travelled over a large number of miles (see Figure 3.9). A radio microphone (see section 4.10) may transmit locally to the outside broadcast vehicle, which may in turn use a microwave radio link to send the programme sound back to the studio centre. The signal may then travel through a length of internal cabling at the studio centre, finally to be connected to the network which distributes signals around the country – to other studio centres and to transmitter sites. This network may be in the form of so-called 'land lines', which are hard-wired connections, equalised to compensate for the losses encountered over the distances involved, or it may be in the form of radio-frequency links, perhaps digitally encoded. From the transmitter site the programme is distributed to the consumer via a further radio-frequency link, with the additional possibility that the signal may be beamed to a satellite for national or international transmission.

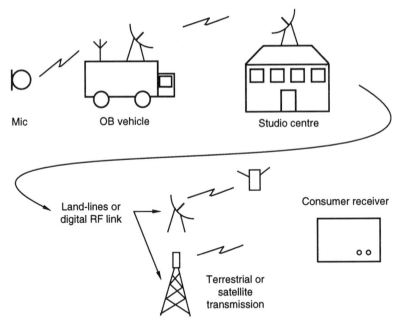

Mic   OB vehicle   Studio centre

Land-lines or   Consumer receiver
digital RF link

Terrestrial or
satellite
transmission

Figure 3.9 A diagrammatic representation of the typical path followed by a television sound signal

3.4 An introduction to signal levels

Looking at the diagram in Figure 3.10 it will be seen that an audio signal passing from a microphone through a mixer to a tape recorder and also to a loudspeaker system encounters a number of changes in overall level (signal voltage). Other potential sources with their nominal output levels are also shown to illustrate the differences involved.

The nominal output voltage of a microphone is usually quite low, depending on the microphone type (see Chapter 4), being of the order of a millivolt or so (1 mV = 0.001 V). Clearly the actual level depends on the audio signal concerned. This is often known as a 'mic level' signal, to distinguish it from a 'line level' signal which has a nominal level of nearer to 1 volt. Line level signals are normally quoted with reference to a level of 0.775 volts – given the designation 0 dBu (see Fact File 1.2). Tape recorders, effects devices, mixers and other studio electrical devices normally have line level outputs. Thus mic level signals are some 60 dB lower in voltage than line level and such signals require amplification to bring them up to line level. Nearly all mixers have microphone amplifiers built into the channels, and a variable gain control is provided to offer amplification of between around 30 and 80 dB, depending on the microphone's output level and the output level of the acoustic source it is facing.

The output of a mixer is at line level, and this is fed to a power amplifier which in turns feeds the loudspeakers. Loudspeakers require a relatively high voltage to drive them sufficiently hard to produce the level of sound required, the precise voltage depending on the efficiency of the loudspeaker and the loudness required. The level of the signal driving the loudspeaker might be, say, around 10 volts nominal.

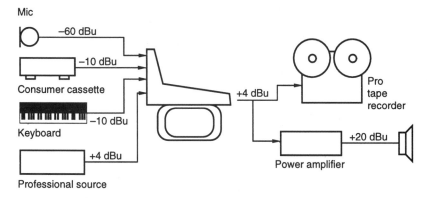

Figure 3.10 Typical signal levels encountered at stages in the recording chain

An analogue tape recorder is driven with a line level electrical signal but converts this to a magnetic signal which it records on the tape (see section 8.2). There are a number of accepted magnetic reference levels, as discussed in Fact File 8.5, and it is important to realise that there is a variable relationship between the magnetic level on tape and the electrical input level to a tape machine which depends on how the machine is aligned.

Consumer sources such as FM tuners, cassette machines, and some cheaper effects devices will have lower output levels than 0 dBu, these being nominally at around 100 mV or sometimes −10 dBV. Line level amplifiers can be obtained to bring such signals up to line level, often at the same time as balancing the unbalanced output of the consumer device (see section 13.2), or it may be possible to connect such signals directly to line level balanced inputs with the possibility for interference and slightly increased noise. Consumer hi-fi amplifiers usually operate with the lower-level input signals, and all of the inputs on such an amplifier except that set aside for the LP turntable (sometimes labelled 'RIAA' or 'Phono') will normally be at this level. Sometimes a special CD/DAT input is provided with a slightly higher nominal input level, since consumer digital equipment tends to produce a higher maximum output voltage than other consumer equipment. If connecting professional equipment outputs to consumer hi-fi equipment it is important not to overload the inputs, and again amplifiers or transformers (see section 13.1) are available to modify the levels accordingly.

LP turntables are somewhat unique, since a typical moving-magnet pickup cartridge produces a similarly low signal level as a microphone, but the signal is designed to be equalised according to the RIAA curve (see section 11.2). A filter is installed in the RIAA pre-amplifier of the system on which the record is replayed. For this reason the LP input on an amplifier should not be used for any other purpose, neither should the output of a turntable be connected to an ordinary mic level input since its frequency balance will be uncorrected.

The meter indications corresponding to different electrical levels in a professional system depend on the type of meter, the operating level of the studio, and any other conventions adhered to, and since this is a large subject it will be covered in greater detail in section 7.5.

Recommended further reading

Historical

Gelatt, R. (1977) *The Fabulous Phonograph*. Cassell and Co., London
Read, O. (1976) *From Tinfoil to Stereo*.

Audio systems and signals (general)

BS 6840. *Audio Equipment*. British Standards Office
CCIR Rec. 574–1 (1982) (re signal levels and dB suffixes). Vol. 13. Green Book

See also *General further reading* at the end of this book.

Chapter 4

Microphones

A microphone is a transducer which converts acoustical sound energy into electrical energy. It performs the opposite function to a loudspeaker which converts electrical energy into acoustical energy. The three most common principles of operation are the moving coil or 'dynamic', the ribbon, and the capacitor or condenser. The principles are described in Fact Files 4.1, 4.2 and 4.3 respectively.

4.1 The moving-coil or dynamic microphone

The moving-coil microphone is widely used in the sound reinforcement industry, its robustness making it particularly suitable for hand-held vocal use. Wire-mesh bulbous wind shields are usually fitted to such models, and contain foam material which attenuates wind noise and 'p-blasting' from the vocalist's mouth. Built-in bass attenuation is also often provided to compensate for the effect known as bass tip-up, a phenomenon whereby sound sources at a distance of less than 50 cm or so are reproduced with accentuated bass if the microphone has a directional response (see Fact File 4.4). The frequency response of the moving-coil mic tends to show a resonant peak of several decibels in the upper-mid frequency or 'presence' range, at around 5 kHz or so, accompanied by a fairly rapid fall-off in response above 8 or 10 kHz. This is due to the fact that the moving mass of the coil–diaphragm structure is sufficient to impede the diaphragm's rapid movement necessary at high frequencies. The shortcomings have actually made the moving coil a good choice for vocalists since the presence peak helps to lift the voice and improve intelligibility. Its robustness has also meant that it is almost exclusively used as a bass drum mic in the rock industry. Its sound quality is restricted by its slightly uneven and limited frequency response, but it is extremely useful in applications such as vocals, drums, and the micing-up of guitar amplifiers.

One or two high-quality moving-coil mics have appeared with an extended and somewhat smoother frequency response, and one way of achieving this has been to use what are effectively two mic capsules in one housing, one covering mid and high frequencies, one covering the bass.

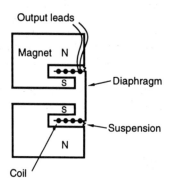

FACT FILE 4.1 Dynamic microphone – principles

The moving-coil microphone functions like a moving-coil speaker in reverse. As shown in the diagram, it consists of a rigid diaphragm, typically 20–30 mm in diameter, which is suspended in front of a magnet. A cylindrical former is attached to the diaphragm on to which is wound a coil of very fine-gauge wire. This sits in the gap of a strong permanent magnet. When the diaphragm is made to vibrate by sound waves the coil in turn moves to and fro in the magnet's gap, and an alternating current flows in the coil, producing the electrical output (see Fact File 3.1). Some models have sufficient windings on the coil to produce a high enough output to be fed directly to the output terminals, whereas other models use fewer windings, the lower output then being fed to a step-up transformer in the microphone casing and then to the output. The resonant frequency of dynamic microphone diaphragms tends to be in the middle frequency region.

The standard output impedance of professional microphones is 200 ohms. This value was chosen because it is high enough to allow useful step-up ratios to be employed in the output transformers, but low enough to allow a microphone to drive long lines of 100 metres or so. It is possible, though, to encounter dynamic microphones with output impedances between 50 and 600 ohms. Some moving-coil models have a transformer that can be wired to give a high-level, high-impedance output suitable for feeding into the lower-sensitivity inputs found on guitar amplifiers and some PA amplifiers. High-impedance outputs can, however, only be used to drive cables of a few metres in length, otherwise severe high-frequency loss results. (This is dealt with fully in Chapter 13.)

4.2 The ribbon microphone

The ribbon microphone at its best is capable of very high-quality results. The comparatively 'floppy' suspension of the ribbon gives it a low-frequency resonance at around 40 Hz, below which its frequency response fairly quickly falls away. At the high-frequency end the frequency response remains smooth. However, the moving mass of the ribbon itself means that it has difficulty in responding to very high frequencies, and there is generally a roll-off above 14 kHz or so. Reducing the size (and therefore the mass) of the ribbon reduces the area for the sound waves to work upon and its electrical output becomes unacceptably low. One manufacturer has adopted a 'double-ribbon' principle which goes some way towards removing this dilemma. Two ribbons, each half the length of a conventional ribbon, are mounted one above the other and are connected in series. They are thus analogous to a conventional ribbon that has been 'clamped' in the centre. Each ribbon now has half the moving mass and thus a better top-end response. Both of them working together still maintain the necessary output.

The ribbon mic is rather more delicate than the moving coil, and it is better suited to applications where its smooth frequency response comes into its own, such as the micing of acoustic instruments and classical ensembles. There are, however, some

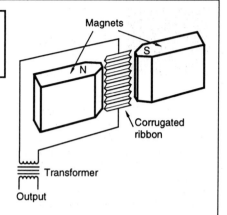

FACT FILE

4.2

Ribbon microphone – principles

The ribbon microphone consists of a long thin strip of conductive metal foil, pleated to give it rigidity and 'spring', lightly tensioned between two end clamps, as shown in the diagram. The opposing magnetic poles create a magnetic field across the ribbon such that when it is excited by sound waves a current is induced into it (see Fact File 3.1). The electrical output of the ribbon is very small, and a transformer is built into the microphone which steps up the output. The step-up ratio of a particular ribbon design is chosen so that the resulting output impedance is the standard 200 ohms, this also giving an electrical output level comparable with that of moving-coil microphones. The resonant frequency of ribbon microphones is normally at the bottom of the audio spectrum.

robust models which look like moving-coil vocal mics and can be interchanged with them. Micing a rock bass drum with one is still probably not a good idea, due to the very high transient sound pressure levels involved.

4.3 The capacitor or condenser microphone

4.3.1 Basic capacitor microphone

The great advantage of the capacitor mic's diaphragm over moving-coil and ribbon types is that it is not attached to a coil and former, and it does not need to be of a shape and size which makes it suitable for positioning along the length of a magnetic field. It therefore consists of an extremely light disk, typically 12–25 mm in diameter, frequently made from polyester coated with an extremely thin vapour-deposited metal layer so as to render it conductive. Sometimes the diaphragm itself is made of a metal such as titanium. The resonant frequency of the diaphragm is typically in the 12–20 kHz range, but the increased output here is rather less prominent than with moving coils due to the diaphragm's very light weight.

Occasionally capacitor microphones are capable of being switched to give a line level output, this being simple to arrange since an amplifier is built into the mic anyway. The high-level output gives the signal rather more immunity to interference when very long cables are employed, and it also removes the need for microphone amplifiers at the mixer or tape recorder. Phantom power does, however, still need to be provided (see section 4.9.1).

4.3.2 Electret designs

A much later development was the so-called 'electret' or 'electret condenser' principle. The need to polarise the diaphragm with 48 volts is dispensed with by

FACT FILE 4.3   Capacitor microphone – principles

The capacitor (or condenser) microphone operates on the principle that if one plate of a capacitor is free to move with respect to the other, then the capacitance (the ability to hold electrical charge) will vary. As shown in the diagram, the capacitor consists of a flexible diaphragm and a rigid back plate, separated by an insulator, the diaphragm being free to move in sympathy with sound waves incident upon it. The 48 volts DC phantom power (see section 4.9) charges the capacitor via a very high resistance. A DC blocking capacitor simply prevents the phantom power from entering the head amplifier, allowing only audio signals to pass.

When sound waves move the diaphragm the capacitance varies, and thus the voltage across the capacitor varies proportionally, since the high resistance only allows very slow leakage of charge from the diaphragm (much slower than the rate of change caused by audio frequencies). This voltage modulation is fed to the head amplifier (via the blocking capacitor) which converts the very high impedance output of the capacitor

capsule to a much lower impedance. The output transformer balances this signal (see section 13.4) and conveys it to the microphone's output terminals. The resonant frequency of a capacitor mic diaphragm is normally at the upper end of the audio spectrum.

The head amplifier consists of a field-effect transistor (FET) which has an almost infinitely high input impedance. Other electronic components are also usually present which perform tasks such as voltage regulation and output stage duties. Earlier capacitor microphones had valves built into the housing, and were somewhat more bulky affairs than their modern counterparts. Additionally, extra wiring had to be incorporated in the mic leads to supply the valves with HT (high-tension) and valve-heater voltages. They were thus not particularly convenient to use, but such is the quality of sound available from capacitor mics that they quickly established themselves. Today, the capacitor microphone is the standard top-quality type, other types being used for relatively specialised applications. The electrical current requirement of capacitor microphones varies from model to model, but generally lies between 0.5 mA and 8 mA, drawn from the phantom power supply.

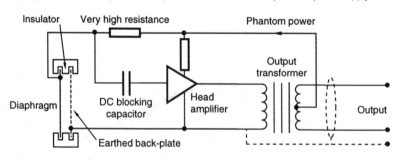

introducing a permanent electrostatic charge into it during manufacture. In order to achieve this the diaphragm has to be of a more substantial mass, and its audio performance is therefore closer to a moving-coil than to a true capacitor type. The power for the head amplifier is supplied either by a small dry-cell battery in the stem of the mic or by phantom power. The electret principle is particularly suited to applications where compact size and light weight are important, such as in small portable cassette machines (all built-in mics are now electrets) and tie-clip microphones which are ubiquitous in television work. They are also made in vast quantities very cheaply.

Later on, the so-called 'back electret' technique was developed. Here, the diaphragm is the same as that of a true capacitor type, the electrostatic charge being induced into the rigid back plate instead. Top-quality examples of back electrets are therefore just as good as conventional capacitor mics with their 48 volts of polarising voltage.

4.3.3 RF capacitor microphone

Still another variation on the theme is the RF (Radio Frequency) capacitor mic, in which the capacitor formed by the diaphragm and back plate forms part of a tuned circuit to generate a steady carrier frequency which is much higher than the highest audio frequency. The sound waves move the diaphragm as before, and this now causes modulation of the tuned frequency. This is then demodulated by a process similar to the process of FM radio reception, and the resulting output is the required audio signal. (It must be understood that the complete process is carried out within the housing of the microphone and it does not in itself have anything to do with radio microphone systems, as discussed in section 4.10.)

4.4 Directional responses and polar diagrams

Microphones are designed to have a specific directional response pattern, described by a so-called 'polar diagram'. The polar diagram is a form of two-dimensional contour map, showing the magnitude of the microphone's output at different angles of incidence of a sound wave. The distance of the polar plot from the centre of the graph (considered as the position of the microphone diaphragm) is usually calibrated in decibels, with a nominal 0 dB being marked for the response at zero degrees at 1 kHz. The further the plot is from the centre, the greater the output of the microphone at that angle.

4.4.1 Omnidirectional pattern

Ideally, an omnidirectional or 'omni' microphone picks up sound equally from all directions. The omni polar response is shown in Figure 4.1, and is achieved by leaving the microphone diaphragm open at the front, but completely enclosing it at the rear, so that it becomes a simple pressure transducer, responding only to the change of air pressure caused by the sound waves. This works extremely well at low and mid frequencies, but at high frequencies the dimensions of the microphone capsule itself begin to be comparable with the wavelength of the sound waves, and a shadowing effect causes high frequencies to be picked up rather less well to the rear and sides of the mic. A pressure increase also results for high-frequency sounds from the front. Coupled with this is the possibility for cancellations to arise when a high-frequency wave, whose wavelength is comparable with the diaphragm diameter, is incident from the side of the diaphragm. In such a case positive and negative peaks of the wave may result in opposing forces on the diaphragm.

Figure 4.2 shows the polar response plot which can be expected from a real omnidirectional microphone with a capsule half an inch (13 mm) in diameter. It is perfectly omnidirectional up to around 2 kHz, but then it begins to lose sensitivity at the rear; at 3 kHz its sensitivity at 180° will typically be 6 dB down compared with

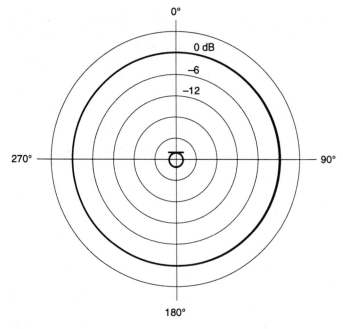

Figure 4.1 Idealised polar diagram of an omnidirectional microphone

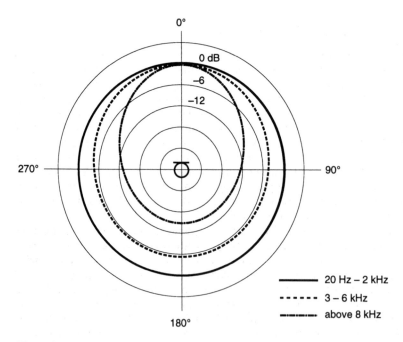

Figure 4.2 Typical polar diagram of an omnidirectional microphone at a number of frequencies

lower frequencies. Above 8 kHz, the 180° response could be as much as 15 dB down, and the response at 90° and 270° could show perhaps a 10 dB loss. As a consequence, sounds which are being picked up significantly off axis from the microphone will be reproduced with considerable treble loss, and will sound dull. It is at its best on axis and up to 45° either side of the front of the microphone.

High-quality omnidirectional microphones are characterised by their wide, smooth frequency response extending both to the lowest bass frequencies and the high treble with minimum resonances or coloration. This is due to the fact that they are basically very simple in design, being just a capsule which is open at the front and completely enclosed at the rear. (In fact a very small opening is provided to the rear of the diaphragm in order to compensate for overall changes in atmospheric pressure which would otherwise distort the diaphragm.) The small tie-clip microphones which one sees in television work are usually omnidirectional electret types which are capable of very good performance. The smaller the dimensions of the mic, the better the polar response at high frequencies, and mics such as these have quarter-inch diaphragms which maintain a very good omnidirectional response right up to 10 kHz.

Omni microphones are usually the most immune to handling and wind noise of all the polar patterns, since they are only sensitive to absolute sound pressure. Patterns such as figure-eight (especially ribbons) and cardioid, described below, are much more susceptible to handling and wind noise than omnis because they are sensitive to the large pressure *difference* created across the capsule by low-frequency movements such as those caused by wind or unwanted diaphragm motion. A pressure-gradient microphone's mechanical impedance (the diaphragm's resistance to motion) is always lower at LF than that of a pressure (omni) microphone, and thus it is more susceptible to unwanted LF disturbances.

4.4.2 Figure-eight or bidirectional pattern

The figure-eight or bidirectional polar response is shown in Figure 4.3. Such a microphone has an output proportional to the mathematical cosine of the angle of incidence (represented by the angle ϑ in the diagram). One can quickly draw a figure-eight plot on a piece of graph paper, using a protractor and a set of cosine tables or pocket calculator. Cos 0° = 1, showing a maximum response on the forward axis (this will be termed the 0 dB reference point). Cos 90° = 0, so at 90° off axis no sound is picked up. Cos 180° is −1, so the output produced by a sound which is picked up by the rear lobe of the microphone will be 180° out of phase compared with an identical sound picked up by the front lobe. The phase is indicated by the + and − signs on the polar diagram. At 45° off axis, the output of the microphone is 3 dB down (cos 45° represents 0.707 or $1/\sqrt{2}$ times the maximum output) compared with the on-axis output.

Traditionally the ribbon microphone has sported a figure-eight polar response, and the ribbon has been left completely open both to the front and to the rear. Such a diaphragm operates on the pressure-gradient principle, responding to the *difference* in pressure between the front and the rear of the microphone. Consider a sound reaching the mic from a direction 90° off axis to it. The sound pressure will be of equal magnitude on both sides of the diaphragm and so no movement will take place, giving no output. When a sound arrives from the 0° direction a phase difference arises between the front and rear of the ribbon, due to the small additional distance travelled by the wave. The resulting difference in pressure produces movement of

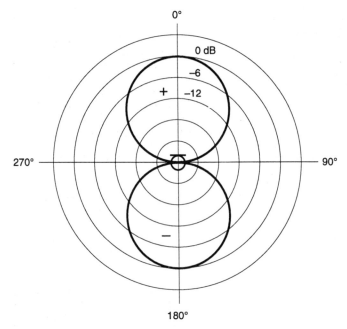

Figure 4.3 Idealised polar diagram of a figure-eight microphone

the diaphragm and an output results.

At very low frequencies, wavelengths are very long and therefore the phase difference between front and rear of the mic is very small, causing a gradual reduction in output as the frequency gets lower. In ribbon microphones this is compensated for by putting the low-frequency resonance of the ribbon to good use, using it to prop up the bass response. Single-diaphragm capacitor mic designs which have a figure-eight polar response do not have this option, since the diaphragm resonance is at a very high frequency, and a gradual roll-off in the bass can be expected unless other means such as electronic frequency correction in the microphone design have been employed. Double-diaphragm switchable types which have a figure-eight capability achieve this by combining a pair of back-to-back cardioids (see section 4.4.3) that are mutually out of phase.

Like the omni, the figure-eight can give very clear uncoloured reproduction. The polar response tends to be very uniform at all frequencies, except for a slight narrowing above 10 kHz or so, but it is worth noting that a ribbon mic has a rather better polar response at high frequencies in the horizontal plane than in the vertical plane, due to the fact that the ribbon is long and thin. A high-frequency sound coming from a direction somewhat above the plane of the microphone will suffer partial cancellation, since at frequencies where the wavelength begins to be comparable with the length of the ribbon the wave arrives partially out of phase at the lower portion compared with the upper portion, therefore reducing the effective acoustical drive of the ribbon compared with mid frequencies. Ribbon figure-eight microphones should therefore be orientated either upright or upside-down with their stems vertical so as to obtain the best polar response in the horizontal plane, vertical polar response usually being less important.

FACT FILE

4.4

Bass tip-up

Pressure-gradient microphones are susceptible to a phenomenon known as bass tip-up, meaning that if a sound source is close to the mic (less than about a metre) the low frequencies become unnaturally exaggerated. In normal operation, the driving force on a pressure-gradient microphone is related almost totally to the phase difference of the sound wave between front and rear of the diaphragm (caused by the extra distance travelled by the wave). For a fixed path-length difference between front and rear, therefore, the phase difference increases with frequency. At LF the phase difference is small and at MF to HF it is larger.

Close to a small source, where the microphone is in a field of roughly spherical waves, sound pressure drops as distance from the source increases (see Fact File 1.3). Thus, in addition to the phase difference between front and rear of the mic's diaphragm, there is a pressure difference due to the natural level-drop with distance from the source. Since the driving force on the diaphragm due to phase difference is small at LF, this pressure drop makes a significant additional contribution, increasing the overall output level at LF. At HF the phase difference is larger, and thus the contribution made by pressure difference is smaller as a proportion of the total driving force.

At greater distances from the source, the sound field approximates more closely to one of plane waves, and the pressure drop over the front–back distance may be considered insignificant as a driving force on the diaphragm, making the mic's output related only to front–back phase difference.

Although the figure-eight picks up sound equally to the front and to the rear, it must be remembered that the rear pickup is out of phase with the front, and so correct orientation of the mic is required.

4.4.3 Cardioid or unidirectional pattern

The cardioid pattern is described mathematically as $1 + \cos\vartheta$, where ϑ is the angle of incidence of the sound. Since the omni has a response of 1 (equal all round) and the figure-eight has a response represented by $\cos\vartheta$, the cardioid may be considered theoretically as a product of these two responses. Figure 4.4(a) illustrates its shape. Figure 4.4(b) shows an omni and a figure-eight superimposed, and one can see that adding the two produces the cardioid shape: at 0°, both polar responses are of equal amplitude and phase, and so they reinforce each other, giving a total output which is actually twice that of either separately. At 180°, however, the two are of equal amplitude but opposite phase, and so complete cancellation occurs and there is no output. At 90° there is no output from the figure-eight, but just the contribution from the omni, so the cardioid response is 6 dB down at 90°. It is 3 dB down at 65° off axis.

One or two early microphone designs actually housed a figure-eight and an omni together in the same casing, electrically combining their outputs to give a resulting cardioid response. This gave a rather bulky mic, and also the two diaphragms could not be placed close enough together to produce a good cardioid response at higher frequencies due to the fact that at these frequencies the wavelength of sound became comparable with the distance between the diaphragms. The designs did, however, obtain a cardioid from first principles.

The cardioid response is now obtained by leaving the diaphragm open at the

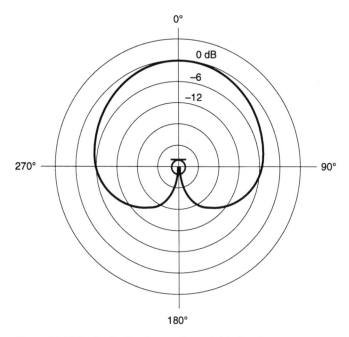

Figure 4.4 (a) Idealised polar diagram of a cardioid microphone

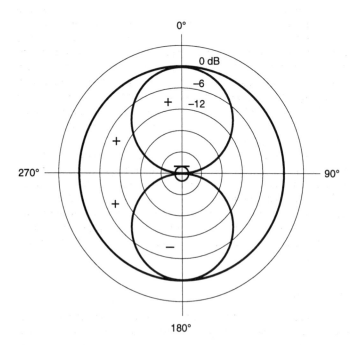

Figure 4.4 (b) A cardioid microphone can be seen to be the mathematical equivalent of an omni and a figure-eight response added together

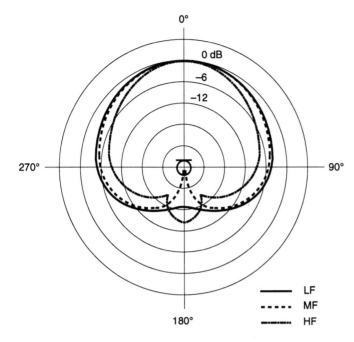

0°

0 dB
−6
−12

270°

90°

180°

LF
MF
HF

Figure 4.5 Typical polar diagram of a cardioid microphone at low, middle and high frequencies

front, but introducing various acoustic labyrinths at the rear which cause sound to reach the back of the diaphragm in various combinations of phase and amplitude to produce a resultant cardioid response. This is difficult to achieve at all frequencies simultaneously, and Figure 4.5 illustrates the polar pattern of a typical cardioid mic with a three-quarter-inch diaphragm. As can be seen, at mid frequencies the polar response is very good. At low frequencies it tends to degenerate towards omni, and at very high frequencies it becomes rather more directional than is desirable. Sound arriving from, say, 45° off axis will be reproduced with treble loss, and sounds arriving from the rear will not be completely attenuated, the low frequencies being picked up quite uniformly.

The above example is very typical of moving-coil cardioids, and they are in fact very useful for vocalists due to the narrow pickup at high frequencies helping to exclude off-axis sounds, and also the relative lack of pressure-gradient component at the bass end helping to combat bass tip-up. High-quality capacitor cardioids with half-inch diaphragms achieve a rather more ideal cardioid response. Owing to the presence of acoustic labyrinths, coloration of the sound is rather more likely, and it is not unusual to find that a relatively cheap electret omni will sound better than a fairly expensive cardioid.

4.4.4 Hypercardioid pattern

The hypercardioid, sometimes called 'cottage loaf' because of its shape, is shown in Figure 4.6. It is described mathematically by the formula $0.5 + \cos\vartheta$, i.e.: it is a combination of an omni attenuated by 6 dB, and a figure-eight. Its response is in between the cardioid and figure-eight patterns, having a relatively small rear lobe

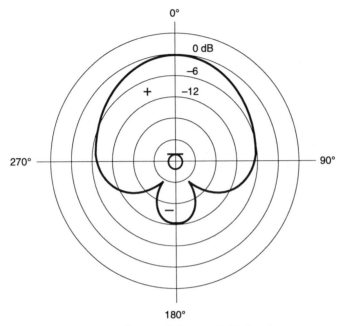

Figure 4.6 Idealised polar diagram of a hypercardioid microphone

which is out of phase with the front lobe. Its sensitivity is 3 dB down at 55° off axis. Like the cardioid, the polar response is obtained by introducing acoustic labyrinths to the rear of the diaphragm. Because of the large pressure-gradient component it too is fairly susceptible to bass tip-up. Practical examples of hypercardioid microphones tend to have polar responses which are tolerably close to the ideal. The hypercardioid has the highest direct-to-reverberant ratio of the patterns described, which means that the ratio between the level of on-axis sound and the level of reflected sounds picked up from other angles is very high, and so it is good for excluding unwanted sounds such as excessive room ambience or unwanted noise.

4.5 Specialised microphone types

4.5.1 Rifle microphone

The rifle microphone is so called because it consists of a long tube of around three-quarters of an inch (1.9 cm) in diameter and perhaps 2 feet (61 cm) in length, and looks rather like a rifle barrel. The design is effectively an ordinary cardioid microphone to which has been attached a long barrel along which slots are cut in such a way that a sound arriving off axis enters the slots along the length of the tube and thus various versions of the sound arrive at the diaphragm at the bottom of the tube in relative phases which tend to result in cancellation. In this way, sounds arriving off axis are greatly attenuated compared with sounds arriving on axis. Figure 4.7 illustrates the characteristic club-shaped polar response. It is an extremely directional device, and is much used by news sound crews where it can be pointed directly at a speaking subject, excluding crowd noise. It is also used for

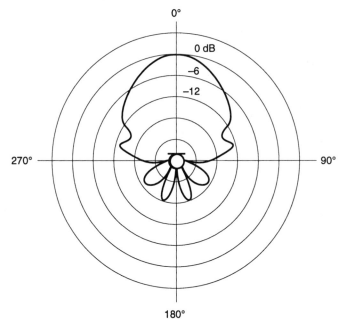

Figure 4.7 Typical polar diagram of a highly-directional microphone

wildlife recording, sports broadcasts, along the front of theatre stages in multiples, and in audience participation discussions where a particular speaker can be picked out. For outside use it is normally completely enclosed in a long, fat wind shield, looking like a very big cigar. Half-length versions are also available which have a polar response mid-way between a club shape and a hypercardioid. All versions, however, tend to have a rather wider pickup at low frequencies.

4.5.2 Parabolic microphone

An alternative method of achieving high directionality is to use a parabolic dish, as shown in Figure 4.8. The dish has a diameter usually of between 0.5 and 1 metre, and a directional microphone is positioned at its focal point. A large 'catchment area' is therefore created in which the sound is concentrated at the head of the mic. An overall gain of around 15 dB is typical, but at the lower frequencies where the wavelength of sound becomes comparable with the diameter of the dish the response falls away. Because this device actually concentrates the sound rather than merely rejecting off-axis sounds, comparatively high outputs are achieved from distant sound sources. They are very useful for capturing bird song, and they are also sometimes employed around the boundaries of cricket pitches. They are, however, rather cumbersome in a crowd, and can also produce a rather coloured sound.

4.5.3 Boundary or 'pressure-zone' microphone

The so-called boundary or pressure-zone microphone (PZM) consists basically of an omnidirectional microphone capsule mounted on a plate usually of around 6 inches (15 cm) square or 6 inches in diameter such that the capsule points directly

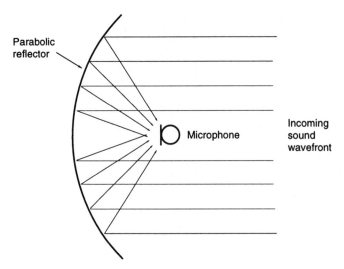

Figure 4.8 A parabolic reflector is sometimes used to 'focus' the incoming sound wavefront at the microphone position, thus making it highly directional

at the plate and is around 2 or 3 millimetres away from it. The plate is intended to be placed on a large flat surface such as a wall or floor, and it can also be placed on the underside of a piano lid for instance. Its polar response is hemispherical. Because the mic capsule is a simple omni, quite good-sounding versions are available with electret capsules fairly cheaply, and so if one wishes to experiment with this unusual type of microphone one can do so without parting with a great deal of money. It is important to remember though that despite its looks it is not a contact mic – the plate itself does not transduce surface vibrations – and it should be used with the awareness that it is equivalent to an ordinary omnidirectional microphone pointing at a flat surface, very close to it. The frequency response of such a microphone is rarely as flat as that of an ordinary omni, but it can be unobtrusive in use.

4.6 Switchable polar patterns

The double-diaphragm capacitor microphone, such as the commercial example shown in Figure 4.9, is a microphone in which two identical diaphragms are employed, placed each side of a central rigid plate in the manner of a sandwich. Perforations in the central plate give both diaphragms an essentially cardioid response. When the polarising voltage on both diaphragms is the same, the electrically combined output gives an omnidirectional response due to the combi-nation of the back-to-back cardioids in phase. When the polarising voltage of one diaphragm is opposite to that of the other, and the potential of the rigid central plate is mid-way between the two, the combined output gives a figure-eight response (back-to-back cardioids mutually out of phase). Intermediate combinations give cardioid and hypercardioid polar responses. In this way the microphone is given a switchable polar response which can be adjusted either by switches on the micro-phone itself or via a remote control box. Some microphones with switchable polar patterns achieve this by employing a conventional single diaphragm around which

Figure 4.9 A typical double-diaphragm condenser microphone with switchable polar pattern: the AKG C414B-ULS. (Courtesy of AKG Acoustics Ltd.)

is placed appropriate mechanical labyrinths which can be switched to give the various patterns.

Another method manufacturers have used is to make the capsule housing on the end of the microphone detachable, so that a cardioid capsule, say, can be unscrewed and removed to be replaced with, say, an omni. This also facilitates the use of extension tubes whereby a long thin pipe of around a metre or so in length with suitably threaded terminations is inserted between the main microphone body and the capsule. The body of the microphone is mounted on a short floor stand and the thin tube now brings the capsule up to the required height, giving a visually unobtrusive form of microphone stand.

4.7 Stereo microphones

Stereo microphones, such as the example shown in Figure 4.10, are available in which two microphones are built into a single casing, one capsule being rotatable with respect to the other so that the angle between the two can be adjusted. Also, each capsule can be switched to give any desired polar response. One can therefore adjust the mic to give a pair of figure-eight microphones angled at, say, 90°, or a pair of cardioids at 120°, and so on. Some stereo mics, such as that pictured in Figure 4.11, are configured in a sum-and-difference arrangement, instead of as a left–right pair, with a 'sum' capsule pointing forwards and a figure-eight 'difference' capsule facing sideways. The sum-and-difference or 'middle and side' (M and S) signals are combined in a matrix box to produce a left–right stereo signal by adding M and S to give the left channel and subtracting M and S to give the right channel. This is discussed in more detail in Fact File 4.5.

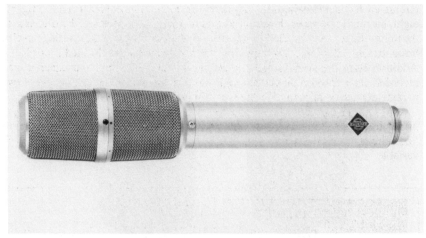

Figure 4.10 A typical stereo microphone: the Neumann SM69. (Courtesy of FWO Bauch Ltd.)

Figure 4.11 A typical 'sum-and-difference' stereo microphone: the Shure VP88. (Courtesy of HW International)

A sophisticated stereo microphone is the AMS/Calrec Soundfield microphone, pictured in Figure 4.12. In this design, four 'subcardioid' capsules (i.e.: between omni and cardioid) are arranged in a tetrahedral array such that their outputs can be combined in various ways to give four outputs, termed 'B format'. The raw output from the four capsules is termed 'A format'. The four B-format signals consist of a forward-facing figure-eight ('X'), a sideways-facing figure-eight ('Y'), an up-and-down-facing figure-eight ('Z'), and an omnidirectional output ('W'). These are then appropriately combined to produce any configuration of stereo microphone output, each channel being fully adjustable from omni through cardioid to figure-eight, the angles between the capsules also being fully adjustable. The tilt angle of the microphone, and also the 'dominance' (the front-to-back pickup ratio) can also be controlled. All of this is achieved electronically by a remotely sited control unit. Additionally, the raw B-format signals can be recorded on a four-channel tape recorder, later to be replayed through the control unit where all of the above parameters can be chosen after the recording session.

AMS has also produced a second-generation microphone based on Soundfield principles, the ST250, designed to be smaller and to be usable either 'end-fire' or 'side-fire' (see Figure 4.13). It can be electronically inverted and the pattern is variable.

FACT FILE
4.5 **Sum and difference processing**

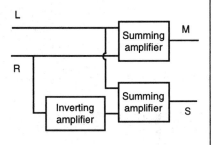

MS signals may be converted to conventional stereo very easily, either using three channels on a mixer, or using an electrical matrix. M is the mono sum of two conventional stereo channels, and S is the difference between them. Thus:

$$M = (L + R) \div 2$$
$$S = (L - R) \div 2$$

and

$$L = (M + S) \div 2$$
$$R = (M - S) \div 2$$

A pair of transformers may be used wired as shown in the diagram to obtain either MS from LR, or vice versa. Alternatively, a pair of summing amplifiers may be used, with the M and S (or L and R) inputs to one being wired

in phase (so that they add) and to the other out of phase (so that they subtract).

The mixer configuration shown in the diagram may also be used. Here the M signal is panned centrally (feeding L and R outputs), whilst the S signal is panned left (M + S=L). A post-fader insertion feed is taken from the S channel to a third channel which is phase reversed to give –S. The gain of this channel is set at 0 dB and is panned right (M – S=R). If the S fader is varied in level, the width of the stereo image and the amount of rear pickup can be varied.

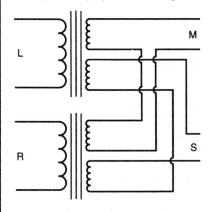

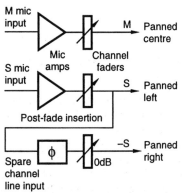

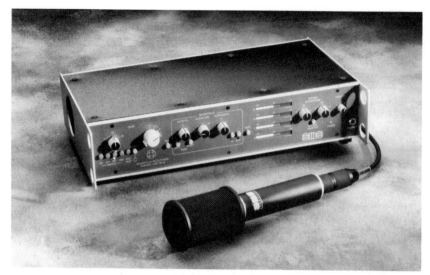

Figure 4.12 The AMS Soundfield microphone and control unit. (Courtesy of AMS Industries PLC)

Figure 4.13 The AMS ST250 microphone is based on soundfield principles, and can be operated either end- or side-fire, or upside-down, using electrical matrixing of the capsules within the control unit

4.8 Microphone performance

Professional microphones have a balanced low-impedance output usually via a three-pin XLR-type plug in their base. The impedance, which is usually around 200 ohms but sometimes rather lower, enables long microphone leads to be used. Also, the balanced configuration, discussed in section 13.4, gives considerable immunity from interference. Other parameters which must be considered are sensitivity (see Fact File 4.6) and noise (see Fact File 4.7).

4.8.1 Microphone sensitivity in practice

The consequence of mics having different sensitivity values is that rather more amplification is needed to bring ribbons and moving coils up to line level than is the case with capacitors. For example, speech may yield, say, 0.15 mV from a ribbon.

FACT FILE 4.6 — Microphone sensitivity

The sensitivity of a microphone is an indication of the electrical output which will be obtained for a given acoustical sound pressure level (SPL). The standard SPL is either 74 dB (= 1 μB) or 94 dB (=1 Pascal or 10 μB) (μB = microbar). One level is simply ten times greater than the other, so it is easy to make comparisons between differently specified models. 74 dB is roughly the level of moderately loud speech at a distance of 1 metre. 94 dB is 20 dB or ten times higher than this, so a microphone yielding 1 mV μB^{-1} will yield 10 mV in a sound field of 94 dB. Other ways of specifying sensitivity include expressing the output as being so many decibels below a certain voltage for a specified SPL. For example, a capacitor mic may have a sensitivity figure of −60 dBV Pa^{-1} meaning that its output level is 60 dB below 1 volt for a 94 dB SPL, which is 1 mV (60 dB = times 1000).

Capacitor microphones are the most sensitive types, giving values in the region of 5–15 mV Pa^{-1}, i.e.: a sound pressure level of 94 dB will give between 5 and 15 millivolts of electrical output. The least sensitive microphones are ribbons, having typical sensitivities of 1–2 mV Pa^{-1}, i.e.: around 15 or 20 dB lower than capacitor types. Moving coils are generally a little more sensitive than ribbons, values being typically 1.5–3 mV Pa^{-1}.

FACT FILE 4.7 — Microphone noise specifications

All microphones inherently generate some noise. The common way of expressing capacitor microphone noise is the 'A'-weighted equivalent self-noise. A typical value of 'A'-weighted self-noise of a high-quality capacitor microphone is around 18 dBA. This means that its output noise voltage is equivalent to the microphone being placed in a sound field with a loudness of 18 dBA. A self-noise in the region of 25 dBA from a microphone is rather poor, and if it were to be used to record speech from a distance of a couple of metres or so the hiss would be noticeable on the recording. The very best capacitor microphones achieve self-noise values of around 12 dBA.

When comparing specifications one must make sure that the noise specification is being given in the same units. Some manufacturers give a variety of figures, all taken using different weighting systems and test meter characteristics, but the 'A'-weighted self-noise discussed will normally be present among them. Also, a signal-to-noise ratio is frequently quoted for a 94 dB reference SPL, being 94 minus the self-noise, so a mic with a self-noise of 18 dBA will have a signal-to-noise ratio of 76 dBA for a 94 dB SPL, which is also a very common way of specifying noise.

To amplify this up to line level (775 mV) requires a gain of around ×5160 or 74 dB. This is a lot, and it taxes the noise performance of the equipment and will also cause considerable amplification of any interference that manages to get into the microphone cables.

Consider now the same speech recording, made using a capacitor microphone of 1 mV μB^{-1} sensitivity. Now only ×775 or 57 dB of gain is needed to bring this up to line level, which means that any interference will have a rather better chance of being unnoticed, and also the noise performance of the mixer will not be so severely taxed. This does not mean that high-output capacitor microphones should always be used, but it illustrates that high-quality mixers and microphone cabling are required to get the best out of low-output mics.

4.8.2 Microphone noise in practice

The noise coming from a capacitor microphone is mainly caused by the head amplifier. Since ribbons and moving coils are purely passive devices one might think that they would therefore be noiseless. This is not the case, since a 200 ohm passive resistance at room temperature generates a noise output between 20 Hz and 20 kHz of 0.26 µV (µV = microvolts). Noise in passive microphones is thus due to thermal excitation of the charge carriers in the microphone ribbon or voice coil, and the output transformer windings. To see what this means in equivalent self-noise terms so that ribbons and moving coils can be compared with capacitors, one must relate this to sensitivity.

Take a moving coil with a sensitivity of 0.2 mV µB^{-1}, which is 2 mV for 94 dB SPL. The noise is 0.26 µV or 0.000 26 mV. The signal-to-noise ratio is given by dividing the sensitivity by the noise:

$$2 \div 0.000\ 26 \approx 7600$$

and then expressing this in decibels:

$$dB = 20 \log 7600 = 77\ dB$$

This is an *unweighted* figure, and 'A' weighting will usually improve it by a couple of decibels. However, the microphone amplifier into which the mic needs to be plugged will add a bit of noise, so it is a good idea to leave this figure as it is to give a fairly good comparison with the capacitor example. (Because the output level of capacitor mics is so much higher than that of moving coils, the noise of a mixer's microphone amplifier does not figure in the noise discussion as far as these are concerned. The noise generated by a capacitor mic is far higher than noise generated by good microphone amplifiers and other types of microphone.)

A 200 ohm moving-coil mic with a sensitivity of 0.2 mV µB^{-1} thus has a signal-to-noise ratio of about 77 dB, and therefore an equivalent self-noise of 94−77 = 17 dB which is comparable with high-quality capacitor types, providing that high-quality microphone amplifiers are also used. A low-output 200 ohm ribbon microphone could have a sensitivity of 0.1 mV µB^{-1}, i.e.: 6 dB less than the above moving-coil example. Because its 200 ohm thermal noise is roughly the same, its equivalent self noise is therefore 6 dB worse, i.e.: 23 dB. This would probably be just acceptable for recording speech and classical music if an ultra-low-noise microphone amplifier were to be used which did not add significantly to this figure.

The discussion of a few decibels here and there may seem a bit pedantic, but in fact self-noises in the low twenties are just on the borderline of being acceptable if one wishes to record speech or the quieter types of classical music. Loud music, and mic positions close to the sound sources such as is the practice with rock music, generate rather higher outputs from the microphones and here noise is rarely a problem. But the high output levels generated by close micing of drums, guitar amps and the like can lead to overload in the microphone amplifiers. For example, if a high-output capacitor microphone is used to pick up a guitarist's amplifier, outputs as high as 150 mV or more can be generated. This would overload some fixed-gain microphone input stages, and an in-line attenuator which reduces the level by an appropriate amount such as 10–20 dB would have to be inserted at the mixer or tape recorder end of the microphone line. Attenuators are available built into a short cylindrical tube which carries an XLR-type plug at one end and a socket at the other end. It is simply inserted between the mixer or tape recorder input and the mic lead connector. It should not be connected at the microphone end because it is best to leave the level of signal along the length of the mic lead high to give it greater immunity from interference.

4.9 Microphone powering options

4.9.1 Phantom power

Consideration of capacitor microphones reveals the need for supplying power to the electronics which are built into the casing, and also the need for a polarising voltage across the diaphragm of many capacitor types. It would obviously be inconvenient and potentially troublesome to incorporate extra wires in the microphone cable to supply this power, and so an ingenious method was devised whereby the existing wires in the cable which carry the audio signal could also be used to carry the DC voltage necessary for the operation of capacitor mics – hence the term 'phantom power', since it is invisibly carried over the audio wires. Furthermore, this system does not preclude the connection of a microphone not requiring power to a powered circuit. The principle is outlined in Fact File 4.8.

FACT FILE

4.8

Phantom powering

The diagram below illustrates the principle of phantom powering. Arrows indicate the path of the phantom power current. (Refer to section 13.4 for details of the balanced line system.) Here 48 volts DC is supplied to the capacitor microphone as follows: the voltage is applied to each of the audio lines in the microphone cable via two equal value resistors, 6800 (6k8) ohms being the standard value. The current then travels along both audio lines and into the microphone. The microphone's output transformer secondary has either a 'centre tap' – that is, a wire connected half-way along the transformer winding, as shown in the diagram – or two resistors as in the arrangement shown at the other end of the line. The current thus travels towards the centre of the winding from each end, and then via the centre tap to the electronic circuit and diaphragm of the microphone. To complete the circuit, the return path for the current is provided by the screening braid of the microphone cable.

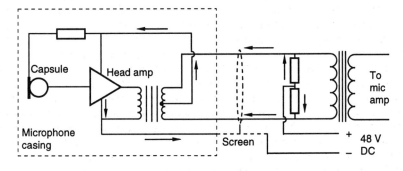

It will be appreciated that if, for instance, a ribbon microphone is connected to the line in place of a capacitor mic, no current will flow into the microphone because there will be no centre tap provided on the microphone's output transformer. Therefore, it is perfectly safe to connect other types of balanced microphone to this line. The two 6k8 resistors are necessary for the system because if they were replaced simply by two wires directly connected to the audio lines, these wires would short-circuit the lines together and so no audio signal would be able to pass. The phantom power could be applied to a centre tap of the input transformer, but if a short circuit were to develop along the cabling between one of the audio wires and the screen, potentially large currents could be drawn through the transformer windings and the phantom power supply, blowing fuses or burning out components. Two 6k8 resistors limit the current to around 14 mA, which should not cause serious problems. The 6k8 value was chosen so as to be high enough not to load the microphone unduly, but low enough for there to be only a small DC voltage drop across them so that the microphone still receives nearly the full 48 volts. Two real-life examples will be chosen to investigate exactly how much voltage drop occurs due to the resistors.

Firstly, the current flows through both resistors equally and so the resistors are effectively 'in parallel'. Two equal-value resistors in parallel behave like a single resistor of half the value, so the two 6k8 resistors can be regarded as a single 3k4 resistor as far as the 48 V phantom power is concerned. Ohm's law (see Fact File 1.1) states that the voltage drop across a resistor is equal to its resistance multiplied by the current passing through it. Now a Calrec 1050C microphone draws 0.5 milliamps (= 0.0005 amps) through the resistors, so the voltage drop is $3400 \times 0.0005 = 1.7$ volts. Therefore the microphone receives $48 - 1.7$ volts, i.e.: 46.3 volts. The Schoeps CMC-5 microphone draws 4 mA so the voltage drop is $3400 \times 0.004 = 13.6$ volts. Therefore the microphone receives $48 - 13.6$ volts, i.e.: 34.4 volts. The manufacturer normally takes this voltage drop into account in the design of the microphone, although examples exist of mics which draw so much current that they load down the phantom voltage of a mixer to a point where it is no longer adequate to power the mics. In such a case some mics become very noisy, some will not work at all, and yet others may produce unusual noises or oscillation. A stand-alone dedicated power supply or internal battery supply may be the solution in difficult cases.

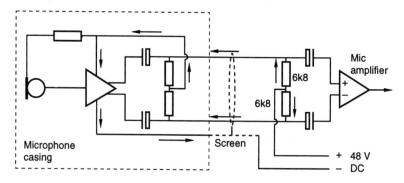

Figure 4.14 A typical 48 volt phantom powering arrangement in an electronically balanced circuit

The universal standard is 48 volts, but some capacitor microphones are designed to operate on a range of voltages down to 9 volts, and this can be advantageous for instance when using battery-powered equipment on location, or out of doors away from a convenient source of mains power.

Figure 4.14 illustrates the situation with phantom powering when electronically balanced circuits are used, as opposed to transformers. Capacitors are used to block the DC voltage from the power supply, but they present a very low impedance to the audio signal.

4.9.2 A–B powering

Another form of powering for capacitor microphones which is sometimes encountered is A–B powering. Figure 4.15 illustrates this system schematically. Here, the power is applied to one of the audio lines via a resistor and is taken to the microphone electronics via another resistor at the microphone end. The return path is provided by the other audio line as the arrows show. The screen is not used for carrying any current. There is a capacitor at the centre of the winding of each transformer. A capacitor does not allow DC to pass, and so these capacitors prevent the current from short-circuiting via the transformer windings. The capacitors have a very low impedance at audio frequencies, so as far as the audio signal is concerned they are not there. The usual voltage used in this system is 12 volts.

Although, like phantom power, the existing microphone lines are used to carry the current, it is dangerous to connect another type of microphone in place of the one illustrated. If, say, a ribbon microphone were to be connected, its output transformer would short-circuit the applied current. Therefore 12 volt A–B powering should be switched off before connecting any other type of microphone, and this is clearly a disadvantage compared with the phantom powering approach. It is encountered most commonly in location film sound recording equipment.

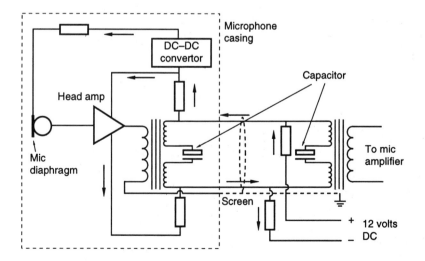

Figure 4.15 A typical 12 volt A–B powering arrangement

4.10 Radio microphones

Radio microphones are widely used in film, broadcasting, theatre and other industries, and it is not difficult to think of circumstances in which freedom from trailing microphone cables can be a considerable advantage in all of the above.

4.10.1 Principles

The radio microphone system consists of a microphone front end (which is no different from an ordinary microphone); an FM (Frequency Modulation) transmitter, either built into the housing of the mic or housed in a separate case into which the mic plugs; a short aerial via which the signal is transmitted; and a receiver which is designed to receive the signal from a particular transmitter. Only one specified transmission frequency is picked up by a given receiver. The audio output of the receiver then feeds a mixer or tape machine in the same manner as any orthodox microphone or line level source would. The principle is illustrated in Figure 4.16.

The transmitter can be built into the stem of the microphone, or it can be housed in a separate case, typically the size of a packet of cigarettes, into which the microphone or other signal source is plugged. A small battery which fits inside the casing of the transmitter provides the power, and this can also supply power to those capacitor mics which are designed to operate at the typical 9 volts of the transmitter battery. The transmitter is of the FM type (see Fact File 4.9), as this offers high-quality audio performance.

Frequently, two or more radio microphones need to be used. Each transmitter must transmit at a different frequency, and the spacing between each adjacent frequency must not be too close otherwise they will interfere with each other. In practice, channels with a minimum spacing of 0.2 MHz are used. Although only one transmitter can be used at a given frequency, any number of receivers can of course be used, as is the case with ordinary radio reception.

4.10.2 Facilities

Transmitters are often fitted with facilities which enable the operator to set the equipment up for optimum performance. A 1 kHz line-up tone is sometimes encountered which sends a continuous tone to the receiver to check continuity. Input gain controls are useful, with an indication of peak input level, so that the transmitter can be used with mics and line level sources of widely different output levels. It is important that the optimum setting is found, as too great an input level may cause a limiter (see section 14.2) to come into action much of the time, which can cause compression and 'pumping' noises as the limiter operates. Too weak a signal gives

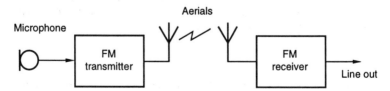

Figure 4.16 A radio microphone incorporates an FM transmitter, resulting in no fixed link between microphone and mixer

FACT FILE

4.9

Frequency modulation

In FM systems the transmitter radiates a high-frequency radio wave (the carrier) whose frequency is modulated by the amplitude of the audio signal. The positive-going part of the audio waveform causes the carrier frequency to deviate upwards, and the negative-going part causes it to deviate downwards. At the receiver, the modulated carrier is demodulated, converting variations in carrier frequency back into variations in the amplitude of an audio signal.

Audio signals typically have a wide dynamic range, and this affects the degree to which the carrier frequency is modulated. The carrier deviation must be kept within certain limits, and manufacturers specify the maximum deviation permitted. The standard figure for a transmitter with a carrier frequency of around 175 MHz is ±75 kHz, meaning that the highest-level audio signal modulates the carrier frequency between 175.075 MHz and 174.925 MHz. The transmitter incorporates a limiter to ensure that these limits are not exceeded.

insufficient drive, and poor signal-to-noise ratios can result.

The receiver will have a signal strength indicator. This can be very useful for locating 'dead spots'; transmitter positions which cause unacceptably low meter readings should be avoided, or the receiving aerial should be moved to a position which gives better results. Another useful facility is an indicator which tells the condition of the battery in the transmitter. When the battery voltage falls below a certain level, the transmitter sends out an inaudible warning signal to the receiver which will then indicate this condition. The operator then has a warning that the battery will soon fail, which is often within 15 minutes of the indication.

4.10.3 Licences

Transmitting equipment usually requires a licence for its operation, and governments normally rigidly control the frequency bands over which a given user can operate. This ensures that local and network radio transmitters do not interfere with police, ambulance and fire brigade equipment, etc. In the UK the frequency band for which radio mics do *not* have to be licensed is between 173.8 MHz and 175 MHz. Each radio mic transmitter needs to be spaced at least 0.2 MHz apart, and commonly used frequencies are 173.8, 174.1, 174.5, 174.8, and 175.0 MHz. An additional requirement is that the frequencies must be crystal controlled, which ensures that they cannot drift outside tightly specified limits. Maximum transmitter power is limited to 10 milliwatts, which gives an effective radiated power (ERP) at the aerial of 2 milliwatts which is very low, but adequate for the short ranges over which radio mics are operated.

Many big musical shows require rather more radio mic channels than are legally available. Television companies are often able to use an additional band of frequencies for radio mics, since they already have licences for certain TV channel frequencies which can be used for radio mics.

4.10.4 Aerials

The dimensions of the transmitting aerial are related to the wavelength of the transmitted frequency. The wavelength (λ) in an electrical conductor at a frequency of 174.5 MHz is approximately 64 inches (160 cm). To translate this into a suitable

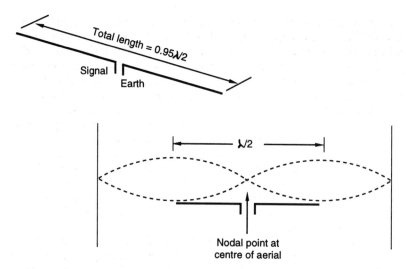

Figure 4.17 A simple dipole aerial configuration

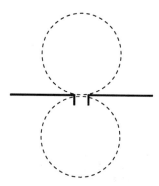

Figure 4.18 The dipole has a figure-eight radiation pattern

aerial length, it is necessary to discuss the way in which a signal resonates in a conductor. It is convenient to consider a simple dipole aerial, as shown in Figure 4.17. This consists of two conducting rods, each a quarter of a wavelength long, fed by the transmitting signal as shown. The centre of the pair is the nodal point and exhibits a characteristic impedance of about 70 ohms. For a radio mic, we need a total length of $\lambda/2$, i.e.: $64/2 = 32$ inches (80 cm).

A 32 inch dipole will therefore allow the standard range of radio mic frequencies to resonate along its length to give efficient radiation, the precise length not being too critical. Consideration also has to be given to the radiated polar response (this is not the same as the microphone's polar response). Figure 4.18 shows the polar response for a dipole. As can be seen, it is a figure-eight with no radiation in the directions the two halves are pointing in. Another factor is polarisation of the signal. Electromagnetic waves consist of an electric wave plus a magnetic wave radiating at right angles to each other, and so if a transmitting aerial is orientated vertically, the receiving aerial should also be orientated vertically. This is termed vertical polarisation.

The radio mic transmitter therefore has a transmitting aerial of about 16 inches long: half of a dipole. The other half is provided by the earth screen of the audio input lead, and will be in practice rather longer than 16 inches. The first-mentioned half is therefore looked upon as being the aerial proper, and it typically hangs vertically downwards. The screened signal input cable will generally be led upwards, but other practical requirements tend to override its function as part of the aerial system.

Another type which is often used for hand-held radio mics is the helical aerial. This is typically rather less than half the length of the 16 inch aerial, and has a diameter of a centimetre or so. It protrudes from the base of the microphone. It consists of a tight coil of springy wire housed in a plastic insulator, and has the advantage of being both smaller and reasonably tolerant of physical abuse. Its radiating efficiency is, however, less good than the 16 inch length of wire. At the receiver, a similar aerial is required. The helical aerial is very common here, and its short stubby form is very convenient for outside broadcast and film crews. A 16 inch length of metal tubing, rather like a short car aerial, can be a bit unwieldy although it is a more efficient receiver.

Other aerial configurations exist, offering higher gain and directionality. In the two-element aerial shown in Figure 4.19 the reflector is slightly larger than the dipole, and is spaced behind it at a distance which causes reflection of signal back

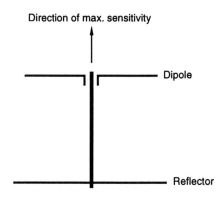

Figure 4.19 A simple two-element aerial incorporates a dipole and a reflector for greater directionality than a dipole

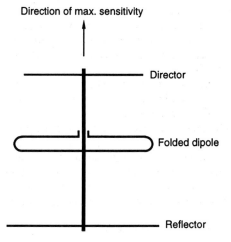

Figure 4.20 The three-element 'Yagi' configuration

on to it. It increases the gain, or strength of signal output, by 3 dB. It also attenuates signals approaching from the rear and sides. The three-element 'Yagi', named after its Japanese inventor and shown in Figure 4.20, uses the presence of a director and reflector to reduce the impedance of a conventional dipole, so a greatly elongated rectangle called a folded dipole is used, which itself has a characteristic impedance of about 300 ohms. The other elements are positioned such that the final impedance is reduced to the standard 50 ohms. The three-element Yagi is even more directional than the dipole, and has increased gain. It can be useful in very difficult reception conditions, or where longer distances are involved such as receiving the signal from a transmitter carried by a rock climber for running commentary! The multi-element, high-gain, highly directional UHF television aerial is of course a familiar sight on our roof-tops.

These aerials can also be used for transmitting, the principles being exactly the same. Their increased directionality also helps to combat multipath problems. The elements should be vertically orientated, because the transmitting aerial will normally be vertical, and the 'direction of maximum sensitivity' arrows on the figures show the direction the aerials should be pointed in.

Another technique for improving the signal-to-noise ratio under difficult reception conditions is noise reduction, which operates as follows. Inside the transmitter there is an additional circuit which compresses the incoming audio signal, thus reducing its overall dynamic range. At the receiver, a reciprocal circuit expands the audio signal, after reception and demodulation, and as it pushes the lower-level audio signals back down to their correct level it also therefore pushes the residual noise level down. Previously unacceptable reception conditions will often yield usable results when such transmitters and receivers are employed. It should be noted though that the system does not increase signal strength, and all the problems of transmission and reception still apply. (Noise reduction systems are covered further in Chapter 9.)

4.10.5 Aerial siting and connection

It is frequently desirable to place the receiving aerial itself closer to the transmitter than the receiver, in order to pick up a strong signal. To do this an aerial is rigged at a convenient position close to the transmitter, for example in the wings of a theatre stage, or on the front of a balcony, and then an aerial lead is run back to the receiver. A helical dipole aerial is frequently employed. In such a situation, characteristic impedance must be considered. As discussed in section 13.9.1, when the wavelength of the electrical signal in a conductor is similar to the length of the conductor, reflections can be set up at the receiving end unless the cable is properly terminated. Therefore, impedance matching must be employed between the aerial and the transmitter or receiver, and additionally the connecting lead needs to have the correct characteristic impedance.

The standard value for radio microphone equipment is 50 ohms, and so the aerial, the transmitter, the receiver, the aerial lead and the connectors must all be rated at this value. This cannot be measured using a simple test meter, but an aerial and cable can be tuned using an SWR (Standing Wave Ratio) meter to detect the level of the reflected signal. The aerial lead should be a good-quality, low-loss type, otherwise the advantage of siting the aerial closer to the transmitter will be wasted by signal loss along the cable. Poor signal reception causes noisy performance, because the receiver has a built-in automatic gain control (AGC), which sets the amplification of the carrier frequency to an appropriate value. Weak signals simply require higher amplification and therefore higher noise levels result.

The use of several radio microphones calls for a complementary number of receivers which all need an aerial feed. It is common practice to use just one aerial which is plugged into the input of an aerial distribution amplifier. This distribution unit has several outputs which can be fed into each receiver. It is not possible simply to connect an aerial to all the inputs in parallel due to the impedance mismatch that this would cause.

Apart from obvious difficulties such as metallic structures between transmitter and receiver, there are two phenomena which cause the reception of the radio signal to be less than perfect. The first phenomenon is known as multipath (see Figure 4.21). When the aerial transmits, the signal reaches the receiving aerial by a number of routes. Firstly, there is the direct path from aerial to aerial. Additionally, signals bounce off the walls of the building and reach the receiving aerial via a longer route. So the receiving aerial is faced with a number of signals of more or less random phase and strength, and these will sometimes combine to cause severe signal cancellation and consequently very poor reception. The movement of the transmitter along with the person wearing it will alter the relationship between these multipath signals, and so 'dead spots' are sometimes encountered where particular combinations of multipath signals cause signal 'drop-out'. The solution is to find out where these dead spots are by trial and error, re-siting the receiving aerial until they are minimised or eliminated. It is generally good practice to site the aerial close to the transmitter so that the direct signal will be correspondingly stronger than many of the signals arriving from the walls. Metal structures should be kept clear of wherever possible due to their ability to reflect and screen RF signals. Aerials can be rigged on metal bars, but at right angles to them, not parallel.

The other phenomenon is signal cancellation from other transmitters when a number of channels are in use simultaneously. Because the transmitting frequencies

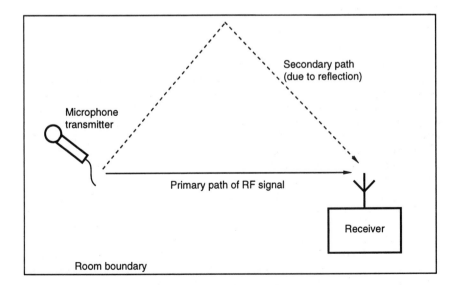

Figure 4.21 Multipath distortion can arise between source and receiver due to reflections

of the radio mics will be quite close together, partial cancellation of all the signals takes place. The received signals are therefore weaker than for a single transmitter on its own. Again, siting the receiving aerial close to the transmitters is a good idea. The 'sharpness' or 'Q' of the frequency tuning of the receivers plays a considerable part in obtaining good reception in the presence of a number of signals. A receiver may give a good performance when only one transmitter is in use, but a poor Q will vastly reduce the reception quality when several are used. This should be checked for when systems are being evaluated, and the testing of one channel on its own will not of course show up these kinds of problems.

4.10.6 Diversity reception

A technique known as 'spaced diversity' goes a good way towards combatting the above problems. In this system, two aerials feed two identical receivers for each radio channel. A circuit continuously monitors the signal strength being received by each receiver and automatically selects the one which is receiving the best signal (see Figure 4.22). When they are both receiving a good signal, the outputs of the two are mixed together. A crossfade is performed between the two as one RF signal fades and the other becomes strong.

The two aerials are placed some distance apart, the distance depending on the frequency involved, so that the multipath relationships between a given transmitter position and each aerial will be somewhat different. A dead spot for one aerial is therefore unlikely to coincide with a dead spot for the other one. A good diversity system overcomes many reception problems, and the considerable increase in reliability of performance is well worth the extra cost. The point at which diversity becomes desirable is when more than two radio microphones are to be used, although good performance from four channels in a non-diversity installation is by no means out of the question. Good radio microphones are very expensive, a single channel of a quality example costing over a thousand pounds today. Cheaper ones exist, but experience suggests that no radio microphone at all is vastly preferable to a cheap one.

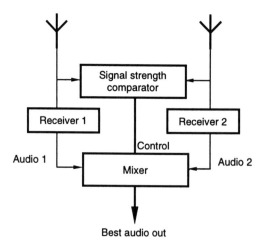

Figure 4.22 A diversity receiver incorporates two aerials spaced apart and two receivers. The signal strength from each aerial is used to determine which output will have the higher quality

Recommended further reading

AES (1979) *Microphones: An Anthology.* Audio Engineering Society
Bartlett, B. (1991) *Stereo Microphone Techniques.* Focal Press
Borwick, J. (1990) *Microphones: Technology and Technique.* Focal Press
Gayford, M. (1994) ed., *Microphone Engineering Handbook.* Focal Press
Huber, D. (1988) *Microphone Manual – Design and Application.* Focal Press
Nisbett, A. (1989) *Use of Microphones.* Focal Press
Robertson, A. E. (1963) *Microphones.* Hayden
Rumsey, F. J. (1989) *Stereo Sound for Television.* Focal Press

See also *General further reading* at the end of this book.

Chapter 5

Loudspeakers

A loudspeaker is a transducer which converts electrical energy into acoustical energy. A loudspeaker must therefore have a diaphragm of some sort which is capable of being energised in such a way that it vibrates to produce sound waves which are recognisably similar to the original sound from which the energising signal was derived. To ask a vibrating plastic loudspeaker cone to reproduce the sound of, say, a violin is to ask a great deal, and it is easy to take for granted how successful the best examples have become. Continuing development and refinement of the loudspeaker has brought about a more or less steady improvement in its general performance, but it is a sobering thought that one very rarely mistakes a sound coming from a speaker for the real sound itself, and that one nevertheless has to use these relatively imperfect devices to assess the results of one's work. Additionally, it is easy to hear significant differences between one model and another. Which is right? It is important not to tailor a sound to suit a particular favourite model. There are several principles by which loudspeakers can function, and the commonly employed ones will be briefly discussed.

A word or two must be said about the loudspeaker enclosure. The box can have as big an influence on the final sound of a speaker system as can the drivers themselves. At first sight surprising, this fact can be more readily appreciated when one remembers that a speaker cone radiates virtually the same amount of sound into the cabinet as out into the room. The same amount of acoustical energy that is radiated is therefore also being concentrated in the cabinet, and the sound escaping through the walls and also back out through the speaker cone has a considerable influence upon the final sound of the system.

5.1 The moving-coil loudspeaker

The moving-coil principle is by far the most widely used, as it can be implemented in very cheap transistor radio speakers, PA (Public Address) systems, and also top-quality studio monitors, plus all performance levels and applications in between. Figure 5.1 illustrates a cutaway view of a typical moving-coil loudspeaker. Such a device is also known as a drive unit or driver, as it is the component of a complete speaker system which actually produces the sound or 'drives' the air. Basically, the speaker consists of a powerful permanent magnet which has an annular gap to accommodate a coil of wire wound around a cylindrical former. This former is attached to the cone or diaphragm which is held in its rest position by a suspension

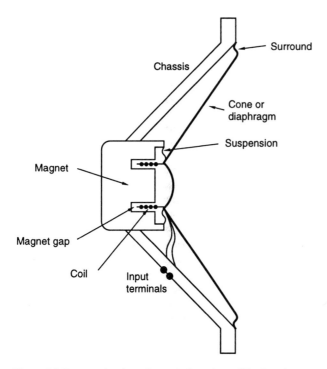

Figure 5.1 Cross-section through a typical moving-coil loudspeaker

system which usually consists of a compliant, corrugated, doped (impregnated) cloth material and a compliant surround around the edge of the cone which can be made of a type of rubber, doped fabric, or it can even be an extension of the cone itself, suitably treated to allow the required amount of movement of the cone.

The chassis usually consists either of pressed steel or a casting, the latter being particularly desirable where large heavy magnets are employed, since the very small clearance between the coil and the magnet gap demands a rigid structure to maintain the alignment, and a pressed steel chassis can sometimes be distorted if the loudspeaker is subject to rough handling as is inevitably the case with portable PA systems and the like. (A properly designed pressed steel chassis should not be overlooked though.) The cone itself can in principle be made of almost any material, common choices being paper pulp (as used in many PA speaker cones for its light weight, giving good efficiency), plastics of various types (as used in many hi-fi speaker cones due to the greater consistency achievable than with paper pulp, and the potentially lower coloration of the sound, usually at the expense of increased weight and therefore lower efficiency which is not crucially important in a domestic loudspeaker), and sometimes metal foil.

The principle of operation is based on the principle of electromagnetic transducers described in Fact File 3.1, and is the exact reverse of the process involved in the moving-coil microphone (see Fact File 4.1). The cone vibration sets up sound waves in the air which are an acoustic analogue of the electrical input signal. Thus in principle the moving-coil speaker is a very crude and simple device, but the results obtained today are incomparably superior to the original 1920s Kellog and Rice design. It is, however, a great tribute to those pioneers that the principle of operation of what is still today's most widely used type of speaker is still theirs.

5.2 Other loudspeaker types

The electrostatic loudspeaker first became commercially viable in the 1950s, and is described in Fact File 5.1. The electrostatic principle is far less commonly employed than is the moving coil, since it is difficult and expensive to manufacture and will not produce the sound levels available from moving-coil speakers. The sound quality of the best examples, such as the Quad ESL 63 pictured in Figure 5.2, is, however, rarely equalled by other types of speaker.

Another technique in producing a panel-type speaker membrane has been to employ a light film on which is attached a series of conductive strips which serve as the equivalent of the coil of a moving-coil cone speaker. The panel is housed within a system of strong permanent magnets, and the drive signal is applied to the conductive strips. Gaps in the magnets allow the sound to radiate. Such systems tend to be large and expensive like the electrostatic models, but again very high-quality results are possible. In order to get adequate bass response and output level from such panel speakers the diaphragm needs to be of considerable area.

The ribbon loudspeaker principle has sometimes been employed in high-frequency applications ('tweeters') and has recently also been employed in large full-range models. Figure 5.3 illustrates the principle. A light corrugated aluminium

---

FACT FILE 5.1   Electrostatic loudspeaker – principles

The electrostatic loudspeaker's drive unit consists of a large, flat diaphragm of extremely light weight, placed between two rigid plates. The diagram shows a side view. There are parallels between this loudspeaker and the capacitor microphone described in Chapter 4. The diaphragm has a very high resistance, and a DC polarising voltage in the kilovolt (kV) range is applied to the centre tap of the secondary of the input transformer, and charges the capacitor formed by the narrow gap between the diaphragm and the plates. The input signal appears (via the transformer) across the two rigid plates and thus modulates the electrostatic field. The diaphragm, being the other plate of the capacitor, thus experiences a force which alters according to the input signal. Being free to move within certain limits with respect to the two rigid plates, it thus vibrates to produce the sound.

There is no cabinet as such to house the speaker, and sound radiates through the holes of both plates. Sound therefore emerges equally from the rear and the front of the speaker, but not from the sides. Its polar response is therefore a figure-eight, similar to a figure-eight microphone with the rear lobe being out of phase with the front lobe.

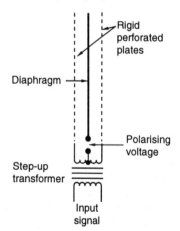

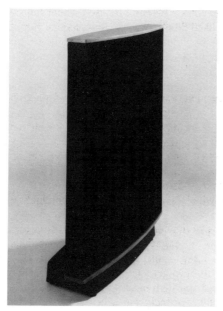

Figure 5.2 The Quad ESL63 electrostatic loudspeaker. (Courtesy of Quad Electroacoustics Ltd.)

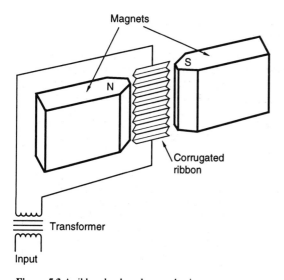

Figure 5.3 A ribbon loudspeaker mechanism

ribbon, clamped at each end, is placed between two magnetic poles, one north, one south. The input signal is applied, via a step-down transformer, to each end of the ribbon. The alternating nature of the signal causes an alternating magnetic field around the ribbon, which behaves like a single turn of a coil in a moving-coil speaker. The magnets each side thus cause the ribbon to vibrate, producing sound waves. The impedance of the ribbon is often extremely low, and an amplifier cannot drive it directly. A transformer is therefore used which steps up the impedance of the ribbon. The ribbon itself produces a very low acoustic output and often has a horn in front of it to improve its acoustical matching with the air, giving a higher output for a given electrical input. Some ribbons are, however, very long – half a metre or more – and drive the air directly.

There are a few other types of speaker in use, but these are sufficiently uncommon for descriptions not to be merited in this brief outline of basic principles.

5.3 Mounting and loading drive units

5.3.1 'Infinite baffle' systems

The moving-coil speaker radiates sound equally in front of and to the rear of the diaphragm or cone. As the cone moves forwards it produces a compression of the air in front of it but a rarefaction behind it, and vice versa. The acoustical waveforms are therefore 180° out of phase with each other and when they meet in the surrounding air they tend to cancel out, particularly at lower frequencies where diffraction around the cone occurs. A cabinet is therefore employed in which the drive unit sits, which has the job of preventing the sound radiated from the rear of the cone from reaching the open air. The simplest form of cabinet is the sealed box (commonly, but wrongly, known as the 'infinite baffle') which will usually have some sound-absorbing material inside it such as plastic foam or fibre wadding. A true 'infinite baffle' would be a very large flat piece of sheet material with a circular hole cut in the middle into which the drive unit would be mounted. Diffraction around the baffle would then only occur at frequencies below that where the wavelength approached the size of the baffle, and thus cancellation of the two mutually out-of-phase signals would not occur over most of the range, but for this to be effective at the lowest frequencies the baffle would have to measure at least 3 or 4 metres square. The only practical means of employing this type of loading is to mount the speaker in the dividing wall between two rooms, but this is rarely encountered for obvious reasons.

5.3.2 Bass reflex systems

Another form of loading is the bass reflex system, as shown in Figure 5.4. A tunnel, or port, is mounted in one of the walls of the cabinet, and the various parameters of cabinet internal volume, speaker cone weight, speaker cone suspension compliance, port dimensions, and thus mass of air inside the port are chosen so that at a specified low frequency the air inside the port will resonate, which reduces the movement of the speaker cone at that frequency. The port thus produces low-frequency output of its own, acting in combination with the driver. In this manner increased low-frequency output, increased efficiency, or a combination of the two can be achieved. However, it is worth remembering that at frequencies lower than the resonant frequency the driver is acoustically unloaded because the port now behaves simply as an open window. If extremely low frequencies from, say, mishandled micro-phones or record player arms reach the speaker they will cause considerable excursion of the speaker cone which can cause damage. The air inside a closed box system, however, provides a mechanical supporting 'spring' right down to the lowest frequencies.

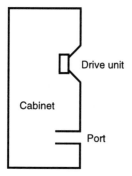

Figure 5.4 A ported bass reflex cabinet construction

FACT FILE
5.2

Transmission line system

extension achievable, but a large cabinet is required for its proper functioning.

A form of bass loading is the acoustic labyrinth or 'transmission line', as shown in the diagram. A large cabinet houses a folded tunnel the length of which is chosen so that resonance occurs at a specified low frequency. Above that frequency, the tunnel, which is filled or partially filled with acoustically absorbent material, gradually absorbs the rear-radiated sound energy along its length. At resonance, the opening, together with the air inside the tunnel, behaves like the port of a bass reflex design. An advantage of this type of loading is the very good bass

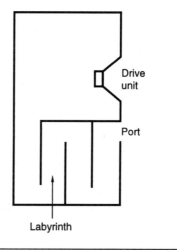

A device known as an auxiliary bass radiator (ABR) is occasionally used as an alternative to a reflex port, and takes the form of a further bass unit without its own magnet and coil. It is thus undriven electrically. Its cone mass acts in the same manner as the air plug in a reflex port, but has the advantage that mid-range frequencies are not emitted, resulting in lower coloration.

A further form of bass loading is described in Fact File 5.2.

5.3.3 Horn loading

Horn loading is a technique commonly employed in large PA loudspeaker systems, as described in Fact File 5.3. Here, a horn is placed in front of the speaker diaphragm.

The so-called 'long-throw' horn tends to beam the sound over an included angle of perhaps 90° horizontally and 40° vertically. The acoustical energy is therefore concentrated principally in the forward direction, and this is one reason for the horn's high efficiency. The sound is beamed forwards towards the rear of the hall

5.3 Horn loudspeaker – principles

A horn is an acoustic transformer, that is it helps to match the air impedance at the throat of the horn (the throat is where the speaker drive unit is) with the air impedance at the mouth. Improved acoustic efficiency is therefore achieved, and for a given electrical input a horn can increase the acoustical output of a driver by 10 dB or more compared with that driver mounted in a conventional cabinet. A horn functions over a relatively limited frequency range, and therefore relatively small horns are used for the high frequencies, larger ones for upper mid frequencies, and so on. This is very worth while where high sound levels need to be generated in large halls, rock concerts and open-air events.

Each design of horn has a natural lower cut-off frequency which is the frequency below which it ceases to load the driver acoustically. Very large horns indeed are needed to reproduce low frequencies, and one technique has been to fold the horn up by building it into a more conventional-looking cabinet. The horn principle is rarely employed at bass frequencies due to the necessarily large size. It is, however, frequently employed at mid and high frequencies, but the higher coloration of the sound it produces tends to rule it out for hi-fi and studio monitoring use other than at high frequencies if high sound levels are required. Horns tend to be more directional than conventional speakers, and this has further advantages in PA applications.

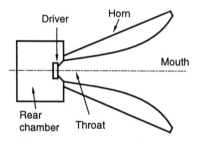

with relatively little sound reaching the side walls. The 'constant directivity' horn aims to achieve a consistent spread of sound throughout the whole of its working frequency range, and this is usually achieved at the expense of an uneven frequency response. Special equalisation is therefore often applied to compensate for this.

The long-throw horn does not do much for those members of an audience who are close to the stage between the speaker stacks, and an acoustic lens is often employed, which, as its name suggests, diffracts the sound, such that the higher frequencies are spread out over a wider angle to give good coverage at the front. Figure 5.5 shows a typical acoustic lens. It consists of a number of metal plates which are shaped and positioned with respect to each other in such a manner as to cause outward diffraction of the high frequencies. The downward slope of the plates is incidental to the design requirements and it is not incorporated to project the sound downwards. Because the available acoustic output is spread out over a wider area than is the case with the long-throw horn, the on-axis sensitivity tends to be lower.

The high efficiency of the horn has also been much exploited in those PA applications which do not require high sound quality, and their use for outdoor events such as fêtes, football matches and the like, as well as on railway station platforms, will have been noticed. Often, a contrivance known as a re-entrant horn is used, as shown in Figure 5.6. It can be seen that the horn has been effectively cut in half, and the half which carries the driver is turned around and placed inside the bell of the other. Quite a long horn is therefore accommodated in a compact structure, and this method of construction is particularly applicable to hand-held loudhailers.

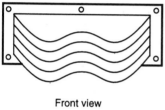

Front view

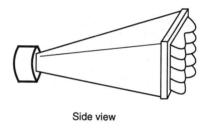

Side view

Figure 5.5 An example of an acoustic lens

Driver

Figure 5.6 A re-entrant horn

The high-frequency horn is driven not by a cone speaker but by a 'compression driver' which consists of a dome-shaped diaphragm usually with a diameter of 1 or 2 inches (2.5 or 5 cm). It resembles a hi-fi dome tweeter but with a flange or thread in front of the dome for fixing on to the horn. The compression driver can easily be damaged if it is driven by frequencies below the cut-off frequency of the horn it is looking into.

5.4 Complete loudspeaker systems

5.4.1 Two-way systems

It is a fact of life that no single drive unit can adequately reproduce the complete frequency spectrum from, say, 30 Hz to 20 kHz. Bass frequencies require large drivers with relatively high cone excursions so that adequate areas of air can be set in motion. Conversely, the same cone could not be expected to vibrate at 15 kHz – 15 000 times a second – to reproduce very high frequencies. A double bass is much larger than a flute, and the strings of a piano which produce the low notes are much fatter and longer than those for the high notes.

The most widely used technique for reproducing virtually the whole frequency spectrum is the so-called two-way speaker system, which is employed at many quality levels from fairly cheap audio packages to very high-quality studio monitors. It consists of a bass/mid driver which handles frequencies up to around 3 kHz, and a high-frequency unit or 'tweeter' which reproduces frequencies from 3 kHz to 20 kHz or more. Figure 5.7 shows a cutaway view of a tweeter. Typically of around 1 inch (2.5 cm) in diameter, the dome is attached to a coil in the same way that a cone

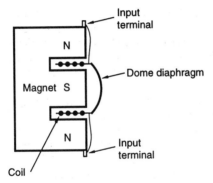

Figure 5.7 Cross-section through a typical dome tweeter

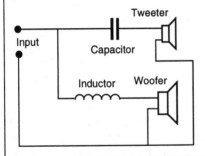

FACT FILE

5.4

A basic crossover network

A frequency-dividing network or 'crossover' is fitted into the speaker enclosure which divides the incoming signal into high frequencies (above about 3 kHz) and lower frequencies, sending the latter to the bass/mid unit or 'woofer' and the former to the tweeter. A simple example of the principle involved is illustrated in the diagram. In practical designs additional account should be taken of the fact that speaker drive units are not pure resistances.

The tweeter is fed by a capacitor. A capacitor has an impedance which is inversely proportional to frequency, that is at high frequencies its impedance is very low and at low frequencies its impedance is relatively high. The typical impedance of a tweeter is 8 ohms, and so for signals below the example of 3 kHz (the 'crossover frequency') a value of capacitor is chosen which exhibits an impedance of 8 ohms also

at 3 kHz, and due to the nature of the voltage/current phase relationship of the signal across a capacitor the power delivered to the tweeter is attenuated by 3 dB at that frequency. It then falls at a rate of 6 dB per octave thereafter (i.e.: the tweeter's output is 9 dB down at 1.5 kHz, 15 dB down at 750 Hz and so on) thus protecting the tweeter from lower frequencies. The formula which contains the value of the capacitor for the chosen 3 kHz frequency is:

$$f = 1/(2\pi RC)$$

where R is the resistance of the tweeter, and C is the value of the capacitor in farads.

The capacitor value will more conveniently be expressed in microfarads (millionths of a farad) and so the final formula becomes:

$$C = 159\ 155 \div (8\ \text{ohms} \times 3000\ \text{Hz})$$
$$= 6.7\ \mu F$$

Turning now to the woofer, it will be seen that an inductor is placed in series with it. An inductor has an impedance which rises with frequency; therefore, a value is chosen that gives an impedance value similar to that of the woofer at the chosen crossover frequency. Again, the typical impedance of a woofer is 8 ohms. The formula which contains the value of the inductor is:

$$f = R/(2\pi L)$$

where L = inductance in henrys, R = speaker resistance, f = crossover frequency. The millihenry (one-thousandth of a henry, mH) is more appropriate, so this gives:

$$L = 8000 \div (2\pi \times 3000) = 0.42\ \text{mH}$$

is in a bass/mid driver. The dome can be made of various materials, 'soft' or 'hard', and metal domes are also frequently employed. A bass/mid driver cannot adequately reproduce high frequencies as has been said. Similarly, such a small dome tweeter would actually be damaged if bass frequencies were fed to it; thus a crossover network is required to feed each drive unit with frequencies in the correct range, as described in Fact File 5.4.

In a basic system the woofer would typically be of around 8 inches (20 cm) in diameter for a medium-sized domestic speaker, mounted in a cabinet having several cubic feet internal volume. Tweeters are usually sealed at the rear, and therefore they are simply mounted in an appropriate hole cut in the front baffle of the enclosure. This type of speaker is commonly encountered at the cheaper end of the price range, but its simplicity makes it well worth study since it nevertheless incorporates the basic features of many much more costly designs. The latter differ in that they make use of more advanced and sophisticated drive units, higher-quality cabinet materials and constructional techniques, and a rather more sophisticated crossover which usually incorporates both inductors and capacitors in the treble and bass sections as well as resistors which together give much steeper filter slopes than our 6 dB/octave example. Also, the overall frequency response can be adjusted by the crossover to take account of, say, a woofer which gives more acoustic output in the mid range than in the bass: some attenuation of the mid range can give a flatter and better-balanced frequency response.

5.4.2 Three-way systems

Numerous three-way loudspeaker systems have also appeared where a separate mid-range driver is incorporated along with additional crossover components to restrict the frequencies feeding it to the mid range, for example between 400 Hz and 4 kHz. It is an attractive technique due to the fact that the important mid frequencies where much of the detail of music and speech resides are reproduced by a dedicated driver designed specially for that job. But the increased cost and complexity does not always bring about a proportional advance in sound quality.

5.5 Active loudspeakers

So far, only 'passive' loudspeakers have been discussed, so named because simple passive components – resistors, capacitors and inductors – are used to divide the frequency range between the various drivers. 'Active' loudspeakers are also encountered, in which the frequency range is divided by active electronic circuitry at line level, after which each frequency band is sent to a separate power amplifier and thence to the appropriate speaker drive unit. The expense and complexity of active systems has tended to restrict the active technique to high-powered professional PA applications where four-, five- and even six-way systems are employed, and to professional studio monitoring speakers, such as the Rogers LS5/8 system pictured in Figure 5.8. Active speakers are still comparatively rare in domestic audio.

Each driver has its own power amplifier, which of course immediately increases the cost and complexity of the speaker system, but the advantages include: lower distortion (due to the fact that the signal is now being split at line level, where only a volt or so at negligible current is involved, as compared with the tens of volts and

Figure 5.8 Rogers LS5/8 high-quality active studio loudspeaker. (Courtesy of Swisstone Electronics Ltd.)

several amps that passive crossovers have to deal with); greater system-design flexibility due to the fact that almost any combination of speaker components can be used because their differing sensitivities, impedances and power requirements can be compensated for by adjusting the gains of the separate power amplifiers or electronic crossover outputs; better control of final frequency response, since it is far easier to incorporate precise compensating circuitry into an electronic crossover design than is the case with a passive crossover; better clarity of sound and firmer bass simply due to the lack of passive components between power amplifiers and drivers; and an improvement in power amplifier performance due to the fact that each amplifier now handles a relatively restricted band of frequencies.

In active systems amplifiers can be better matched to loudspeakers, and the system can be designed as a whole, without the problems which arise when an unpredictable load is attached to a power amplifier. In passive systems, the designer has little or no control over which type of loudspeaker is connected to which type of amplifier, and thus the design of each is usually a compromise between adaptability and performance. Some active speakers have the electronics built into the speaker cabinet which simplifies installation.

5.6 Subwoofers

Good bass response from a loudspeaker requires a large internal cabinet volume so that the resonant frequency of the system can be correspondingly low, the response of a given speaker normally falling away below this resonant point. This implies the use of two large enclosures which are likely to be visually obtrusive in a living room for instance. A way around this problem is to incorporate a so-called 'subwoofer' system. A separate speaker cabinet is employed which handles only the deep bass frequencies, and it is usually driven by its own power amplifier. The signal to drive the power amp comes from an electronic crossover which subtracts the low bass frequencies from the feed to the main stereo amplifier and speakers, and sends the mono sum of the deep bass to the subwoofer system.

Freed from the need to reproduce deep bass, the main stereo speakers can now be small high-quality systems; the subwoofer can be positioned anywhere in the room according to the manufacturers of such systems since it only radiates frequencies below around 100 Hz or so, where sources tend to radiate only omnidirectionally anyway. Degradation of the stereo image has sometimes been noted when the subwoofer is a long way from the stereo pair, and a position close to one of these is probably a good idea.

Subwoofers are also employed in concert and theatre sound systems. It is difficult to achieve both high efficiency and a good bass response at the same time from a speaker intended for public address use, and quite large and loud examples often have little output below 70 Hz or so. Subwoofer systems, if properly integrated into the system as a whole, can make a large difference to the weight and scale of live sound.

5.7 Loudspeaker performance

5.7.1 Impedance

The great majority of loudspeaker drive units and systems are labelled 'Impedance = 8 ohms'. This is, however, a nominal figure, the impedance in practice varying widely with frequency (see section 1.8). A speaker system may indeed have an 8 ohm impedance at, say, 150 Hz, but at 50 Hz it may well be 30 ohms, and at 10 kHz it could be 4 ohms. Figure 5.9 shows the impedance plot of a typical two-way, sealed box, domestic hi-fi speaker.

The steep rise in impedance at a certain low frequency is indicative of the low-frequency resonance of the system. Other undulations are indicative of the reactive nature of the speaker due to capacitive and inductive elements in the crossover components and the drive units themselves. Also, the driver/box interface has an effect, the most obvious place being at the already-mentioned LF resonant frequency.

Figure 5.10 shows an impedance plot of a bass reflex design. Here we see the characteristic 'double hump' at the bass end. The high peak at about 70 Hz is the bass driver/cabinet resonance point. The trough at about 40 Hz is the resonant frequency of the bass reflex port where maximum LF sound energy is radiated from the port itself and minimum energy is radiated from the bass driver. The low peak at about 20 Hz is virtually equal to the free-air resonance of the bass driver itself because at very low frequencies the driver is acoustically unloaded by the cabinet due to the presence of the port opening. A transmission-line design exhibits a similar impedance characteristic.

The DC resistance of an 8 ohm driver or speaker system tends to lie around 7 ohms, and this simple measurement is a good guide if the impedance of an unlabelled speaker is to be estimated. Other impedances encountered include

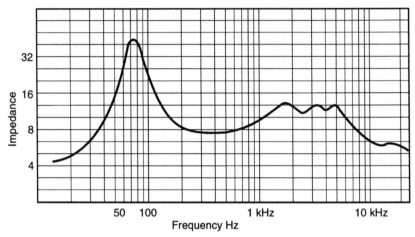

Figure 5.9 Impedance plot of a typical two-way sealed-box domestic loudspeaker

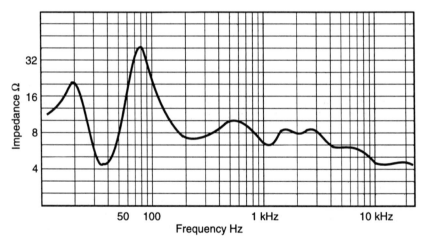

Figure 5.10 Impedance plot of a typical bass reflex design

15 ohm and 4 ohm models. The 4 ohm speakers are harder to drive because for a given amplifier output voltage they draw twice as much current. The 15 ohm speaker is an easy load, but its higher impedance means that less current is drawn from the amplifier and so the power (volts × amps) driving the speaker will be correspondingly less. So a power amplifier may not be able to deliver its full rated power into this higher impedance. Thus 8 ohms has become virtually standard, and competently designed amplifiers can normally be expected to drive competently designed speakers. Higher-powered professional power amplifiers can also be expected to drive two 8 ohm speakers in parallel, giving a resultant nominal impedance of 4 ohms.

5.7.2 Sensitivity

A loudspeaker's sensitivity is a measure of how efficiently it converts electrical sound energy into acoustical sound energy. The principles are described in Fact File 5.5. Loudspeakers are very inefficient devices indeed. A typical high-quality domestic speaker system has an efficiency of less than 1%, and therefore if 20 watts is fed into it the resulting acoustic output will be less than 0.2 acoustical watts. Almost all of the rest of the power is dissipated as heat in the voice coils of the drivers. Horn-loaded systems can achieve a much better efficiency, figures of around 10% being typical. An efficiency figure is not in itself a very helpful thing to know, parameters such as sensitivity and power handling being much more useful. But it is as well to be aware that most of the power fed into a speaker has to be dissipated as heat, and prolonged high-level drive causes high voice-coil temperatures.

It has been suggested that sensitivity is not an indication of quality. In fact, it is often found that lower-sensitivity models tend to produce a better sound. This is because refinements in sound quality usually come at the expense of reduced acoustical output for a given input, and PA speaker designers generally have to sacrifice absolute sound quality in order to achieve the high sensitivity and sound output levels necessary for the intended purpose.

5.7.3 Distortion

Distortion in loudspeaker systems is generally an order of magnitude or more higher than in other audio equipment. Much of it tends to be second-harmonic distortion (see section A1.3) whereby the loudspeaker will add frequencies an octave above the legitimate input signal. This is especially manifest at low frequencies where speaker diaphragms have to move comparatively large distances to reproduce them. When output levels of greater than 90 dB for domestic systems and 105 dB or so for high-sensitivity systems are being produced, low-frequency distortion of around

---

5.5 Loudspeaker sensitivity

Sensitivity is defined as the acoustic sound output for a given voltage input. The standard conditions are an input of 2.83 volts (corresponding to 1 watt into 8 ohms) and an acoustic SPL measurement at a distance of 1 metre in front of the speaker. The input signal is pink noise which contains equal sound energy per octave (see section 1.6). A single frequency may correspond with a peak or dip in the speaker's response, leading to an inaccurate overall assessment. For example, a domestic speaker may have a quoted sensitivity of 86 dB W^{-1}, that is 1 watt of input will produce 86 dB output at 1 metre.

Sensitivities of various speakers differ quite widely and this is not an indication of the sound quality. A high-level professional monitor speaker may have a sensitivity of 98 dB W^{-1} suggesting that it will be very much louder than its domestic cousin, and this will indeed be the case. High-frequency PA horns sometimes achieve a value of 118 dB for just 1 watt input. Sensitivity is thus a useful guide when considering which types of speaker to choose for a given application. A small speaker having a quoted sensitivity of 84 dB W^{-1} and 40 watts power handling will not fill a large hall with sound. The high sound level capability of large professional models will be wasted in a living room.

10% is quite common, this consisting mainly of second-harmonic and partly of third-harmonic distortion.

At mid and high frequencies distortion is generally below 1%, this being confined to relatively narrow bands of frequencies which correspond to areas such as crossover frequencies or driver resonances. Fortunately, distortion of this magnitude in a speaker does not indicate impending damage, and it is just that these transducers are inherently non-linear to this extent. Much of the distortion is at low frequencies where the ear is comparatively insensitive to it, and also the predominantly second-harmonic character is subjectively innocuous to the ear. Distortion levels of 10–15% are fairly common in the throats of high-frequency horns.

5.7.4 Frequency response

The frequency response of a speaker also indicates how linear it is. Ideally, a speaker would respond equally well to all frequencies, producing a smooth 'flat' output response to an input signal sweeping from the lowest to the highest frequencies at a constant amplitude. In practice, only the largest speakers produce a significant output down to 20 Hz or so, but even the smallest speaker systems can respond to 20 kHz. The 'flatness' of the response, i.e.: how evenly a speaker responds to all frequencies, is a rather different matter. High-quality systems achieve a response that is within 6 dB of the 1 kHz level from 80 Hz to 20 kHz, and such a frequency response might look like Figure 5.11(a). Figure 5.11(b) is an example of a rather lower-quality speaker which has a considerably more ragged response and an earlier bass roll-off.

The frequency response can be measured using a variety of different methods, some manufacturers taking readings under the most favourable conditions to hide inadequacies. Others simply quote something like '±3 dB from 100 Hz to 15 kHz'. This does at least give a fairly good idea of the smoothness of the response. These specifications do not, however, tell you how a system will sound, and they must be used only as a guide. They tell nothing of coloration levels, or the ability to reproduce good stereo depth, or the smoothness of the treble, or the 'tightness' of the bass.

5.7.5 Power handling

Power handling is the number of watts a speaker can handle before unacceptable amounts of distortion ensue. It goes hand in hand with sensitivity in determining the maximum sound level a speaker can deliver. For example, a domestic speaker may be rated at 30 watts and have a sensitivity of 86 dB W^{-1}. The decibel increase of 30 watts over 1 watt is given by:

dB increase $= 10 \log 30 \approx 15$ dB

Therefore, the maximum output level of this speaker is $86 + 15 = 101$ dB at 1 metre for 30 watts input. This is loud, and quite adequate for domestic use. Consider now a PA speaker with a quoted sensitivity of 99 dB W^{-1}. 30 watts input now produces $99 + 15 = 114$ dB, some 13 dB more than with the previous example for the same power input. To get 114 dB out of the 86 dB W^{-1} speaker one would need to drive it with no less than 500 watts, which would of course be way beyond its capabilities. This dramatically demonstrates the need to be aware of the implications of sensitivity and power handling.

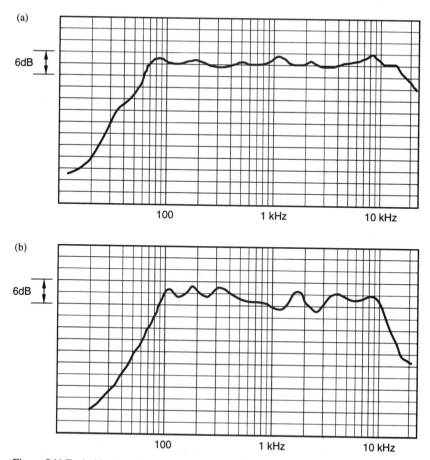

(a)

6dB

100   1 kHz   10 kHz

(b)

6dB

100   1 kHz   10 kHz

Figure 5.11 Typical loudspeaker frequency response plots. (a) A high-quality unit. (b) A lower-quality unit

A 30 watt speaker can, however, safely be driven even by a 500 watt amplifier providing that sensible precautions are taken with respect to how hard the amplifier is driven. Occasional peaks of more than 30 watts will be quite happily tolerated; it is sustained high-level drive which will damage a speaker. It is perfectly all right to drive a high-power speaker with a low-power amplifier, but care must be taken that the latter is not overdriven otherwise the harsh distortion products can easily damage high-frequency horns and tweeters even though the speaker system may have a quoted power handling well in excess of the amplifier. The golden rule is to listen carefully. If the sound is clean and unstressed, all will be well.

5.8 Setting up loudspeakers

5.8.1 Phase

Phase is a very important consideration when wiring up speakers. A positive-going voltage will cause a speaker cone to move in a certain direction, which is usually forwards, although at least two American and two British manufacturers have unfortunately adopted the opposite convention. It is essential that both speakers of a stereo pair, or all of the speakers of a particular type in a complete sound rig, are 'in phase', that is all the cones are moving in the same direction at any one time when an identical signal is applied. If two stereo speakers are wired up out of phase, this produces vague 'swimming' sound images in stereo, and cancellation of bass frequencies. This can easily be demonstrated by temporarily connecting *one* speaker in opposite phase and then listening to a mono signal source – speech from the radio is a good test. The voice will seem to come from nowhere in particular, and small movements of the head produce sudden large shifts in apparent sound source location. Now reconnect the speakers in phase and the voice will come from a definite position in between the speakers. It will also be quite stable when you move a few feet to the left or to the right.

Occasionally it is not possible to check the phase of an unknown speaker by listening. An alternative method is to connect a 1.5 V battery across the input terminals and watch which way the cone of the bass driver moves. If it moves forwards, then the positive terminal of the battery corresponds to the positive input terminal of the speaker. If it moves backwards as the battery is connected, then the positive terminal of the battery is touching the negative input terminal of the speaker. The terminals can then be labelled + and –.

5.8.2 Positioning

Loudspeaker positioning has a significant effect upon the performance. In smaller spaces such as control rooms and living rooms the speakers are likely to be positioned close to the walls, and 'room gain' comes into effect whereby the low frequencies are reinforced. This happens because at these frequencies the speaker is virtually omnidirectional, i.e.: it radiates sound equally in all directions. The rear- and side-radiated sound is therefore reflected off the walls and back into the room to add more bass power. As we move higher in frequency, a point is reached whereby the wavelength of lower mid frequencies starts to become comparable with the distance between the speaker and a nearby wall. At half wavelengths the reflected sound is out of phase with the original sound from the speaker and some cancellation of sound is caused. Additionally, high-frequency 'splash' is often caused by nearby hard surfaces, this often being the case in control rooms where large consoles, tape machines, outboard processing gear, etc. can be in close proximity to the speakers. Phantom stereo images can thus be generated which distort the perspective of the legitimate sound. A loudspeaker which has an encouragingly flat frequency response can therefore often sound far from neutral in a real listening environment. It is therefore essential to give consideration to loudspeaker placement, and a position such that the speakers are at head height when viewed from the listening position (high-frequency dispersion is much narrower than at lower frequencies, and therefore a speaker should be listened to on axis) and also away from room boundaries will give the most tonally accurate sound.

Some speakers, however, are designed to give of their best when mounted directly against a wall, the gain in bass response from such a position being allowed for in the design. A number of professional studio monitors are designed to be let into a wall such that their drivers are then level with the wall's surface. The manufacturer's instructions should be heeded, in conjunction with experimentation and listening tests. Speech is a good test signal. Male speech is good for revealing

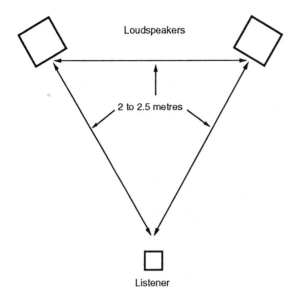

Figure 5.12 Recommended layout of loudspeakers for optimum stereo listening

boominess in a speaker, and female speech reveals treble splash from hard-surfaced objects nearby. Electronic music is probably the least helpful since it has no real-life reference by which to assess the reproduced sound. It is worth emphasising that the speaker is the means by which the results of previous endeavour are judged, and that time spent in both choosing and siting is time well spent.

Speakers are of course used in audio–visual work, and one frequently finds that it is desirable to place a speaker next to a video monitor screen. But the magnetic field from the magnets can affect the picture quality by pulling the internal electron beams off course. Some speakers are specially magnetically screened so as to avoid this.

For stereo reproduction, the optimum listening position is normally considered to be just rear of the apex of an equilateral triangle, formed by the loudspeakers and the listener, as shown in Figure 5.12. Depending on the directional properties of the loudspeakers in question, the stereo image can collapse into the nearest speaker for fairly small movements of the listener away from the centre.

Recommended further reading

Borwick. J. (1995) ed. *Loudspeaker and Headphone Handbook*, 2nd Edition. Focal Press

Colloms, M. (1991) *High Performance Loudspeakers*, 4th Edition. Pentech Press/Wiley

Earl, J. (1973) *Pickups and Loudspeakers*. Fountain Press

Gayford, M. (1970) *Loudspeakers*. Newnes–Butterworth

Sinclair, I. (1993) ed. *Audio and Hi-Fi Handbook*, 2nd Edition; Chapter 15: *Loudspeakers*, by Stan Kelly. Butterworth-Heinemann

Chapter 6

Mixers 1

In its simplest form an audio mixer combines several incoming signals into a single output signal. This cannot be achieved simply by connecting all the incoming signals in parallel and then feeding them into a single input because they may influence each other. The signals need to be isolated from each other. Individual control of at least the level of each signal is also required.

In practice, mixers also do rather more things than simply mix. They can provide phantom power for capacitor microphones (see section 4.3); pan control (whereby each signal can be placed in any desired position in a stereo image); filtering and equalisation; routing facilities; and monitoring facilities, whereby one of a number of sources can be routed to a pair of loudspeakers for listening, often without affecting the mixer's main output.

6.1 A simple six-channel mixer

6.1.1 Overview

By way of example, a simple six-channel mixer will be considered, having six inputs and two outputs (for stereo). Figure 6.1 illustrates such a notional six-into-two mixer with basic facilities. It also illustrates the back panel. The inputs illustrated are via XLR-type three-pin latching connectors, and are of a balanced configuration. Separate inputs are provided for microphone and line level signals, although it is possible to encounter systems which simply use one socket switchable to be either mic or line. Many cheap mixers have unbalanced inputs via quarter-inch jack sockets, or even 'phono' sockets such as are found on hi-fi amplifiers. Some mixers employ balanced XLR inputs for microphones, but unbalanced jack or phono inputs for line level signals, since the higher-level line signal is less susceptible to noise and interference, and will probably have travelled a shorter distance.

On some larger mixers a relatively small number of multipin connectors are provided, and multicore cables link these to a large jackfield which consists of rows of jack sockets mounted in a rack, each being individually labelled. All inputs and outputs will appear on this jackfield, and patch cords of a metre or so in length with GPO-type jack plugs at each end enable the inputs and outputs to be interfaced with other equipment and tie-lines in any appropriate combination. (The jackfield is more fully described in sections 6.4.9 and 13.12.)

The outputs are also on three-pin XLR-type connectors. The convention for

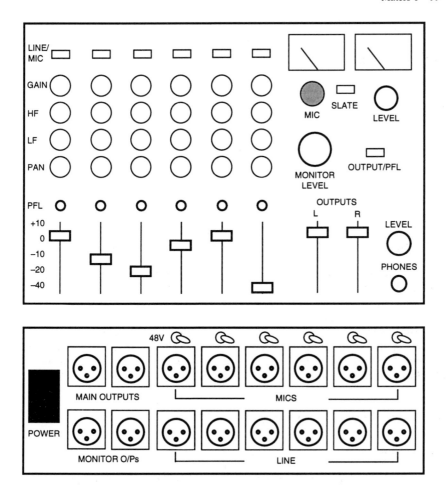

Figure 6.1 Front panel and rear connectors of a typical simple six-channel mixer

these audio connections is that inputs have sockets or holes, outputs have pins. This means that the pins of the connectors 'point' in the direction of the signal, and therefore one should never be confused as to which connectors are inputs and which are outputs. The microphone inputs also have a switch each for supplying 48 V phantom power to the microphones if required. Sometimes this switch is found on the input module itself, or sometimes on the power supply, switching 48 V for all the inputs at once.

6.1.2 Input channels

All the input channels in this example are identical, and so only one will be described. The first control in the signal chain is input gain or sensitivity. This control adjusts the degree of amplification provided by the input amplifier, and is often labelled in decibels, either in detented steps or continuously variable. Inputs are normally switchable between mic and line. In 'mic' position, depending on the

FACT FILE 6.1   Fader facts

Fader law

Channel and output faders, and also rotary level controls, can have one of two laws: linear or logarithmic (the latter sometimes also termed 'audio taper'). A linear law means that a control will alter the level of a signal (or the degree of cut and boost in a tone control circuit) in a linear fashion: that is, a control setting mid-way between maximum and minimum will attenuate a signal by half its voltage, i.e.: −6 dB. But this is not a very good law for an audio level control because a 6 dB drop in level does not produce a subjective halving of loudness. Additionally, the rest of the scaling (−10 dB, −20 dB, −30 dB and so on) has to be accommodated within the lower half of the control's travel, so the top half gives control over a mere 6 dB, the bottom half all the rest.

For level control, therefore, the logarithmic or 'log' law is used whereby a non-linear voltage relationship is employed in order to produce an approximately even spacing when the control is calibrated in decibels, since the decibel scale is logarithmic. A log fader will therefore attenuate a signal by 10 dB at a point approximately a quarter of the way down from the top of its travel. Equal dB increments will then be fairly evenly spaced below this point. A rotary log pot ('pot' is short for potentiometer) will have its maximum level usually set at the 5 o'clock

position, and the −10 dB point will be around the 2 o'clock position. An even subjective attenuation of volume level is therefore produced by the log law as the control is gradually turned down. A linear law causes very little to happen subjectively until one reaches the lowest quarter of the range, at which point most of the effect takes place.

The linear law is, however, used where a symmetrical effect is required about the central position; for example, the cut and boost control of a tone control section will have a central zero position about which the signal is cut and boosted to an equal extent either side of this.

Electrical quality

There are two types of electrical track in use, along which a conductive 'wiper' runs as the fader is moved to vary its resistance. One type of track consists of a carbon element, and is cheap to manufacture. The quality of such carbon tracks is, however, not very consistent and the 'feel' of the fader is often scrapy or grainy, and as it is moved the sound tends to jump from one level to another in a series of tiny stages rather than in a continuous manner. The carbon track wears out rather quickly, and can become unreliable.

The second type employs a conductive plastic track. Here, an electrically conductive material is diffused into a strip of plastic in a controlled manner to give the desired resistance value and law (linear or log). Much more expensive than the carbon track, the conductive plastic track gives smooth, continuous operation and maintains this standard over a long period of time. It is the only serious choice for professional-quality equipment.

---

output level of the microphone connected to the channel (see section 4.8), the input gain is adjusted to raise the signal to a suitable line level, and up to 80 dB or so of gain is usually available here (see section 6.4.1, below). In 'line' position little amplification is used and the gain control normally provides adjustment either side of unity gain (0 dB), perhaps ±20 dB either way, allowing the connection of high-level signals from such devices as CD players, tape machines and musical keyboards.

The equalisation or EQ section which follows (see section 6.4.4) has only two bands in this example – treble and bass – and these provide boost and cut of around ±12 dB over broad low- and high-frequency bands (e.g.: centred on 100 Hz and 10 kHz). This section can be used like the tone controls on a hi-fi amplifier to adjust

Pan control

The pan control on a mixer is used for positioning a signal somewhere between left and right in the stereo mix image. It does this by splitting a *single* signal from the output of a fader into two signals (left and right), setting the position in the image by varying the level difference between left and right channels. It is thus not the same as the balance control on a stereo amplifier, which takes in a *stereo* signal and simply varies the relative levels between the two channels. A typical pan-pot law would look similar to that shown in the diagram, and ensures a roughly constant perceived level of sound as the source is panned from left to right in stereo. The output of the pan-pot usually feeds the left and right channels of the stereo mix bus (the two main summation lines which combine the outputs of all channels on the mixer), although on mixers with more than two mix buses the

pan-pot's output may be switched to pan between any pair of buses, or perhaps simply between odd and even groups (see Fact File 6.4).

On some older consoles, four way routing is provided to a quadraphonic mix bus, with a left–right pot and a front–back pot. These are rare now. Many stereo pan-pots use a dual-gang variable resistor which follows a law giving a 4.5 dB level drop to each channel when panned centrally, compared with the level sent to either channel at the extremes. The 4.5 dB figure is a compromise between the –3 dB and –6 dB laws. Pan-pots which only drop the level by 3 dB in the centre cause a rise in level of any centrally panned signal if a mono sum is derived from the left and right outputs of that channel, since two identical signals summed together will give a rise in level of 6 dB. A pot which gives a 6 dB drop in the centre results in no level rise for centrally panned signals in the mono sum. Unfortunately, the 3 dB drop works best for stereo reproduction, resulting in no perceived level rise for centrally panned signals.

Only about 18 dB of level difference is actually required between left and right channels to give the impression that a source is either fully left or fully right in a loudspeaker stereo image, but most pan-pots are designed to provide full attenuation of one channel when rotated fully towards the other. This allows for the two buses between which signals are panned to be treated independently, such as when a pan control is used to route a signal either to odd or even channels of a multitrack bus (see section 6.4.2).

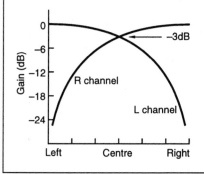

The spectral balance of the signal. The fader controls the overall level of the channel, usually offering a small amount of gain (up to 12 dB) and infinite attenuation. The law of the fader is specially designed for audio purposes (see Fact File 6.1). The pan control divides the mono input signal between left and right mixer outputs, in order to position the signal in a virtual stereo sound stage (see Fact File 6.2).

6.1.3 Output section

The two main output faders (left and right) control the overall level of all the channel signals which have been summed on the left and right mix buses, as shown in the block diagram (Figure 6.2). The outputs of these faders (often called the group

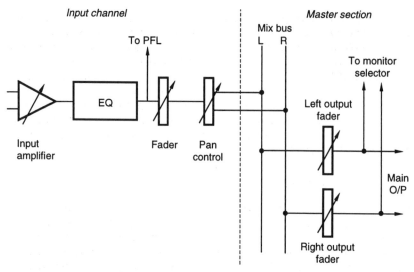

Figure 6.2 Block diagram of a typical signal path from channel input to main output on a simple mixer

FACT FILE

Pre-fade listen (PFL)

6.3

Pre-fade listen, or PFL, is a facility which enables a signal to be monitored without routing it to the main outputs of the mixer. It also provides a means for listening to a signal in isolation in order to adjust its level or EQ.

Normally, a separate mono mixing bus runs the length of the console picking up PFL outputs from each channel. A PFL switch on each channel routes the signal

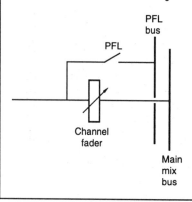

from before the fader of that channel to the PFL bus (see diagram), sometimes at the same time as activating internal logic which switches the mixer's monitor outputs to monitor the PFL bus. If no such logic exists, the mixer's monitor selector will allow for the selection of PFL, in which position the monitors will reproduce any channel currently with its PFL button pressed. On some broadcast and live consoles a separate small PFL loudspeaker is provided on the mixer itself, or perhaps on a separate output, in order that selected sources can be checked without affecting the main monitors.

Sometimes PFL is selected by 'over-pressing' the channel fader concerned at the bottom of its travel (i.e.: pushing it further down). This activates a microswitch which performs the same functions as above. PFL has great advantages in live work and broadcasting, since it allows the engineer to listen to sources before they are faded up (and thus routed to the main outputs which would be carrying the live programme). It can also be used in studio recording to isolate sources from all the others without cutting all the other channels, in order to adjust equalisation and other processing with greater ease.

outputs) feed the main output connectors on the rear panel, and an internal feed is taken from the main outputs to the monitor selector. The monitor selector on this simple example can be switched to route either the main outputs or the PFL bus (see Fact File 6.3) to the loudspeakers. The monitor gain control adjusts the loudspeaker output level *without affecting* the main line output level, but of course any changes made to the main fader gain will affect the monitor output.

The slate facility on this example allows for a small microphone mounted in the mixer to be routed to the main outputs, so that comments from the engineer (such as take numbers) can be recorded on a tape machine connected to the main outputs. A rotary control adjusts the slate level.

6.1.4 Miscellaneous features

Professional-quality microphones have an output impedance of around 200 ohms, and the balanced microphone inputs will have an input impedance of between 1 and 2000 ohms ('2 kΩ', k = thousand). The outputs should have an impedance of around 200 ohms or lower. The headphone output impedance will typically be 100 ohms or so. Small mixers usually have a separate power supply which plugs into the mains. This typically contains a mains transformer, rectifiers and regulating circuitry, and it supplies the mixer with relatively low DC voltages. The main advantage of a separate power supply is that the mains transformer can be sited well away from the mixer, since the alternating 50 Hz mains field around the former can

Figure 6.3 A typical simple stereo mixer: the Seem 'Seeport'. (Courtesy Seem Audio A/S)

be induced into the audio circuits. This manifests itself as 'mains hum' which is only really effectively dealt with by increasing the distance between the mixer and the transformer. Large mixers usually have separate rack-mounting power supply.

The above-described mixer is very simple, offering few facilities, but it provides a good basis for the understanding of more complex models. A typical commercial example of a basic mixer is shown in Figure 6.3.

6.2 A multitrack mixer

6.2.1 Overview

The stereo mixer outlined in the previous section only forms half the story in a multitrack recording environment. Conventionally, popular music recording involves at least two distinct stages: the 'track-laying' phase, and the 'mixdown' phase. In the former, musical tracks are layed down on a multitrack tape recorder in stages, with backing tracks and rhythm tracks being recorded first, followed by lead tracks and vocals. In the mixdown phase, all the previously recorded tracks from the tape recorder are played back through the mixer and combined into stereo to form the finished product which goes to be made into a commercial release (see section 3.2). More recently, with the widespread adoption of electronic instruments and MIDI equipment (see Chapter 15), the multitrack tape recorder has begun to play a smaller rôle in some recording studios, because MIDI-sequenced sound sources are now played directly into the mix in the second stage.

For these reasons, as well as requiring mixdown signal paths from many inputs to a stereo bus the mixer also requires signal paths for routing many input signals to a multitrack tape recorder. Often it will be necessary to perform both of these functions simultaneously — that is, recording microphone signals to multitrack tape whilst also mixing the return from tape into stereo, so that the engineer and producer can hear what the finished result will sound like, and so that any musicians who may be overdubbing additional tracks can be given a mixed feed of any previously recorded tracks in headphones. The latter is known as the *monitor mix* and this often forms the basis for the stereo mixdown when the track-laying job is finished.

So there are two signal paths in this case: one from the microphone or line source

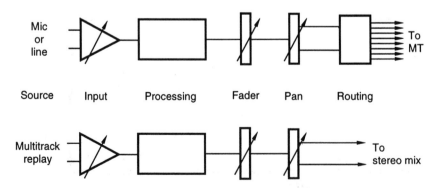

Figure 6.4 In multitrack recording two signal paths are needed – one from mic or line input to the multitrack recorder, and one returning from the recorder to contribute to a 'monitor' mix

to the multitrack tape recorder, and one from the multitrack recorder back to the stereo mix, as shown in Figure 6.4. The path from the microphone input which usually feeds the multitrack machine will be termed the *channel path,* whilst the path from the line input or tape return which usually feeds the stereo mix will be termed the *monitor path.*

It is likely that some basic signal processing such as equalisation will be required in the feed to the multitrack recorder (see below), but the more comprehensive signal processing features are usually applied in the mixdown path. The situation is somewhat different in the American market where there is a greater tendency to record on multitrack 'wet', that is with all effects and EQ, rather than applying the effects on mixdown.

6.2.2 In-line and split configurations

As can be seen from Figure 6.4, there are two complete signal paths, two faders, two sets of EQ, and so on. This takes up space, and there are two ways of arranging this physically, one known as the *split-monitoring,* or *European-style* console, the other as the *in-line* console. The split console is the more obvious of the two, and its physical layout is shown in Figure 6.5. It contains the input channels on one side (usually the left), a master control section in the middle, and the monitor mixer on the other side. So it really is two consoles in one frame. It is necessary to have as many monitor channels as there are tracks on the tape, and these channels are likely to need some signal processing. The monitor mixer is used during track laying for mixing a stereo version of the material that is being recorded, so that everyone can hear a rough mix of what the end result will sound like, and then on mixdown every input to the console can be routed to the stereo mix bus so as to increase the number of inputs for outboard effects, etc. and so that the comprehensive facilities provided perhaps only on the left side of the console are available for the tape returns.

This layout has advantages in that it is easily assimilated in operation, and it makes the channel module less cluttered than the in-line design (described below), but it can make the console very large when a lot of tracks are involved. It can also

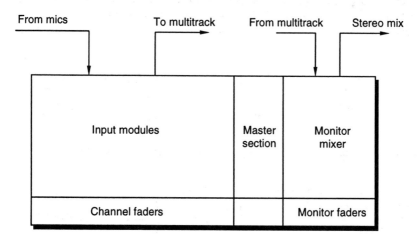

Figure 6.5 A typical 'split' or 'European-style' multitrack mixer has input modules on one side and monitor modules on the other: two separate mixers in effect

increase the build cost of the console because of the near doubling in facilities and metalwork required, and it lacks flexibility, especially when switching over from track laying to remixing.

The in-line layout involves the translation of everything from the right-hand side of the split console (the monitor section) into the left side, rather as if the console were sawn in half and the right side merged with the left, as shown in Figure 6.6. In this process a complete monitor signal path is fitted into the same module as the same-numbered channel path, making it no more than a matter of a few switches to enable facilities to be shared between the two paths. In such a design each module will contain two faders (one for each signal path), but usually only one EQ section, one set of auxiliary sends (see below), one dynamics control section, and so on, with switches to swap facilities between paths. (A simple example showing only the switching needed to swap one block of processing is shown in Figure 6.7.) Usually

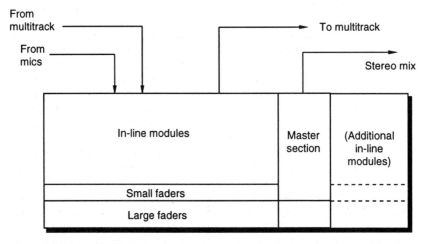

Figure 6.6 A typical 'in-line' mixer incorporates two signal paths in one module, providing two faders per module (one per path). This has the effect of reducing the size of the mixer for a given number of channels, when compared with a split design

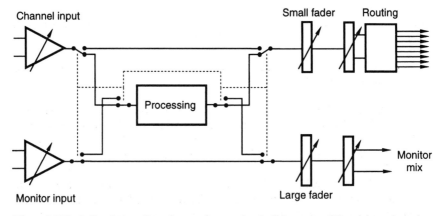

Figure 6.7 The in-line design allows for sound processing facilities such as EQ and dynamics to be shared or switched between the signal paths

this means that it is not possible to have EQ in both the multitrack recording path *and* the stereo mix path, but some more recent designs have made it possible to split the equaliser so that some frequency-band controls are in the channel path whilst others are in the monitor path. The band ranges are then made to overlap considerably which makes the arrangement quite flexible.

6.2.3 Further aspects of the in-line design

It has already been stated that there will be two main faders associated with each channel module in an in-line console: one to control the gain of each signal path. Sometimes the small fader is not a linear slider but a rotary knob. It is not uniformly agreed as to whether the large fader at the bottom of the channel module should normally control the *monitor* level of the like-numbered tape track or whether it should control the *channel* output level to multitrack tape. Convention has it that American consoles make the large fader 'the monitor fader' in normal operation, while British consoles tend to make it 'the channel fader'. Normally their functions may be swapped over, depending on whether one is mixing down or track laying, either globally (for the whole console), in which case the fader swap will probably happen automatically when switching the console from 'recording' to 'remix' mode, or on individual channels, in which case the operation is usually performed using a control labelled something like 'fader flip', 'fader reverse' or 'changeover'. The process of fader swapping is mostly used for convenience, since more precise control can be exercised over a large fader near the operator than over a small fader which is further away, and thus the large fader is assigned to the function which is being used most in the current operation. This is coupled with the fact that in an automated console, it is almost invariably the large fader which is automated, and the automation is required most in the mixdown process.

Confusion can arise when operating in-line mixers, such as when a microphone signal is fed into, say, mic input 1 and is routed to track 13 on the tape, because the operator will control the monitor level of that track (and therefore the level of that microphone's signal in the stereo mix) on monitor fader 13, whilst the channel fader on module 1 will control the multitrack record level for that mic signal.

If a 24 track tape machine is in use with the mixer, then monitor faders higher than number 24 will not normally carry a tape return, but will be free for other sources. Remember that more than one microphone signal can be routed to each track on the tape, and so there will be a number of level controls which affect each source's level in the monitor mix, each of which has a different purpose:

- **MIC LEVEL TRIM** – adjusts the gain of the microphone pre-amplifier at the channel input. Usually located at the top of the module.

- **CHANNEL FADER** – comes next in the chain and controls the individual level of the mic (or line) signal connected to that module's input before it goes to tape. Located on the same-numbered module as the input. (May be switched to be either the large or small fader, depending on configuration.)

- **BUS TRIM or TRACK SUBGROUP** – will affect the overall level of all signals routed to a particular tape track. Usually located with the track routing buttons at the top of the module. Sometimes a channel fader can be made to act as a subgroup master.

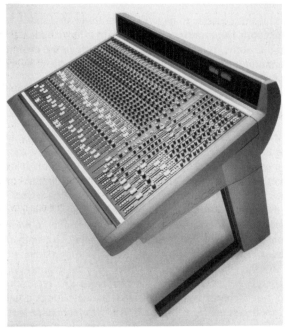

Figure 6.8 A typical in-line mixer: the Soundcraft 'Sapphyre'. (Courtesy of Soundcraft Electronics Ltd.)

- **MONITOR FADER** – is located in the return path from the multitrack recorder to the stereo mix. Does not affect the recorded level on the multitrack tape, but affects the level of this track in the mix. (May be switched to be either the large or small fader, depending on configuration.)

A typical in-line multitrack mixer is shown in the photograph in Figure 6.8.

6.3 Channel grouping

Grouping is a term which refers to the simultaneous control of more than one signal at a time. It usually means that one fader controls the levels of a number of slave channels. Two types of channel grouping are currently common: *audio grouping* and *'control' grouping*. The latter is often called VCA grouping, but there are other means of control grouping that are not quite the same as the direct VCA control method. The two approaches have very different results, although initially they may appear to be very similar due to the fact that one fader appears to control a number of signal levels. The primary reason for adopting group faders of any kind is in order to reduce the number of faders which the engineer has to handle at a time, and is feasible in a situation where a number of channels are carrying audio signals which can be faded up and down together. These signals do not all have to be at the same initial level, and indeed one is still free to adjust levels individually within a group. A collection of channels carrying drum sounds, or carrying an orchestral string section, would be examples of suitable groups. The two approaches are described in Fact Files 6.4 and 6.5.

FACT FILE 6.4   Audio groups

Audio groups are so called because they create a single audio output which is the sum of a number of channels. A single fader controls the level of the summed signal, and there will be a group output from the console which is effectively a mix of the audio signals in that group, as shown in the diagram. The audio signals from each input to the group are fed via equal-value resistors to the input of a summing or virtual-earth amplifier.

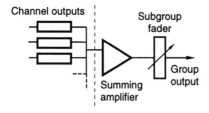

The stereo mix outputs from an in-line console are effectively audio groups, one for the left, one for the right, as they constitute a sum of all the signals routed to the stereo output and include overall level control. In the same way, the multitrack routing buses on an in-line console are also audio groups, as they are sums of all the channels routed to their respective tracks. More obviously, some smaller or older consoles will have routing buttons on each channel module for, say, four audio group destinations, these being really the only way of routing channels to the main outputs.

The master faders for audio groups will often be in the form of four or eight faders in the central section of the console. They may be arranged such that one may pan a channel between odd and even groups, and it would be common for two of these groups (an odd and an even one) to be used as the stereo output in mixdown. It is also common for perhaps eight audio group faders to be used as 'subgroups', themselves having routing to the stereo mix, so that channel signals can be made more easily manageable by routing them to a subgroup (or panning between two subgroups) and thence to the main mix via a single level control (the subgroup fader), as shown in the diagram. (Only four subgroups are shown in the diagram, without pan controls. Subgroups 1 and 3 feed the left mix bus, and 2 and 4 feed the right mix bus. Sometimes subgroup outputs can be panned between left and right main outputs.)

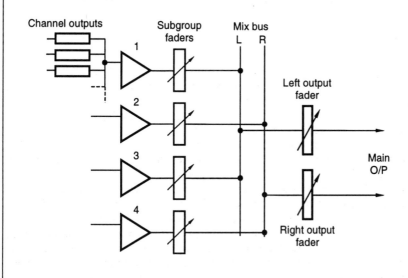

FACT FILE
6.5

Control groups

Control grouping differs from audio grouping primarily because it does not give rise to a single summed audio output for the group: the levels of the faders in the group are controlled from one fader, but their outputs remain separate. Such grouping can be imagined as similar in its effect to a large hand moving many faders at the same time, each fader maintaining its level in relation to the others.

The most common way of achieving control grouping is to use VCAs (Voltage-Controlled Amplifiers), whose gain can be controlled by a DC voltage applied to a control pin. In the VCA fader, audio is not passed through the fader itself but is routed through a VCA, whose gain is controlled by a DC voltage derived from the fader position, as shown in the diagram. So the fader now carries DC instead of audio, and the audio level is controlled indirectly.

Indirect gain control opens up all sorts of new possibilities. The gain of the channel could be controlled externally from a variety of sources, either by combining the voltage from an external controller in an appropriate way with the fader's voltage so that it would still be possible to set the relative level of the channel, or by breaking the direct connection between the DC fader and the VCA so that an automation system could intervene, as discussed in section 7.2. It becomes possible to see that group faders could be DC controls which could be connected to a number of channel VCAs such that their gains would go up and down together. Further to this, a channel VCA could be assigned to any of the available groups simply by selecting the appropriate DC path: this is often achieved by means of thumbwheel switches on each fader, as shown in the diagram.

Normally, there are dedicated VCA group master faders in a non-automated system. They usually reside in the central section of a mixer and will control the overall levels of any channel faders assigned to them by the thumbwheels by the faders. In such a system, the channel audio outputs would normally be routed to the main mix directly, the grouping affecting the levels of the individual channels in this mix.

In an automated system grouping may be achieved via the automation processor which will allow any fader to be designated as the group master for a particular group. This is possible because the automation processor reads the levels of all the faders, and can use the position of the designated master to modify the data sent back to the other faders in the group (see section 7.2).

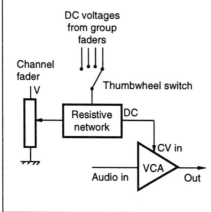

6.4 An overview of typical mixer facilities

Most mixing consoles provide a degree of sound signal processing on board, as well as routing to external processing devices. The very least of these facilities is some form of equalisation (a means of controlling the gain at various frequencies), and there are few consoles which do not include this. As well as signal processing, there will be a number of switches which make changes to the signal path or operational mode of the console. These may operate on individual channels, or they may function globally (affecting the whole console at once). The following section is a

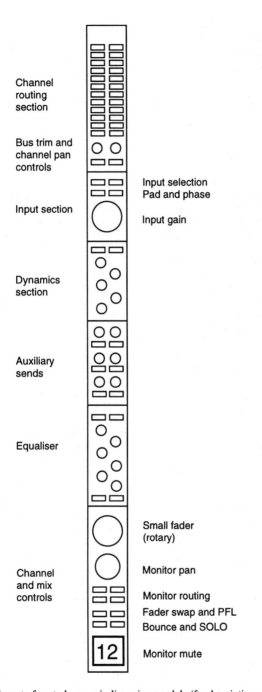

Figure 6.9 Typical layout of controls on an in-line mixer module (for description see text)

guide to the facilities commonly found on multitrack consoles. Figure 6.9 shows the typical location of these sections on an in-line console module.

6.4.1 Input section

- *Input gain control*
Sets the microphone or line input amplifier gain to match the level of the incoming signal. This control is often a coarse control in 10 dB steps, sometimes accompanied by a fine trim. Opinion varies as to whether this control should be in detented steps or continuous. Detented steps of 5 or 10 dB make for easy reset of the control to an exact gain setting, and precise gain matching of channels.

- *Phantom power*
Many professional mics require 48 volts phantom powering (see section 4.9). There is sometimes a switch on the module to turn it on or off, although most balanced mics which do not use phantom power will not be damaged if it is accidentally left on. Occasionally this switch is on the rear of the console, by the mic input socket, or it may be in a central assignable switch panel. Other methods exist: for example, one console requires that the mic gain control is pulled out to turn on the phantom power.

- *MIC/LINE switch*
Switches between the channel's mic input and line input. The line input could be the playback output from a tape machine, or another line level signal such as a synth or effects device.

- *PAD*
Usually used for attenuating the mic input signal by something like 20 dB, for situations when the mic is in a field of high sound pressure. If the mic is in front of a kick drum, for example, its output may be so high as to cause the mic input to clip. Also, capacitor mics tend to produce a higher output level than dynamic mics, requiring that the pad be used on some occasions.

- *Phase reverse or 'ø'*
Sometimes located after the mic input for reversing the phase of the signal, to compensate for a reversed directional mic, a miswired lead, or to create an effect. This is often left until later in the signal path.

- *HPF/LPF*
Filters can sometimes be switched in at the input stage, which will usually just be basic high- and low-pass filters which are either in or out, with no frequency adjustment. These can be used to filter out unwanted rumble or perhaps hiss from noisy signals. Filtering rumble at this stage can be an advantage because it saves clipping later in the chain.

6.4.2 Routing section

- *Track routing switches*
The number of routing switches depends on the console: some will have 24, some 32 and some 48. The switches route the channel path signal to the multitrack machine, and it is possible to route a signal to more than one track. The track assignment is often arranged as pairs of tracks, so that odd and even tracks can be assigned together, with a pan-pot used to pan between them as a stereo

pair, e.g.: tracks 3 and 4 could be a stereo pair for background vocals, and each background vocal mic could be routed to 3 and 4, panned to the relevant place in the image. In an assignable console these controls may be removed to a central assignable routing section.

It is common for there to be fewer routing switches than there are tracks, so as to save space, resulting in a number of means of assigning tracks. Examples are rotary knobs to select the track, one button per pair of tracks with 'odd/even/both' switch, and a 'shift' function to select tracks higher than a certain number. The multitrack routing may be used to route signals to effects devices during mixdown, when the track outputs are not being used for recording. In this case one would patch into the track output on the patchfield (see below) and take the relevant signal to an effects input somewhere else on the patchfield. In order to route monitor path signals to the track routing buses it may be necessary to use a switch which links the output of the monitor fader to the track assignment matrix.

In theatre sound mixers it is common for output routing to be changed very frequently, and thus routing switches may be located close to the channel fader, rather than at the top of the module as in a music mixer. On some recent mixers, track routing is carried out on a matrix which resides in the central section above the main faders. This removes unnecessary clutter from the channel modules and reduces the total number of switches required. It may also allow the storing of routing configurations in memory for later recall.

- *Mix routing switches*
  Sometimes there is a facility for routing the channel path output signal to the main monitor mix, or to one of perhaps four output groups, and these switches will often be located along with the track routing.

- *Channel pan*
  Used for panning channel signals between odd and even tracks of the multitrack, in conjunction with the routing switches.

- *Bus trim*
  Used for trimming the overall level of the send to multitrack for a particular bus. It will normally trim the level sent to the track which corresponds to the number of the module.

- *Odd/Even/Both*
  Occasionally found when fewer routing buttons are used than there are tracks. When one routing button is for a pair of tracks, this switch will determine whether the signal is sent to the odd channel only, the even channel only, or to both (in which case the pan control is operative).

- *DIRECT*
  Used for routing the channel output directly to the corresponding track on the multitrack machine without going via the summing buses. This can reduce the noise level from the console since the summing procedure used for combining a number of channel outputs to a track bus can add noise. If a channel is routed directly to a track, no other signals can be routed to that track.

6.4.3 Dynamics section

Some advanced consoles incorporate dynamics control on every module, so that each signal can be treated without resorting to external devices. The functions available on the best designs rival the best external devices, incorporating compressor and expander sections which can act as limiters and gates respectively if required. One system allows the EQ to be placed in the side-chain of the dynamics unit, providing frequency-sensitive limiting, among other things, and it is usually possible to link the action of one channel's dynamics to the next in order to 'gang' stereo channels so that the image does not shift when one channel has a sudden change in level while the other does not.

When dynamics are used on stereo signals it is important that left and right channels have the same settings, otherwise the image may be affected. If dynamics control is not available on every module, it is sometimes offered on the central section with inputs and outputs on the patchbay. Dynamics control will not be covered further here, but is discussed in more detail in section 14.2.

6.4.4 Equaliser section

The EQ section is usually split into three or four sections, each operating on a different frequency band. As each band tends to have similar functions these will be described in general. The principles of EQ are described in greater detail in section 6.5.

- *HF, MID 1, MID 2, LF*
  A high-frequency band, two mid-frequency bands, and a low-frequency band are often provided. If the EQ is *parametric* these bands will allow continuous variation of frequency (over a certain range), 'Q', and boost/cut. If it is not parametric, then there may be a few switched frequencies for the mid band, and perhaps a fixed frequency for the LF and HF bands.

- *Peaking/shelving or BELL*
  Often provided on the upper and lower bands for determining whether the filter will provide boost/cut over a fixed band (whose width will be determined by the Q), or whether it will act as a shelf, with the response rising or rolling off above or below a certain frequency (see Figure 6.14).

- *Q*
  The Q of a filter is defined as its centre frequency divided by its bandwidth (the distance between frequencies where the output of the filter is 3 dB lower than the peak output). In practice this affects the 'sharpness' of the filter peak or notch, high Q giving the sharpest response, and low Q giving a very broad response. Low Q would be used when boost or cut over a relatively wide range of frequencies is required, while high Q is used to boost or cut one specific region (see Fact File 6.6).

- *Frequency control*
  Sets the centre frequency of a peaking filter, or the turnover frequency of a shelf.

FACT FILE

6.6

Variable Q

Some EQ sections provide an additional control whereby the Q of the filter can be adjusted. This type of EQ section is termed a parametric EQ since all parameters, cut/boost, frequency, and Q can be adjusted. The diagram below illustrates the effect of varying the Q of an EQ section. High Q settings affect very narrow bands of frequencies, low Q settings affect wider bands. The low Q settings sound 'warmer' because they have gentle slopes and therefore have a more gradual and natural effect on the sound. High Q slopes are good for a rather more overt emphasis of a particular narrow band, which of course can be just as useful in the appropriate situation. Some EQ sections are labelled parametric even though the Q is not variable. This is a misuse of the term, and it is wise to check whether or not an EQ section is truly parametric even though it may be labelled as such.

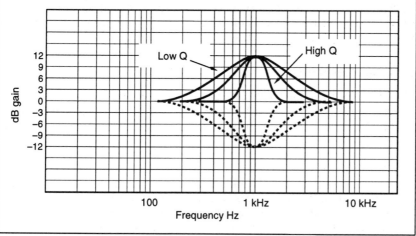

- *Boost/cut*
  Determines the amount of boost or cut applied to the selected band, usually up to a maximum of around ±15 dB.

- *HPF/LPF*
  Sometimes the high- and low-pass filters are located here instead of at the input, or perhaps in addition. They normally have a fixed frequency turnover point and a fixed roll-off of either 12 or 18 dB per octave. Often these will operate even if the EQ is switched out.

- *CHANNEL*
  The American convention is for the main equaliser to reside normally in the monitor path, but it can be switched so that it is in the channel path. Normally the whole EQ block is switched at once, but on some recent models a section of the EQ can be switched separately. This would be used to equalise the signal which is being *recorded* on multitrack tape. If the EQ is in the monitor path then it will only affect the replayed signal. The traditional European convention is for EQ to reside normally in the *channel* path, so as to allow recording with EQ.

- *IN/OUT*
  Switches the EQ in or out of circuit. Equalisation circuits can introduce noise and phase distortion, so they are best switched out when not required.

6.4.5 Channel and mix controls

- *Pan*
  See Fact File 6.2.

- *Fader reverse*
  Swaps the faders between mix and channel paths, so that the large fader can be made to control either the mix level or the channel level. Some systems defeat any fader automation when the large fader is put in the channel path. Fader reverse can often be switched globally, and may occur when the console mode is changed from recording to mixdown.

- *Line/Tape or Bus/Tape*
  Switches the source of the input to the monitor path between the line *output* of the same-numbered channel and the return from multitrack tape. Again it is possible that this may be switched globally. In 'line' or 'bus' mode the monitor paths are effectively 'listening to' the line output of the console's track assignment buses, while in 'tape' mode the monitor paths are listening to the off-tape signal (unless the tape machine's monitoring is switched to monitor the line input of the tape machine, in which case 'line' and 'tape' will effectively be the same thing!). If a problem is suspected with the tape machine, switching to monitor 'line' will bypass the tape machine entirely and allow the operator to check if the console is actually sending anything.

- *Broadcast, or 'mic to mix', or 'simulcast'*
  Used for routing the mic signal to both the channel and monitor paths simultaneously, so that a multitrack recording can be made while a stereo mix is being recorded or broadcasted. The configuration means that any alterations made to the channel path will not affect the stereo mix, which is important when the mix output is live (see Figure 6.10).

- *BUS or 'monitor-to-bus'*
  Routes the output of the monitor fader to the input of the channel path (or the channel fader) so that the channel path can be used as a post-fader effects send to any one of the multitrack buses (used in this case as aux sends), as shown in Figure 6.11. If a BUS TRIM control is provided on each multitrack output this can be used as the master effects-send level control.

- *DUMP*
  Incorporated (rarely) on some consoles to route the stereo panned mix output of a track (i.e.: after the monitor path pan-pot) to the multitrack assignment switches. In this way, the mixed version of a group of tracks can be 'bounced down' to two tracks on the multitrack, panned and level-set as in the monitor mix (see Figure 6.11).

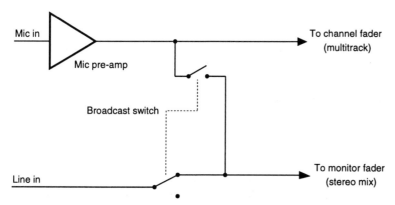

Figure 6.10 A 'broadcast mode' switch in an in-line console allows the microphone input to be routed to both signal paths, such that a live stereo mix may be made independent of any changes to multitrack recording levels

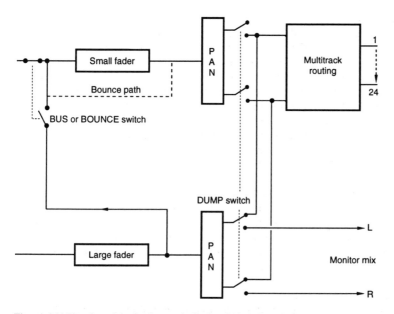

Figure 6.11 Signal routings for 'bounce', 'bus' and 'dump' modes (see text)

- *BOUNCE*
  A facility for routing the output of the monitor fader to the multitrack assignment matrix, before the pan control, in order that tracks can be 'bounced down' so as to free tracks for more recording by mixing a group of tracks on to a lower number of tracks. BOUNCE is like a mono version of DUMP (see Figure 6.11).

- *MUTE or CUT*
  Cuts the selected track from the mix. There may be two of these switches, one for cutting the channel signal from the multitrack send, the other for cutting the mix signal from the mix.

- *PFL*
  See Fact File 6.3.

- *AFL*
  After-fade listen is similar to PFL, except that it is taken from after the fader. This is sometimes referred to as *SOLO,* which routes a panned version of the track to the main monitors, cutting everything else. These functions are useful for isolating signals when setting up and spotting faults. On many consoles the AFL bus will be stereo. Solo functions are useful when applying effects and EQ, in order that one may hear the isolated sound and treat it individually without hearing the rest of the mix. Often a light is provided to show that a solo mode is selected, because there are times when nothing can be heard from the loudspeakers due to a solo button being down with no signal on that track. A *solo safe* control may be provided centrally, which prevents this feature from being activated.

- *In-place solo*
  On some consoles, solo functions as an 'in-place' solo, which means that it actually changes the mix output, muting all tracks which are not solo'ed and picking out all the solo'ed tracks. This may be preferable to AFL as it reproduces the exact contribution of each channel to the mix, at the presently set master mix level. Automation systems often allow the solo functions to be automated in groups, so that a whole section can be isolated in the mix. In certain designs, the function of the automated mute button on the monitor fader may be reversed so that it becomes solo.

6.4.6 Auxiliary sends

The number of aux sends depends on the console, but there can be up to ten on an ordinary console, and sometimes more on assignable models. Aux sends are 'take-off points' for signals from either the channel or mix paths, and they appear as outputs from the console which can be used for foldback to musicians, effects sends, cues, and so on. Each module will be able to send to auxiliaries, and each numbered auxiliary output is made up of all the signals routed to that aux send. So they are really additional mix buses. Each aux will have a master gain control, usually in the centre of the console for adjusting the overall gain of the signal sent from the console, and may have basic EQ. Aux sends are often a combination of mono and stereo buses. Mono sends are usually used as routes to effects, while stereo sends may have one level control and a pan control per channel for mixing a foldback source.

- *Aux sends 1–n*
  Controls for the level of each individual channel in the numbered aux mix.

- *Pre/post*
  Determines whether the send is taken off before or after the fader. If it is before then the send will still be live even when the fader is down. Generally, 'cue' feeds will be pre-fade, so that a mix can be sent to foldback which is independent of the monitor mix. Effects sends will normally be taken post-fade, in order that the effect follows a track's mix level.

- *Mix/channel*
  Determines whether the send is taken from the mix or channel paths. It will often be sensible to take the send from the channel path when effects are to be recorded on to multitrack rather than on to the mix. This function has been labelled 'WET' on some designs.

- *MUTE*
  Cuts the numbered send from the aux mix.

6.4.7 Master control section

The master control section usually resides in the middle of the console, or near the right-hand end. It will contain some or all of the following facilities:

- *Monitor selection*
  A set of switches for selecting the source to be monitored. These will include tape machines (stereo), aux sends, the main stereo mix, and perhaps some miscellaneous external sources like CD players, cassette machines, etc. They only select the signal going to the loudspeakers, not the mix outputs. This may be duplicated to some extent for a set of additional studio loudspeakers, which will have a separate gain control.

- *DIM*
  Reduces the level sent to the monitor loudspeakers by a considerable amount (usually around 40 dB), for quick silencing of the room.

- *MONO*
  Sums the left and right outputs to the monitors into mono so that mono compatibility can be checked.

- *Monitor phase reverse*
  Phase reverses one channel of the monitoring so that a quick check on suspected phase reversals can be made.

- *TAPE/LINE*
  Usually a global facility for switching the inputs to the mix path between the tape returns and the console track outputs. Can be reversed individually on modules.

- *FADER REVERSE*
  Global swapping of small and large faders between mix and channel paths.

- *Record/Overdub/Mixdown*
  Usually globally configures mic/line input switching, large and small faders and auxiliary sends depending on mode of operation. (Can be overridden on individual channels.)

- *Auxiliary level controls*
  Master controls for setting the overall level of each aux send output.

- *Foldback and Talkback*
  There is often a facility for selecting which signals are routed to the stereo foldback which the musicians hear on their headphones. Sometimes this is as comprehensive as a cue mixer which allows mixing of aux sends in various amounts to various stereo cues, while often it is more a matter of selecting whether foldback consists of the stereo mix, or one of the aux sends. Foldback level is controllable, and it is sometimes possible to route left and right foldback signals from different sources. *Talkback* is usually achieved using a small microphone built into the console, which can be routed to a number of destinations. These destinations will often be aux sends, multitrack buses, mix bus, studio loudspeakers and foldback.

- *Oscillator*
  Built-in sine-wave oscillators vary in quality and sophistication, some providing only one or two fixed frequencies, while others allow the generation of a whole range. If the built-in oscillator is good it can be used for lining up the tape machine, as it normally can be routed to the mix bus or the multitrack outputs. The absolute minimum requirement is for accurate 1 kHz and 10 kHz tones, the 10 kHz being particularly important for setting the bias of an analogue tape machine. The oscillator will have an output level control.

- *Slate*
  Provides a feed from the console talkback mic to the stereo output, often superimposing a low-frequency tone (around 50 Hz) so that the slate points can be heard when winding a tape at high speed. Slate would be used for recording take information on to tape.

- *Master faders*
  There may be either one stereo fader or left and right faders to control the overall mix output level. Often the group master faders will reside in this section.

6.4.8 Effects returns

Effects returns are used as extra inputs to the mixer, supplied specifically for inputs from external devices such as reverberation units. These are often located in the central section of the console and may be laid out like reduced-facility input channels. Returns sometimes have EQ, perhaps more basic than on channels, and they may have aux sends. Normally they will feed the mix, although sometimes facilities are provided to feed one or more returns to the multitrack via assignment switches. A small fader or rotary level control is provided, as well as a pan-pot for a mono return. Occasionally, automated faders may be assigned to the return channels so as to allow automated control of their levels in the mix.

6.4.9 Patchfield or jackfield

Most large consoles employ a built-in jackfield or patchbay for routing signals in ways which the console switching does not allow, and for sending signals to and from external devices. Just about every input and output on every module in the console comes up on the patchbay, allowing signals to be cross-connected in virtually any configuration. The jackfield is usually arranged in horizontal rows, each row having an equal number of jacks. Vertically, it tries to follow the signal path of the console as closely as possible, so the mic inputs are at the top and the multitrack outputs are nearer the bottom. In between these there are often *insert points* which allow the engineer to 'break into' the signal path, often before or after the EQ, to insert an effects device, compressor, or other external signal processor. Insert points usually consist of two rows, one which physically breaks the signal chain when a jack is inserted, and one which does not. Normally it is the lower row which breaks the chain, and should be used as inputs. The upper row is used as an output or send. *Normalling* is usually applied at insert points, which means that unless a jack is inserted the signal will flow directly from the upper row to the lower.

At the bottom of the patchfield will be all the master inputs and outputs, playback returns, perhaps some parallel jacks, and sometimes some spare rows for connection of one's own devices. Some consoles bring the microphone signals up to the patchbay, but there are some manufacturers who would rather not do this unless absolutely necessary as it is more likely to introduce noise, and phantom power may be present on the jackfield. Jackfields are covered in further detail in section 13.12.

6.5 EQ explained

The tone control or EQ (= equalisation) section provides mid-frequency controls in addition to bass and treble. A typical comprehensive EQ section may have firstly an HF (High-Frequency) control similar to a treble control but operating only at the highest frequencies. Next would come a hi-mid control, affecting frequencies from around 1 kHz to 10 kHz, the centre frequency being adjusted by a separate control. Lo-mid controls would come next, similar to the hi-mid but operating over a range of say 200 Hz to 2 kHz. Then would come an LF (Low-Frequency) control. Additionally, high- and low-frequency filters can be provided. The complete EQ section looks something like Figure 6.12. An EQ section takes up quite a bit of space, and so it is quite common for dual concentric controls to be used. For instance, the cut/boost controls of the hi- and lo-mid sections can be surrounded by annular skirts which select the frequency. Console area is therefore saved.

6.5.1 Principal EQ bands

The HF section affects the highest frequencies and provides up to 12 dB of boost or cut. This type of curve is called a shelving curve because it gently boosts or cuts the frequency range towards a shelf where the level remains relatively constant (see Figure 6.13(a)). Next comes the hi-mid section. Two controls are provided here, one to give cut or boost, the other to select the desired centre frequency. The latter is commonly referred to as a 'swept mid' because one can sweep across the frequency range.

Figure 6.13(b) shows the result produced when the frequency setting is at the 1 kHz position, termed the centre frequency. Maximum boost and cut affects this frequency the most, and the slopes of the curve are considerably steeper than those of the previous shelving curves. This is often referred to as a 'bell' curve due to the upper portion's resemblance to the shape of a bell. It has a fairly high 'Q', that is its

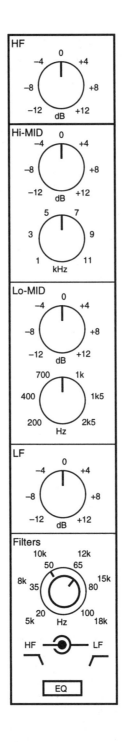

Figure 6.12 Typical layout of an EQ section

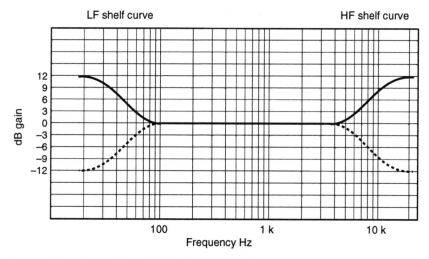

Figure 6.13 (a) Typical HF and LF shelf EQ characteristics shown at maximum settings

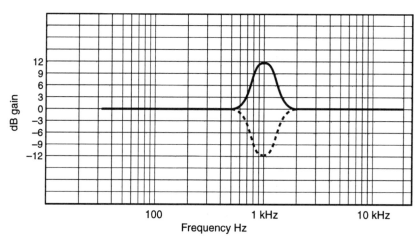

Figure 6.13 (b) Typical MF peaking filter characteristic

sides are steep. Q is defined as:

Q = centre frequency ÷ bandwidth

where the bandwidth is the distance in hertz between the two points at which the response of the filter is 3 dB lower than that at the centre frequency. In the example shown the centre frequency is 1 kHz and the bandwidth is 400 Hz, giving Q = 2.5.

MF EQ controls are often used to hunt for troublespots; if a particular instrument (or microphone) has an emphasis in its spectrum somewhere which does not sound very nice, some mid cut can be introduced, and the frequency control can be used to search for the precise area in the frequency spectrum where the trouble lies. Similarly, a dull sound can be given a lift in an appropriate part of the spectrum which will bring it to life in the overall mix. Figure 6.13(c) shows the maximum cut

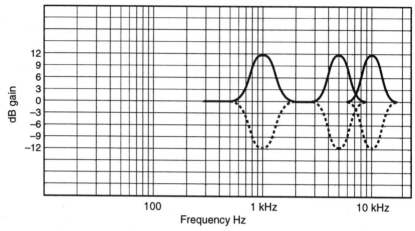

Figure 6.13 (c) MF peaking filter characteristics at 1, 5 and 10 kHz

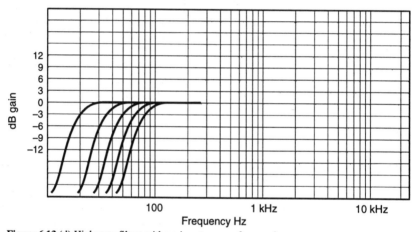

Figure 6.13 (d) High-pass filters with various turnover frequencies

and boost curves obtained with the frequency selector at either of the three settings of 1, 5 and 10 kHz. The high Q of the filters enables relatively narrow bands to be affected. Q may be varied in some cases, as described in Fact File 6.6.

The lo-mid section is the same as the hi-mid section except that it covers a lower band of frequencies. Note though that the highest frequency setting overlaps the lowest setting of the hi-mid section. This is quite common, and ensures that no 'gaps' in the frequency spectrum are left uncovered.

6.5.2 Filters

High- and low-cut filters provide fixed attenuation slopes at various frequencies. Figure 6.13(d) shows the responses at LF settings of 80, 65, 50, 35 and 20 Hz. The slopes are somewhat steeper than is the case with the HF and LF shelving curves, and slope rates of 18 or 24 dB per octave are typical. This enables just the lowest, or highest, frequencies to be rapidly attenuated with minimal effect on the mid band. Very low traffic rumble could be removed by selecting the 20 or 35 Hz setting. More serious low-frequency noise may require the use of one of the higher turnover frequencies. High-frequency hiss from, say, a noisy guitar amplifier or air escaping from a pipe organ bellows can be dealt with by selecting the turnover frequency of the HF section which attenuates just sufficient HF noise without unduly curtailing the HF content of the wanted sound.

6.6 Stereo line input modules

In broadcast situations it is common to require a number of inputs to be dedicated to stereo line level sources, such as cart machines, CD players, tapes, etc. Such modules are sometimes offered as an option for multitrack consoles, acting as replacements for conventional I/O modules and allowing two signals to be faded up and down together with one fader. Often the EQ on such modules is more limited, but the module may provide for the selection of more than one stereo source, and routing to the main mix as well as the multitrack. It is common to require that stereo modules always reside in special slots on the console, as they may require special wiring. Such modules may also provide facilities for handling LP turntable outputs, offering RIAA equalisation (see section 11.2).

With the advent of stereo television, the need for stereo *microphone* inputs is also becoming important, with the option for MS (middle and side) format signals as well as AB (conventional left and right) format (see section 4.7).

6.7 Dedicated monitor mixer

A dedicated monitor mixer is often used in live sound reinforcement work to provide a separate monitor mix for each musician, in order that each artist may specify his or her precise monitoring requirements. A comprehensive design will have, say, 24 inputs containing similar facilities to a conventional mixer, except that below the EQ section there will be a row of rotary or short-throw faders which individually send the signal from that channel to the group outputs, in any combination of relative levels. Each group output will then provide a separate monitor mix to be fed to headphones or amplifier racks.

6.8 Basic operational techniques

6.8.1 Level setting

If one is using a microphone to record speech or classical music then normally a fairly high input gain setting will be required. If the microphone is placed up against a guitar amplifier then the mic's output will be high and a much lower input gain setting can be used. There are essentially three ways of setting the gain control to the optimum position. Firstly, using PFL or pre-fade listen (see Fact File 6.3).

PFL is pressed, or the fader overpressed (i.e.: pressed beyond the bottom of its travel against a sprung microswitch), on the input module concerned and the level read on either a separate PFL meter or with the main meters switched to monitor the PFL bus. The channel input gain should be adjusted to give a meter reading of, say, PPM 5, or 0 VU (although bearing in mind that VUs tend to give a lower reading than PPMs on material with a large transient content, thus making it easy to overrecord when using VUs, as discussed in section 7.5). This gain-setting procedure must be carried out at a realistic input level from the source. It is frequently the case during rehearsals that vocalists and guitarists will produce a level that is rather lower than that which they will use when they actually begin to play.

The pan control should be set next (see Fact File 6.2) to place the source in the stereo image. The main output faders will normally be set to 0 dB on their calibration, which is usually at the top. The channel faders can then be set to give both the desired subjective sound balance and appropriate output meter readings.

The second way of setting the gain is a good way in its own right, and it has to be used if PFL facilities are not provided. First of all both the channel fader and the output faders need to be positioned to the 0 dB point. This will be either at the top of the faders' travels or at a position about a quarter of the way down from the top of their travel. If no 0 dB position is indicated then the latter position should be set. After the pan control and faders have been positioned, the input gain may then be adjusted to give the desired reading on the output level meters. When several incoming signals need to be balanced the gain controls should all be positioned to give both the desired sound balance between them and the appropriate meter readings – normally PPM 6 or just over 0 VU during the loudest passages.

These two gain-setting methods differ in that with the former method the channel fader positions will show a correspondence to the subjective contribution each channel is making towards the overall mix, whereas the latter method places all the channel faders at roughly the same level.

The third way is similar to the second way, but one channel at a time is set up, placing channel and output faders at 0 dB and adjusting the gain for a peak meter reading. That channel fader is then turned completely down and the next channel is set up in a similar way. When all the channels which are to be used have been set up, the channel faders can then be advanced to give both the desired subjective balance and peak meter readings.

Use of the EQ controls often necessitates the resetting of the channel's input gain. For example, if a particular instrument requires a bit of bass boost, applying this will also increase the level of signal and so the gain will often need to be reduced a little to compensate. Applying bass or treble cut will sometimes require a small gain increase.

6.8.2 Using auxiliary sends

Aux facilities were described in section 6.4.6. The auxiliaries are configured either 'pre-fade' or 'post-fade'. Pre-fade aux sends are useful for providing a monitor mix for musicians, since this balance will be unaffected by movements of the faders which control the main mix. The engineer then retains the freedom to experiment in the control room without disturbing the continuity of feed to the musicians.

Post-fade sends are affected by the channel fader position. These are used to send signals to effects devices and other destinations where it is desirable to have the aux level under the overall control of the channel fader. For example, the engineer may wish to add a little echo to a voice. Aux 2, set to post-fade, is used to send the signal to an echo device, probably positioning the aux 2 control around the number 6 position and the aux 2 master at maximum. The output of the echo device is returned to another input channel or an echo return channel, and this fader can be adjusted to set the amount of echo. The level of echo will then rise and fall with the fader setting for the voice.

The post-fade aux could also be used simply as an additional output to drive separate amplifiers and speakers in another part of a hall, for example.

6.8.3 Using audio groups

The group outputs (see section 6.3) or multitrack routing buses (see section 6.4.2) can be used for overall control of various separate groups of instruments, depending on whether mixing down or track laying. For example, a drum kit may have eight microphones on it. These eight input channels can be routed to groups 1 and 2 with appropriate stereo pan settings. Groups 1 and 2 would then be routed to stereo outputs left and right respectively. Overall control of the drum kit level is now achieved simply by moving group faders 1 and 2.

When feeding a multitrack tape machine it is normally desirable to use the highest possible recording level on every track regardless of the final required balance, in order to achieve the best noise performance, and each multitrack group output will usually have an output level meter to facilitate this.

Recommended further reading

See *General further reading* at the end of this book.

Chapter 7

Mixers 2

7.1 Technical specifications of analogue mixers

7.1.1 Input noise

The output from a microphone is in the millivolt range, and so needs considerable amplification to bring it up to line level. Amplification of the signal also brings with it amplification of the microphone's own noise output (discussed in section 4.8.2), which one can do nothing about, and amplification of the mixer's own input noise. The latter must therefore be as low as possible so as not to compromise the noise performance unduly. A 200 ohm source resistance on its own generates 0.26 μV of noise (20 kHz bandwidth). Referred to the standard line level of 775 mV (0 dBu) this is −129.6 dBu. A microphone amplifier will add its own noise to this, and so manufacturers quote an 'equivalent input noise' (EIN) value which should be measured with a 200 ohm source resistance across the input.

An amplifier with a noise contribution equal to that of the 200 ohm resistor will degrade the theoretically 'perfect' noise level by 3 dB, and so the quoted equivalent input noise will be −129.6 + 3 = −126.6 dBm. (Because noise contributions from various sources sum according to their power content, not their voltage levels, dBm is traditionally used to express input noise level.) This value is quite respectable, and good-quality mixers should not be noisier than this. Values of around −128 dBm are sometimes encountered which are excellent, indicating that the input resistance is generating more noise than the amplifier is. Make sure that the EIN is quoted with a 200 ohm source, and a bandwidth up to 20 kHz, unweighted. A 150 ohm source, sometimes specified, will give an apparently better EIN simply because this resistor is itself quieter than a 200 ohm one, resistor noise being proportional to ohmic value. Also, weighting gives a flattering result, so one always has to check the measuring conditions. Make sure that EIN is quoted in dBm or dBu. Some manufacturers quote EIN in dBV (i.e.: ref. 1 volt) which gives a result 2.2 dB better.

An input should have high common mode rejection as well as low noise, as discussed in Fact File 7.1.

7.1.2 Output noise

The output residual noise of a mixer, with all faders at minimum, should be at least −90 dBu. There is no point in having a very quiet microphone amplifier if a noisy output stage ruins it. With all channels routed to the output, and all faders at the

FACT FILE

7.1

Common mode rejection

As discussed in section 13.4, common mode rejection is the ability of a balanced input to reject interference which can be induced into the signal lines. A microphone input should have a CMRR (Common Mode Rejection Ratio) of 70 dB or more; i.e.: it should attenuate the interference by 70 dB. But look at how this measurement is made. It is relatively easy to achieve 70 dB at, say, 500 Hz, but rejection is needed most at high frequencies – between 5 and 20 kHz – and so a quoted CMRR of '70 dB at 15 kHz' or '70 dB between 100 Hz and 10 kHz' should be sought. Line level CMRR can be allowed to be rather lower since the signal voltage level is a lot higher than in microphone cabling. CMRRs of as low as 30 dB at 10 kHz are deemed to be adequate.

Common mode rejection is a property of a balanced input, and so it is not applicable to a balanced output. However, output balance is sometimes quoted which gives an indication of how closely the two legs of a balanced output are matched. If the two legs were to be combined in antiphase total cancellation would ideally be achieved. In practice, around 70 dB of attenuation should be looked for.

'zero' position, output noise (or 'mixing' noise) should be at least −80 dBu with the channel inputs switched to 'line' and set for unity gain. Switching these to 'mic' inevitably increases noise levels because this increases the gain of the input amplifier. It underlines the reason why all unused channels should be switched out, and their faders brought down to a minimum. Make sure that the aux outputs have a similarly good output noise level.

7.1.3 Impedance

A microphone input should have a minimum impedance of 1 kΩ. A lower value than this degrades the performance of many microphones. A line level input should have a minimum impedance of 10 kΩ. Whether it is balanced or unbalanced should be clearly stated, and consideration of the type of line level equipment that the mixer will be partnered with will determine the importance of balanced line inputs. All outputs should have a low impedance, below 200 ohms, balanced (600 ohms sounds nice and professional, but it is much too high, as described in section 13.9). Check that the aux outputs are also of very low impedance. Sometimes they are not. If insert points are provided on the input channels and/or outputs, these also should have very low output and high input impedances.

7.1.4 Frequency response

A frequency response which is within 0.2 dB between 20 Hz and 20 kHz for all combinations of input and output is desirable. The performance of audio transformers varies slightly with different source and load impedances, and a specification should state the range of loads between which a 'flat' frequency response will be obtained. Above 20 kHz, and probably below 15 Hz or so, the frequency response should fall away so that unwanted out-of-band frequencies are not amplified, for example radio-frequency breakthrough or subsonic interference.

7.1.5 Distortion

Distortion should be quoted at maximum gain through the mixer and a healthy output level of, say, +10 dBu or more. This will produce a typical worst case, and should normally be less than 0.1% THD (Total Harmonic Distortion). The distortion of the low-gain line level inputs to outputs can be expected to be lower: around 0.01%. The outputs should be loaded with a fairly low impedance which will require more current from the output stages than a high impedance will, this helping to reveal any shortcomings. A typical value is 600 ohms.

Clipping and overload margins are discussed in Fact File 7.2.

7.1.6 Crosstalk

A signal from one input may induce a small signal in another channel, and this is termed 'crosstalk'. Crosstalk from adjacent channels should be well below the level of the legitimate output signal, and a figure of –80 dB or more should be looked for. Crosstalk performance tends to deteriorate at high frequencies due to capacitive coupling in wiring harnesses for instance, but a crosstalk of at least –60 dB at 15 kHz should still be sought. Similarly, very low-frequency crosstalk often deteriorates due to the power supply source impedance rising here, and a figure of –50 dB at 20 Hz is reasonable.

Ensure that crosstalk between all combinations of input and output is of a similarly good level. Sometimes crosstalk between channel auxiliaries is rather poorer than that between the main outputs.

FACT FILE | **Clipping**

7.2

A good mixer will be designed to provide a maximum electrical output level of at least +20 dBu. Many will provide +24 dBu. Above this electrical level clipping will occur, where the top and bottom of the audio waveform are chopped off, producing sudden and excessive distortion (see diagram). Since the nominal

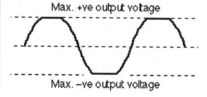

Max. +ve output voltage

Max. –ve output voltage

reference level of 0 dBu usually corresponds to a meter indication of PPM 4 or –4 VU, it is very difficult to clip the output stages of a mixer. The maximum meter indication on a PPM would correspond in this case to an electrical output of around +12 dBu, and thus one would have to be severely bending the meter needles to cause clipping.

Clipping, though, may occur at other points in the signal chain, especially when large amounts of EQ boost have been added. If, say, 12 dB of boost has been applied on a channel, and the fader is set well above the 0 dB mark, clipping on the mix bus may occur, depending on overload margins here. Large amounts of EQ boost should not normally be used without a corresponding overall gain reduction of the channel for this reason.

An input pad or attenuator is often provided to prevent the clipping of mic inputs in the presence of high-level signals (see section 6.4.1).

7.2 Metering systems

Metering systems are provided on audio mixers to indicate the levels of audio signals entering and leaving the mixer. Careful use of metering is vital to the recording of the correct audio level on tape (and thus the avoidance of distortion and noise). In this section the merits of different metering systems are examined.

7.2.1 Mechanical metering

Two primary types of mechanical meters are in existence today: the VU (Volume Unit) meter (Figure 7.1) and the PPM (Peak Program Meter), as shown in Figure 7.2. These are very different to each other, the only real similarity being that they both have swinging needles. The British, or BBC-type, PPM is distinctive in styling in that it is black with numbers ranging from 1 to 7 equally spaced across its scale, there being 4 dB level difference between each gradation, except between 1 and 2 where there is usually a 6 dB change in level. The EBU PPM (also shown in Figure 7.2) has a scale calibrated in decibels. The VU, on the other hand, is usually white or cream, with a scale running from –20 dB up to +3 dB, ranged around a zero point which is usually the studio's electrical reference level.

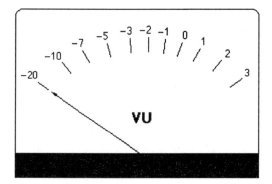

Figure 7.1 Typical VU meter scale

Figure 7.2 (a) BBC-type peak programme meter (PPM). (b) European-type PPM

FACT FILE

7.3

Metering and tape distortion

Within a studio there is usually a 'house reference level' and a 'house peak recording level'. The reference level usually relates to the level at which the 1 kHz line-up tone from a test tape should play back on the console's meters, and may correspond to PPM 4 or PPM 5 on a BBC-type PPM, which may in turn correspond to either –4 dB or 0 dB on a VU meter (the relationship between VUs and PPMs depends on the standard in use, as described in section 7.5). PPM 4 usually corresponds to an electrical level of 0 dBu at the console main output. The tone on the test tape will have been recorded at a specific magnetic flux level (usually one of 320, 250 or 200 nanowebers per metre) depending on the recording standard in use, and thus a relationship is set up between the meter reading and the magnetic recording level. This is covered in greater detail in Chapter 8.

It is the magnetic recording level which is the most important factor influencing distortion and compression on any analogue

tape recorder, and any distortion introduced by the electronics will be minimal in comparison unless the level is excessively high (above, say, +20 dBu, which would be bending the needles on most normal console meters). All modern professional tape has specifications quoted for it, as will the tape machine in use, and these will include figures for the maximum output level (MOL), which is usually the point at which third-harmonic distortion reaches 3% of the total output when recording a 1kHz tone (see Appendix 1). This is a generally accepted peak recording level above which it is not a good idea to venture unless a particular effect is required, and tends to lie between 8 and 12 dB above 320 nWb m^{-1} on modern tape. Thus if PPM 4 or –4 VU is set to correspond to a magnetic level of 320 nWb m^{-1}, PPM 6 will be 8 dB above this reference, and PPM 7 will be 12 dB above, and somewhere between these two should be assumed to be the peak recording level. Naturally, the setup needs to be tailored to the studio's individual requirements, but these are general guidelines. Typically, in a broadcast environment, peak level is PPM 6 in order to avoid overmodulating transmitters.

Although this section is not intended to be a tutorial on tape machine line-up and reference levels, it is impossible to cover the subject of metering without reference to such topics, as they are inextricably intertwined. It is important to know how meter readings relate to the line-up standard in use in a particular environment, and to understand that these standards may vary between establishments. Fact File 7.3 discusses the relationship between meter indication and recording level on a tape recorder.

7.2.2 Problems with mechanical meters

PPMs respond well to signal peaks, that is they have a fast rise-time, whereas VUs are quite the opposite: they have a very slow rise-time. This means that VUs do not give a true representation of the peak level going on to tape, especially in cases when a signal with a high transient content, such as a harpsichord, is being recorded, often showing as much as 10–15 dB lower than a peak-reading meter. This can result in overmodulation of the recording, especially with digital recorders where the system is very sensitive to peak overload. None the less, many people are used to working with VUs, and have learned to interpret them. They are good for measuring continuous signals such as tones, but their value for monitoring programme material is dubious in the age of digital recording.

VUs have no control over the fall-time of the needle, which is much the same as the rise-time, whereas PPMs are engineered to have a fast rise-time and a longer fall-time, which tends to be more subjectively useful. The PPM was designed to indicate peaks which would cause audible distortion, but does not measure the absolute peak level of a signal. Mechanical meters take up a lot of space on a console, and it can be impossible to find space for one meter per channel in the case of a multitrack console. In this case there are often only meters on the main outputs, and perhaps measuring some auxiliary signals, these being complemented on more expensive consoles by electronic bargraph meters, usually consisting of LED or liquid crystal displays, or some form of 'plasma' display.

7.2.3 Electronic bargraph metering

Unlike mechanical meters, electronic bargraphs have no mechanical inertia to be overcome, so they can effectively have an infinitely fast rise-time although this may not be the ideal in practice. Cheaper bargraphs are made out of a row of LEDs (Light Emitting Diodes), and the resolution accuracy depends on the number of LEDs used. This type of display is sometimes adequate, but unless there are a lot of gradations it is difficult to use them for line-up purposes. Plasma and liquid crystal displays look almost continuous from top to bottom, and do not tend to have the glare of LEDs, being thus more comfortable to work with for any period of time. Such displays often cover a dynamic range far greater than any mechanical meter, perhaps from –50 dB up to +12 dB, and so can be very useful in showing the presence of signals which would not show up on a mechanical PPM. Such a meter is illustrated in Figure 7.3.

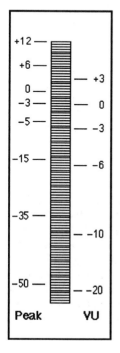

Figure 7.3 Typical peak-reading bargraph meter with optional VU scale

There may be a facility provided to switch the peak response of these meters from PEAK to VU mode, where they will imitate the scale and ballistic response of a VU meter. On more up-market designs it may be possible to use the multitrack bargraphs as a spectrum analyser display, indicating perhaps a one-third octave frequency-band analysis of the signal fed to it. Occasionally, bargraph displays incorporate a peak-hold facility. A major advantage of these vertical bargraphs is that they take up very little horizontal space on a meter bridge and can thus be used for providing one meter for every channel of the console: useful for monitoring the record levels on a multitrack tape machine. In this case, the feed to the meter is usually taken off at the input to the monitor path of an in-line module.

Miscellaneous meters may also be provided on the aux send outputs for giving some indication of the level being sent to auxiliary devices such as effects. These are commonly smaller than the main meters, or may consist of LED bargraphs with lower resolution. A phase meter or correlation meter is another option often available, this usually being connected between the left and right main monitor outputs to indicate the degree of phase correlation between these signals. This can be either mechanical or electronic. In broadcast environments, sum and difference (or M and S) meters may be provided to show the level of the mono-compatible and stereo difference signals in stereo broadcasting. These often reside alongside a stereo meter for left and right output levels.

7.2.4 Correlation between different metering standards

The correlation between meter indication and electrical output level varies depending on the type of meter and the part of the world concerned. Figure 7.4 shows a number of common meter scales and the relationship between these scales and electrical output level of the mixer. As introduced in Fact File 7.3, there is a further correlation to be concerned with, this being the relationship between the electrical output level of the mixer and the recording level on an analogue or digital tape machine. This is discussed in greater detail in Fact File 8.5.

7.2.5 Meter take-off point

Output level meter-driving circuits should normally be connected directly across the outputs so that they register the real output levels of the mixer. This may seem self-evident but there are certain models in which this is not the case, the meter circuit taking its drive from a place in the circuit just before the output amplifiers. In such configurations, if a faulty lead or piece of equipment the mixer is connected to places, say, a short-circuit across the output the meter will nevertheless read normal levels, and lack of signal reaching a destination will be attributed to other causes. The schematic circuit diagrams of the mixer can be consulted to ascertain whether such an arrangement has been employed. If it is not clear, a steady test tone can be sent to the mixer's output, giving a high meter reading. Then a short-circuit can be deliberately applied across the output (the output amplifier will not normally be harmed by several seconds of short-circuit) and the meter watched. If the indicated level drastically reduces then the meter is correctly registering the real output. If it stays high then the meter is taking its feed from elsewhere.

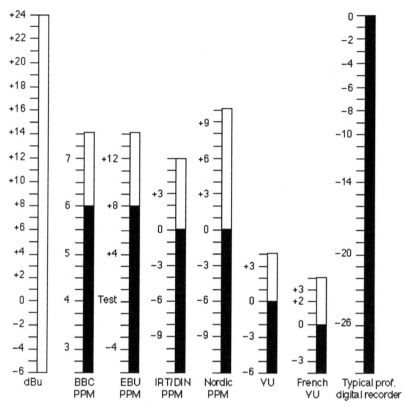

Figure 7.4 Graphical comparison of commonly encountered meter scalings and electrical levels in dBu. (After David Pope, with permission)

7.3 Automation

7.3.1 Background

The original, and still most common form of mixer automation is a means of storing fader positions dynamically against time for reiteration at a later point in time, synchronous with recorded material. The aim of automation has been to assist an engineer in mixdown when the number of faders that need to be handled at once become too great for one person. Fader automation has resulted in engineers being able to concentrate on sub-areas of a mix at each pass, gradually building up the finished product and refining it.

MCI first introduced VCA (voltage controlled amplifier) automation for their JH-500 series of mixing consoles in the mid-seventies, and this was soon followed by imitations with various changes from other manufacturers. Moving fader automation systems, such as Neve's NECAM, were introduced slightly later and tended to be more expensive than VCA systems. During the mid 1980s, largely because of the falling cost of microprocessor hardware, console automation enjoyed further advances resulting in developments such as snapshot storage, total dynamic automation, retrofit automation packages, and MIDI-based automation. It is now possible to install basic fader automation on a console for only a few thousand pounds, whereas previously one might have been contemplating tens of thousands. The rise of digital mixers and digitally controlled analogue mixers with integral automation is likely to continue the trend towards total automation of most mixer controls as a standard feature of new products.

In the following sections a number of different approaches to console automation will be presented and discussed.

7.3.2 Fader automation

There are two common means of memorising and controlling the gain of a channel: one which stores the positions of the fader and uses this data to control the gain of a VCA or digital attenuator, the other which also stores fader movements but uses this information actually to drive the fader's position using a motor. The former is cheaper to implement than the latter, but is not so ergonomically satisfactory because the fader's physical position may not always correspond to the gain of the channel.

It is possible to combine elements of the two approaches in order that gain control can be performed by a VCA but with the fader being moved mechanically to display the gain. This allows for rapid changes in level which might be impossible using physical fader movements, and also allows for dynamic gain offsets of a stored mix whilst retaining the previous gain profile (see below). In the following discussion the term 'VCA faders' may be taken to refer to any approach where indirect gain control of the channel is employed.

With VCA faders it is possible to break the connection between a fader and the corresponding VCA, as was described in Fact File 6.5. It is across this break point that an automation system will normally be connected. The automation processor then reads a digital value corresponding to the position of the fader and can return a value to the VCA to control the gain of the channel (see Figure 7.5). The information sent back to the VCA would depend on the operational mode of the system at the time, and might or might not correspond directly to the fader position. Common operational modes are:

- WRITE: VCA gain corresponds directly to the fader position
- READ: VCA gain controlled by data derived from a previously stored mix
- UPDATE: VCA gain controlled by a combination of previously stored mix data and current fader position
- GROUP: VCA gain controlled by a combination of the channel fader's position and that of a group master

The fader position is measured by an analogue-to-digital convertor (see Chapter 10), which turns the DC value from the fader into a binary number (usually eight or ten bits) which the microprocessor can read. An eight bit value suggests that the fader's position can be represented by one of 256 discrete values, which is usually enough to give the impression of continuous movements, although professional systems tend to use ten bit representation for more precise control (1024 steps). The automation computer 'scans' the faders many times a second and reads their values. Each fader has a unique address and the information obtained from each address is stored in a different temporary memory location by the computer. A generalised block diagram of a typical system is shown in Figure 7.6.

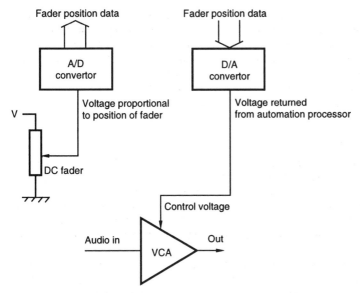

Figure 7.5 Fader position is encoded so that it can be read by an automation computer. Data returned from the computer is used to control a VCA through which the audio signal flows

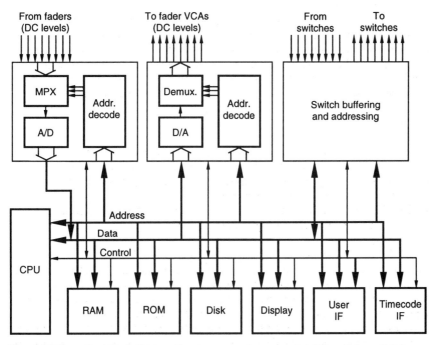

Figure 7.6 Generalised block diagram of a mixer automation system handling switches and fader positions. The fader interfaces incorporate a multiplexer (MPX) and demultiplexer (Demux) to allow one convertor to be shared between a number of faders. RAM is used for temporary mix data storage, ROM may hold the operating software program. The CPU is the controlling microprocessor

The disadvantage of such a system is that it is not easy to see what the level of the channel is. During a read or update pass the automation computer is in control of the channel gain, rather than the fader. The fader could be half way to the bottom of its travel while the gain of the VCA was near the top. Sometimes a mixer's bargraph meters can be used to display the value of the DC control voltage which is being fed from the automation to the VCA, and a switch is sometimes provided to change their function to this mode. Alternatively a separate display is provided for the automation computer, indicating fader position with one marker and channel gain with another.

VCA faders are commonly provided with 'null' LEDs: little lights on the fader package which point in the direction that the fader must be moved to make its position correspond to the gain of the VCA. When the lights go out (or when they are both on), the fader position is correct. This can sometimes be necessary when modifying a section of the mix by writing over the original data. If the data fed to the VCA from the automation is different to the position of the fader, then when the mode is switched from read to write there will be a jump in level as the fader position takes over from the stored data. The null lights allow the user to move the fader towards the position dictated by the stored data, and most systems only switch from read to write when the null point is crossed, to ensure a smooth transition. The same procedure is followed when coming out of rewrite mode, although it can be bypassed in favour of a sudden jump in level.

Update mode involves using the *relative* position of the fader to modify the stored data. In this mode, the fader's absolute position is not important because the system assumes that its starting position is a point of unity gain, thereafter adding the *changes* in the fader's position to the stored data. So if a channel was placed in update mode and the fader moved up by 3 dB, the overall level of the updated passage would be increased by 3 dB (see Figure 7.7). For fine changes in gain the fader can be preset near the top of its range before entering update mode, whereas larger changes can be introduced nearer the bottom (because of the gain law of typical faders).

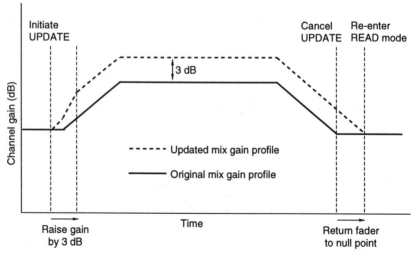

Figure 7.7 Graphical illustration of stages involved in entering and leaving an UPDATE or RELATIVE mode on an automated VCA fader

Some systems make these modes relatively invisible, anticipating which mode is most appropriate in certain situations. For example, WRITE mode is required for the first pass of a new mix, where the absolute fader positions are stored, whereas subsequent passes might require all the faders to be in UPDATE.

A moving fader system works in a similar fashion, except that the data which is returned to the fader is used to set the position of a drive mechanism which physically moves the fader to the position in which it was when the mix was written. This has the advantage that the fader is its own means of visual feedback from the automation system and will always represent the gain of the channel.

If the fader was permanently driven, there would be a problem when both the engineer and the automation system wanted to control the gain. Clutches or other forms of control are employed to remove the danger of a fight between fader and engineer in such a situation, and the fader is usually made touch-sensitive to detect the presence of a hand on it.

Such faders are, in effect, permanently in update mode, as they can at any time be touched and the channel gain modified, but there is usually some form of relative mode which can be used for offsetting a complete section by a certain amount. The problem with relative offsets and moving faders is that if there is a sudden change in the stored mix data while the engineer is holding the fader, it will not be executed. The engineer must let go for the system to take control again. This is where a combination of moving fader and VCA-type control comes into its own.

7.3.3 Grouping automated faders

Conventional control grouping (Fact File 6.5) is normally achieved by using dedicated VCA master faders. In an automated console it may be possible to do things differently. The automation computer has access to data representing the positions of all the main faders on the console, so it may allow any fader to be designated a group master for a group of faders assigned to it. It can do this by allowing the user to set up a fader as a group master (either by pressing a button on the fader panel, or from a central control panel). It will then use the level from this fader to modify the data sent back to all the other VCAs in that group, taking into account their individual positions as well. This idea means that a master fader can reside physically within the group of faders to which it applies, although this may not always be the most desirable way of working.

Sometimes the computer will store automation data relating to groups in terms of the motions of the individual channels in the group, without storing the fact that a certain fader was the master, whereas other systems will store the data from the master fader, remembering the fact that it was a master originally.

7.3.4 Mute automation

Mutes are easier to automate than faders because they only have two states. Mute switches associated with each fader are also scanned by the automation computer, although only a single bit of data is required to represent the state of each switch. A simple electronic switch can be used to effect the mute, and this often takes the form of a FET (field effect transistor) in the signal path, which has very high attenuation in its 'closed' position (see Figure 7.8). Alternatively, some more basic systems effect mutes by a sudden change in VCA gain, pulling it down to maximum attenuation.

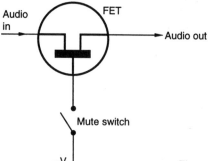

Figure 7.8 Typical implementation of a FET mute switch

7.3.5 Storing the automation data

Early systems converted the data representing the fader positions and mute switches into a modulated serial data stream which could be recorded alongside the audio to which it related on a multitrack tape. In order to allow updates of the data, at least two tracks were required: one to play back the old data, and one to record the updated data, these usually being the two outside tracks of the tape (1 and 24 in the case of a 24 track machine). This was limiting, in that only the two most recent mixes were ever available for comparison (unless more tracks were set aside for automation), whole tracks had to be mixed at a time (because otherwise the updated track would be incomplete), and at least two audio tracks were lost on the tape. Yet it meant that the mix data was always available alongside the music, eliminating the possibility of losing a disk with the mix data stored separately.

More recent systems use computer hardware to store mix data, in RAM and on disks (either hard or floppy). Data is synchronised to the audio by recording timecode on one track of the tape which uniquely identifies any point in time, this being read by the automation system and used to relate tape position to stored data. This method gives almost limitless flexibility in the modification of a mix, allowing one to store many versions, of which sections can be joined together 'off-line' (that is, without the tape running) or on-line, to form the finished product. The finished mix can be dumped to a disk for more permanent storage, and this disk could contain a number of versions of the mix.

It is becoming quite common for cheaper automation systems to use MIDI for the transmission of fader data. A basic automation computer associated with the mixer converts fader positions into MIDI information using a device known as a UART which generates and decodes serial data at the appropriate rate for the MIDI standard, as shown in Figure 7.9. MIDI data can then be stored on a conventional sequencer or using dedicated software. This proves adequate for a small number of channels or for uncomplicated mixes, especially if a dedicated sequencer is used for the storage of automation data, but it is possible that a single MIDI interface would be overloaded with information if large mixes with much action were attempted. One recent system overcomes these limitations by using a multiport MIDI interface (see Chapter 15) and a non-standard implementation of MIDI to carry ten bit fader data for a large number of channels.

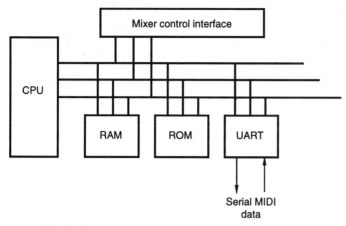

Figure 7.9 A UART is used to route MIDI data to and from the automation computer

7.3.6 Integrating machine control

Control of a tape machine or machines is a common feature of modern mixers. It may only involve transport remotes being mounted in the centre panel somewhere, or it may involve a totally integrated autolocator/synchroniser associated with the rest of the automation system. On top-flight desks, controls are provided on the channel modules for putting the relevant tape track into record-ready mode, coupled with the record function of the transport remotes. This requires careful interfacing between the console and the tape machine, but means that it is not necessary to work with a separate tape machine remote unit by the console.

It is very useful to be able to address the automation in terms of the mix in progress: in other words, 'go back to the second chorus', should mean something to the system, even if abbreviated. The alternative is to have to address the system in terms of timecode locations. Often, keys are provided which allow the engineer to return to various points in the mix, both from a mix data point-of-view and from the tape machines' point-of-view, so that the automation system locates the tape machine to the position described in the command, ready to play. This facility is often provided using an integral synchroniser with which the automation computer can communicate, and this can be a semi-customised commercial synchroniser from another manufacturer.

MIDI Machine Control (MMC) is also becoming a popular means of remote control for modular multitrack recording equipment, and is another way of inter-facing a tape recorder to an automation system.

7.3.6 Retrofitting automation

Automation can usually be retrofitted into existing consoles which do not have any automation. These systems usually control only the faders and the mutes, as anything else requires considerable modification of the console's electronics, but the relatively low price of some systems makes them attractive, even on a modest budget. Fitting normally involves a modification or replacement of the fader package, to incorporate VCAs in consoles which don't have them, or to break into the control path between fader and VCA in systems which do. This job can normally be achieved in a day. It is also possible to retrofit moving fader automation.

A separate control panel may be provided, with buttons to control the modes of operation, as well as some form of display to show things like VCA gains, editing modes, and set up data. The faders will be interfaced to a processor rack which would reside either in a remote bay, or under the console, and this will normally contain a disk drive to store the final mixes. Alternatively a standard desktop computer will be used as the control interface.

7.3.7 Total automation systems

This title means different things to different people. SSL originally coined the term 'Total Recall' for its system, which was not in fact a means of completely resetting every control on the console; rather it was a means of telling the operator where the controls should be and leaving him to reset them himself. None the less, this saved an enormous amount of time in the resetting of the console in between sessions, because it saved having to write down the positions of every knob and button.

True Total Reset is quite a different proposition and requires an interface between the automation system and every control on the console, with some means of measuring the position of the control, some means of resetting it, and some means of displaying what is going on. A number of options exist, for example one could:

- motorise all the rotary pots
- make all the pots continuously rotating and provide a display
- make the pots into up/down-type incrementers with display
- provide assignable controls with larger displays

Of these, the first is impractical in most cases due to the space that motorised pots would take up, the reliability problem, and the cost, although it does solve the problem of display. The second would work, but again there is the problem that a continuously rotating pot would not have a pointer because it would merely be a means of incrementing the level from wherever it was at the time, so extra display would be required and this takes up space. None the less some ingenious solutions have been developed, including incorporating the display in the head of rotary controls (see Figure 7.10). The third is not ergonomically very desirable, as the human prefers analogue interfaces rather than digital ones, and there is no room on a conventional console for all the controls to be of this type with their associated displays. Most of the designs which have implemented total automation have adopted a version of the fourth option: that is to use fewer controls than there are functions, and to provide larger displays.

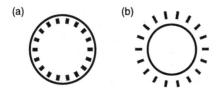

Figure 7.10 Two possible options for positional display with continuously rotating knobs in automated systems. (a) Lights around the rim of the knob itself; (b) lights around the knob's base

The concept of total automation is inherent in the principles of an assignable mixing console. In such a console, few of the controls carry audio directly as they are only interfaces to the control system, so one knob may control the HF EQ for any channel to which it is assigned, for example. Because of this indirect control, usually via a microprocessor, it is relatively easy to implement a means of storing the switch closures and settings in memory for reiteration at a later date.

7.3.8 Dynamic and static systems

Many analogue assignable consoles use the modern equivalent of a VCA: the digitally controlled attenuator, to control the levels of various functions such as EQ, aux sends, and so on. If every function on the desk is to be dynamically memorised (that is, changes stored and replayed in real time, and continuously) then a lot of data is produced and must be dealt with quickly to ensure suitably seamless audible effects. Complex software and fast processors are required, along with a greater requirement for memory space.

Static systems exist which do not aim to store the continuous changes of all the functions, but they will store 'snapshots' of the positions of controls which can be recalled either manually or with respect to timecode. This can often be performed quite regularly (many times a second) and in these cases we approach the dynamic situation, but in others the reset may take a second or two which precludes the use of it during mixing. Changes must be silent to be useful during mixing.

Other snapshot systems merely store the settings of switch positions, without storing the variable controls, and this uses much less processing time and memory. Automated routing is of particular use in theatre work where sound effects may need to be routed to a complex combination of destinations. A static memory of the required information is employed so that a single command from the operator will reset all the routing ready for the next set of sound cues.

7.4 Digital mixers

In a digital mixer incoming analogue signals are converted to a digital signal as early as possible so that all the functions are performed entirely in the digital domain. Digital inputs and outputs can be provided to connect recording devices and other digital equipment (see Chapter 10) without conversion to analogue. The advantage of this is that once the signal is in the digital domain it is inherently more robust than its analogue counterpart: it is virtually immune from crosstalk, and is unaffected by lead capacitance, electromagnetic fields from mains wiring, additional circuit distortion and noise, and other forms of interference.

Functions such as gain, EQ, delay, phase, routing, and effects such as echo, reverb, compression and limiting, can all be carried out in the digital domain precisely and repeatably using digital signal processing as described in Chapter 10. The microphone amplifiers can be sited remotely from the mixer, close to the microphones themselves, the gain still being adjusted from the mixer, so that their line level outputs can now be converted to digital format before being input to the mixer. Operationally a digital mixer can remain similar to its analogue counterpart, although the first commercial examples have tended at least partially to follow the assignable route described above.

Digital mixing has now reached the point where it can be implemented cost

Figure 7.11 Yamaha 02R digital mixing console (Courtesy of Sound PR)

Figure 7.12 Sony OXF-R3 (Courtesy of Sony Broadcast and Professional Europe)

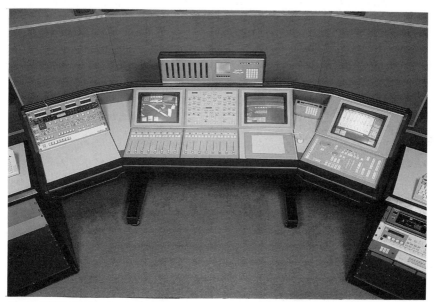

Figure 7.13 Solid State Logic *OmniMix*

effectively, and Yamaha has released a number of low cost digital mixers with full automation. An example is pictured in Figure 7.11. At the other end of the scale companies are manufacturing large-scale studio mixers with an emphasis on ultra-high sound quality and an ergonomically appropriate control interface. An example of such a mixer is pictured in Figure 7.12. There is also the possibility for multitrack digital audio storage and editing to be integrated with a digital mixer, as demonstrated by Solid State Logic in their audio-for-video post production products such as *OmniMix* and *Scenaria*, one of which is shown in Figure 7.13. At the low cost end of the scale, digital mixers are implemented within computer-based workstations and represented graphically on the computer display. Faders and other controls are moved using a mouse.

Recommended further reading

See *General further reading* at the end of this book.

Chapter 8

Analogue tape recording

Analogue sound recording is the process by which a continuously variable electrical waveform, representing an acoustic sound signal (see Chapter 1), is converted into an analogous pattern of modulation on a physical medium, which may be stored and subsequently replayed. In tape recording, this pattern of modulation is magnetic, such that the pattern of magnetisation (strength and direction of field) on the tape after recording is analogous to the sound waveform which was recorded. Analogue recording differs from digital recording (see Chapter 10) in that the recorded signal is physically and temporally continuous, whereas in digital recording it is time discrete (sampled) and stored in the form of data. A magnified analysis of the pattern of modulation of an analogue recording would show up clear similarities with the causatory sound waveform, whereas one would be unlikely to see any such similarity in a visual analysis of the recorded signal on a digital tape.

8.1 Magnetic tape

8.1.1 Structure

Magnetic tape consists of a length of plastic material which is given a surface coating capable of retaining magnetic flux rather in the manner that, say, an iron rod is capable of being magnetised (see Figure 8.1). The earliest recorders actually used a length of iron wire as the recording medium. In practice all modern tape has a polyester base which was chosen, after various trials with other formulations which proved either too brittle (they snapped easily) or too plastic (they stretched), for its good strength and dimensional stability. It is used throughout the tape industry from the dictation microcassette to the 2 inch (5 cm) multitrack variety. The coating is of a metal oxide, or metal alloy particles.

The most common coating used is of gamma-ferric oxide, a kind of purified rust with specially shaped particles (signified by 'gamma'). This formulation is used in cassettes along with the alternative formulations of chromium dioxide and its substitutes, and metal particles. It is also almost exclusively used for open-reel

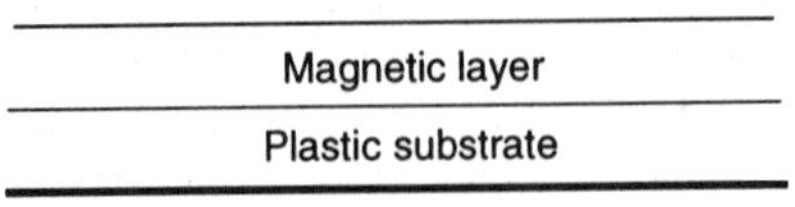

Magnetic layer

Plastic substrate

Figure 8.1 Cross-section through a magnetic recording tape

analogue master tapes of all widths – quarter inch, half inch, 1 inch, 2 inch – although a particular brand and formulation may not always be available in all widths.

8.1.2 Recent history

The BASF company of Germany introduced a chromium dioxide formulation in the early 1970s, claiming better high-frequency performance and improved signal-to-noise ratios. The patent on this formulation spurred other manufacturers to experiment with alternatives, and a substitute which consisted of cobalt-enriched ferric oxide emerged. Then so-called 'ferri-chrome' tape was introduced which consisted of a layer of ferric oxide over which was applied a layer of chromium dioxide or a substitute. High frequencies tend to be recorded close to the surface of an oxide layer, lower frequencies being recorded more deeply. The dual layer was therefore intended to exploit this, taking advantage of the top layer's good performance at high frequencies together with the lower layer's strengths in the areas of distortion and good lower-frequency output capability. This type of formulation is infrequently used today, partially because it tended to display appalling middle-frequency distortion characteristics, and also because of improvements in standard ferric and chrome formulations.

During the 1980s pure metal tape was developed for the cassette medium. Initially this posed great difficulties for the manufacturers due to the fact that very tiny particles of pure iron or iron alloy of the size required for recording tape tended to oxidise rapidly – they caught fire! This problem was, however, overcome and metal cassettes were launched which were rather more expensive than the oxide types and additionally required specially equipped cassette machines with record heads and circuitry capable of effectively magnetising the high-coercivity metal tape. Many cheaper cassette machines, although intended to cope with this tape and labelled as such, could not fully exploit it. The coercivity of the tape – its willingness to accept and retain magnetic flux – was such that cheap tape heads often saturated magnetically before the tape did, due to the extremely high bias currents required to magnetise it (see section 8.2.3). Improvements in distortion, output level and signal-to-noise ratios were obtainable with metal tape from good cassette machines, at a cost.

The metal formulation came into its own in a modified form when its high recording density capability proved ideal for digital tape recording, in special cassette form for R-DAT (see section 10.5.3) and also for hand-held domestic camcorder machines.

8.1.3 Cassette tape

Cassette tape comes in various lengths to give appropriate playing times for given applications – 'C5' gives 2.5 minutes playing time per side, C90 gives 45 minutes per side and so on. The longer-playing tapes are thinner in order that more tape can be loaded on to the spools. The C120 cassette has very thin tape which causes problems with many machines since their transports are unable to handle the delicate and very flimsy tape adequately. The tape will tend to stick on to the rubber pinch roller as it passes between the roller and the capstan, causing the transport to chew the tape up in bad cases, wrapping it around the capstan and pinch roller. Excessively thin tape does not sit comfortably in the tape guides in the cassette housing, and it is also not pulled across the tape heads very evenly, causing poor head-to-tape contact with loss of output and frequency response. It would be difficult to find a cassette machine manufacturer who actually sanctioned the use of C120 cassettes.

8.1.4 Open-reel tape

Open-reel quarter-inch tape intended for analogue recorders has been available in a variety of thicknesses. Standard Play tape has an overall thickness of 50 microns (micrometres), and a playing time (at 15 inches (38 cm) per second) of 33 minutes is obtained from a 10 inch (25 cm) reel. Long Play tape has an overall thickness of 35 microns giving a corresponding 48 minutes of playing time, which is very useful for live recording work. In the past 'Double Play' and even 'Triple Play' thicknesses have been available, these being aimed at the domestic open-reel market. These formulations are prone to snapping or stretching, as well as offering slightly poorer sound quality, and should not really be considered for professional use.

Standard Play tape is almost always 'back coated'. A rough coating is applied to the back of the tape during manufacture which produces neater and more even winding on a tape machine, by providing a certain amount of friction between layers which holds the tape in place. Also, the rough surface helps prevent air being trapped between layers during fast spooling which can contribute to uneven winding. Long Play tape is also available with a back coating, but as often as not it will be absent. It is worth noting that the flanges of a tape spool should only serve to protect the tape from damage. The 'pancake' of tape on the spool should not touch these flanges. Metal spools are better than plastic spools because they are more rigid and they do not warp. Professional open-reel tape can be purchased either on spools or in 'pancake' form on hubs without flanges. The latter is of course cheaper, but considerable care is needed in its handling so that spillage of the unprotected tape does not occur. Such pancakes are either spooled on to empty reels before use, or they can be placed on top of a special reel with only a lower flange. Professional tape machines are invariably operated with their decks horizontal. Half inch, 1 inch and 2 inch tape intended for multitrack recorders always comes on spools, is always of Standard Play thickness, and is always back coated.

Open-reel tape should have a batch number printed on the box. This number ensures that a given delivery of tape was all manufactured at the same time and so will have virtually identical magnetic properties throughout the batch. Different batches can have slightly different properties and a studio may wish to realign its machines when a new batch is started. It can be said, however, that variations from batch to batch are minimal these days.

8.2 The magnetic recording process

8.2.1 Introduction

Since tape is magnetic, the recording process must convert an electrical audio signal into a magnetic form. On replay the recorded magnetic signal must be converted back into electrical form. The process is outlined in Fact File 8.1. Normally a professional tape recorder has three heads, as shown in Figure 8.2, in the order erase–record–replay. This allows for the tape to be first erased, then re-recorded,

FACT FILE 8.1   A magnetic recording head

When an electrical current flows through a coil of wire a magnetic field is created. If the current only flows in one direction (DC) the electromagnet thus formed will have a north pole at one end and a south pole at the other (see diagram). The audio signal to be recorded on to tape is alternating current (AC), and when this is passed through a similar coil the result is an alternating magnetic field whose direction changes according to the amplitude and phase of the audio signal.

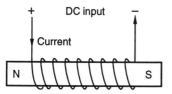

Magnetic flux is rather like the magnetic equivalent of electrical current, in that it flows from one pole of the magnet to the other in invisible 'lines of flux'. For sound recording it is desirable that the tape is magnetised with a pattern of flux representing the sound signal. A recording head is used which is basically an electromagnet with a small gap in it. The tape passes across the gap, as shown in the diagram. The electrical audio signal is applied across the coil and an alternating magnetic field is created across the gap. Since the gap is filled with a non-magnetic material it appears as a very high 'resistance' to magnetic flux, but the tape represents a very low resistance in comparison and thus the flux flows across the gap via the tape, leaving it magnetised.

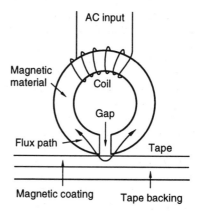

On replay, the magnetised tape moves across the head gap of a similar or identical head to that used during recording, but this time the magnetic flux on the tape flows through the head and thus *induces* a current in the coil, providing an electrical output.

and then monitored by the third head. The structure of the three heads is similar, but the gap of the replay head is normally smaller than that of the record head. It is possible to use the same head for both purposes, but usually with a compromise in performance. Such a two-head arrangement is often found in cheaper cassette machines which do not allow off-tape monitoring whilst recording. A simplified block diagram of a typical tape recorder is shown in Figure 8.3.

The magnetisation characteristics of tape are by no means linear, and therefore a high-frequency signal known as *bias* is added to the audio signal at the record head, generally a sine wave of between 100 and 200 kHz, which biases the tape towards a more linear part of its operating range. Without bias the tape retains very little magnetisation and distortion is excessive. The bias signal is of too high a frequency to be retained by the tape, so does not appear on the output during replay. Different types of tape require different levels of bias for optimum recording conditions to be achieved, and this will be discussed in section 8.2.3.

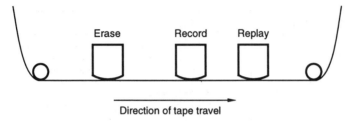

Direction of tape travel

Figure 8.2 Order of heads on a professional analogue tape recorder

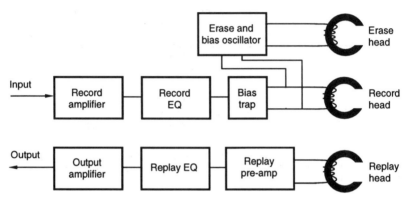

Figure 8.3 Simplified block diagram of a typical analogue tape recorder. The bias trap is a filter which prevents the HF bias signal feeding back into an earlier stage

8.2.2 Equalisation

'Pre-equalisation' is applied to the audio signal before recording. This equalisation is set in such a way that the replayed short-circuit flux in an ideal head follows a standard frequency response curve (see Figure 8.4). A number of standards exist for different tape speeds, whose time constants are the same as those quoted for replay EQ in Table 8.1. Although the replayed *flux level* must conform to these curves, the electrical pre-EQ may be very different, since this depends on the individual head and tape characteristics. Replay equalisation (see Figure 8.5) is used to ensure that a flat response is available at the tape machine's output. It compensates for losses incurred in the magnetic recording/replay process, the rising output of the replay head with frequency, the recorded flux characteristic, and the fall-off in HF response where the recorded wavelength approaches the head gap width (see Fact File 8.2). Table 8.1 shows the time constants corresponding to the turnover frequencies of replay equalisers at a number of tape speeds. Again a number of standards exist. Time constant (normally quoted in microseconds) is the product of resistance and capacitance (RC) in the equivalent equalising filter, and the turnover frequency corresponding to a particular time constant can be calculated using:

$$f = 1/(2\pi RC)$$

The LF time constant of 3180 µs was introduced in the American NAB standard to reduce hum in early tape recorders, and has remained. HF time constants resulting in low turnover frequencies tend to result in greater replay noise, since HF is boosted

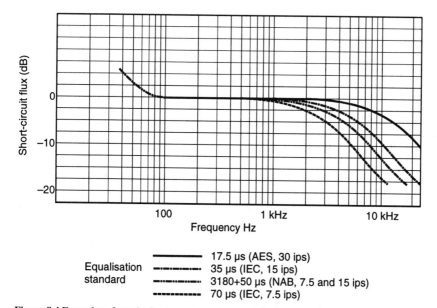

Figure 8.4 Examples of standardised recording characteristics for short-circuit flux. (N.B: this is not equivalent to the electrical equalisation required in the record chain, but represents the resulting flux level replayed from tape, measured using an ideal head)

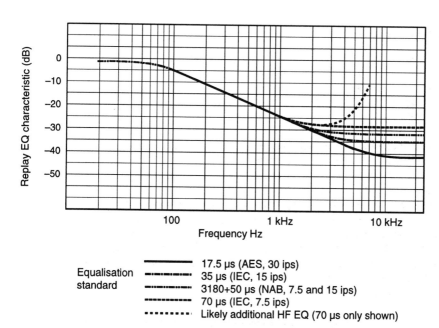

Figure 8.5 Examples of replay equalisation required to correct for the recording characteristic (see Figure 8.4), replay-head losses, and the rising output of the replay head with frequency

FACT FILE 8.2 Replay head effects

The output level of the replay head coil is proportional to the *rate of change of flux*, and thus the output level increases by 6 dB per octave as frequency rises (assuming a constant flux recording). Replay equalisation is used to correct for this slope.

At high frequencies the recorded wavelength on tape is very short (in other words the distance between magnetic flux reversals is very short). The higher the tape speed, the longer the recorded wavelength. At a certain high frequency the recorded wavelength will equal the replay-head gap width (see diagram) and the net flux in the

upper cut-off frequency on replay (the extinction frequency), which is engineered to be as high as possible. Gap effects are noticeable below the cut-off frequency, resulting in a gradual roll-off in the frequency response as the wavelength approaches the gap length. Clearly, at low tape speeds (in which case the recorded wavelength is short) the cut-off frequency will be lower than at high tape speeds for a given gap width.

At low frequencies, the recorded wavelength approaches the dimensions of the length of tape in contact with the head, and various additive and cancellation effects occur when not all of the flux from the tape passes through the head, or when flux takes a 'short-circuit' path through the head. This results in low-frequency 'head bumps' or 'woodles' in the frequency response. The diagram below summarises these effects on the output of the replay head.

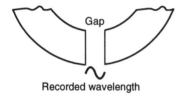

head will be zero, thus no current will be induced. The result of this is that there is an

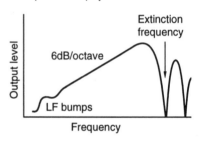

Table 8.1 Replay equalisation time constants

Tape speed ips (cm/s)	Standard	Time constants (µs) HF	LF
30 (76)	AES/IEC	17.5	–
15 (38)	IEC/CCIR	35	–
15 (38)	NAB	50	3180
7.5 (19)	IEC/CCIR	70	–
7.5 (19)	NAB	50	3180
3.75 (9.5)	All	90	3180
1.875 (4.75)	DIN (Type I)	120	3180
1.875 (4.75)	DIN (Type II or IV)	70	3180

over a wider band on replay, thus amplifying tape noise considerably. This is mainly why Type I cassette tapes (120 µs EQ) sound noisier than Type II tapes (70 µs EQ). Most professional tape recorders have switchable EQ to allow the replay of NAB- and IEC/CCIR-recorded tapes. EQ switches automatically with tape speed in most machines.

Additional adjustable HF and LF EQ is provided on many tape machines, so that the recorder's frequency response may be optimised for a variety of operational conditions, bias levels and tape types.

8.2.3 Bias requirements

Higher bias levels enable the audio signal to be recorded more deeply into the oxide layer than would be the case with lower bias levels. High-coercivity tapes require a higher bias than low-coercivity tapes in order to magnetise the tape suitably. Such tapes can, however, display slightly poorer print-through performance (see below). The correct setting of bias level is vital for obtaining optimum performance from an analogue tape, and in professional systems the setting of bias is part of the day-to-day alignment of the tape machine (see section 8.7).

Bias requirements vary from tape to tape as has been said, and cassette tapes are grouped into four 'slots': Type I (ferric), Type II ('chrome' or CrO_2), Type III (ferri-chrome), and Type IV (metal). Cassette formulations are expected to conform closely to one of the above groups with respect to required bias level so that good results are achieved when various different brands are used with a given machine. This is important because variations in bias level cause significant variations in high-frequency performance. The domestic user cannot be expected to optimise his or her cassette machine for a particular type of tape as is routine in professional open-reel usage, although some cassette machines do offer fine bias adjustment, either manual or automatic. It is usually found that a given cassette machine will perform better with one brand of tape than with another depending on exactly how the manufacturer or distributor has aligned the machine, and it is worth experimenting with different types in order to find the one which gives the best performance on that machine.

8.2.4 Print-through

Print-through is caused by a modulated area of tape inducing its magnetism into the adjacent layer of tape on the spool during storage, rather in the manner that a pin, when left stuck to a magnet, will itself become magnetised. This manifests itself as pre-echo or post-echo, depending on which way the tape is wound off the machine, meaning that one can hear the beginning of a movement or programme after a pause at a very low level a second or so before it actually starts. At the end of the programme, faint traces of the last second or so can be heard repeated, although this is usually effectively masked by decaying reverberation or, if no significant reverberation is present, some leader tape will be spliced on immediately after the programme material has ceased.

The storing of master tapes 'tail out' – standard industry practice, indicated by red leader tape which should always be present on the end of every master tape – helps to avoid pre-echo because now a silent section will lie adjacent to the previous programme material rather than adjacent to the beginning of the following section. Tail-out storage of master tapes is also desirable because tape machines invariably give neater spooling on replay than on fast wind or rewind, and so a tape which has just been played and is now tail out will be stored in this neat state which helps to maintain the tape in good condition. Post-echo is preferable because it is at least more natural to hear a faint echo of what one has just heard than to hear a preview of a following section. Dying reverberation also helps to mask it as has been said.

Noise reduction (see Chapter 9) helps to reduce the consequences of print-through because on replay the decoding process pushes low-level signals further down in level as part of the expansion process. Since print-through signals are introduced in between noise reduction encoding and decoding they are therefore reduced in level.

8.3 The tape recorder

8.3.1 Studio recorder

Professional open-reel recorders fall into two categories: console mounted and portable. The stereo console recorder, intended for permanent or semi-permanent installation in a recording studio, outside broadcast truck or whatever generally sports rather few facilities, but has balanced inputs and outputs at line level (no microphone inputs), transport controls, editing modes, possibly a headphone socket, a tape counter (often in real time rather than in arbitrary numbers or revs), tape speed selector, reel size selector, and probably (though not always) a pair of level meters. It is deliberately simple because its job is to accept a signal, store it as faithfully as possible, and then reproduce it on call. It is also robustly built, stays aligned for long periods without the need for frequent adjustment, and will be expected to perform reliably for long periods. A typical example is pictured in Figure 8.6.

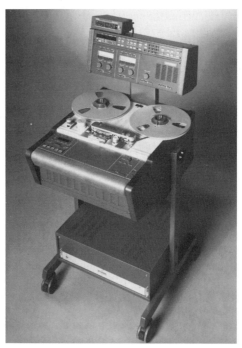

Figure 8.6 A typical professional open-reel two-track analogue tape recorder: the Studer A807-TC. (Courtesy of FWO Bauch Ltd.)

The inputs of such a machine will be capable of accepting high electrical levels – up to at least +20 dBu or around 8 volts – so that there is virtually no possibility of electrical input overload. The input impedance will be at least 10 kΩ. The outputs will be capable of driving impedances down to 600 ohms, and will have a source impedance of below 100 ohms. A facility will be provided for connecting a remote control unit so that the transport can be controlled from the mixing console, for instance. Often the real-time tape counter can also be remotely displayed so that the machine itself can virtually be ignored during a recording session. A noise reduction system can often be located within the housing of the machine, and the record button can send a DC control voltage to this which will automatically switch the noise reduction to encode during recording. Input and output level, bias and EQ controls will be provided, but tucked away so that they cannot be accidently misaligned. Often, small screwdriver holes are employed for these.

Depending on the machine, record 'safe' and 'ready' switches are normally provided for each track, as well as the ability to switch between sync replay and normal replay if it is a multitrack machine (see below). Varispeed controls often exist too, being a means of finely adjusting the speed of the machine either side of a standard play speed. There may also be the option to lock the machine to an external speed reference for synchronisation purposes.

8.3.2 Semi-professional recorder

Its semi-professional counterpart will be capable at its best of a performance that is little inferior, and in addition to being smaller and lighter will sport rather more facilities such as microphone inputs and various alternative input and output options. Headphone outlets will be provided along with record-level meters, source/

Figure 8.7 A typical semi-professional two-track recorder: the Revox PR99. (Courtesy of Revox UK Ltd.)

tape monitor switching, variable output level, and perhaps 'sound on sound'-type facilities for simple overdub work. A typical example is shown in Figure 8.7. The semi-professional machine will not usually be as robustly constructed, this being of particular concern for machines which are to be transported since rough treatment can easily send a chassis askew, causing misalignment of the tape transport system which will be virtually impossible to correct. Some chassis are constructed of pressed steel which is not very rigid. A casting is much better.

Input and output connectors on a semi-pro machine will generally be of the domestic type – phono and unbalanced jack – and voltage levels may well be referenced around –10 dBV instead of +4 dBu. Tape guides may well not include the rollers and low-friction ball races of the professional machine. Facilities for interfacing with noise reduction systems will probably be absent. Such a machine is considerably cheaper than its fully professional counterpart, but the best examples of such machines do, however, return excellent performance consistently and reliably.

8.3.3 The portable machine

The professional portable tape machine, unlike its console equivalent, needs to offer a wide range of facilities since it will be required to provide such things as balanced outputs and inputs, both at line and microphone level, phantom and A–B mic powering, metering, battery operation which allows usefully long recording times, the facility to record timecode and pilot tone for use in TV and film work, illumination of the important controls and meters, and possibly even basic mixing facilities. It must be robust to stand up to professional field use, and small enough to be carried easily. Nevertheless, it should also be capable of accepting professional 10 inch (25 cm) reels, and adaptors are usually available to facilitate this. A lot has to be provided in a small package, and the miniaturisation necessary does not come cheap. The audio performance of such machines is at least as good as that of a studio recorder. A typical commercial example is pictured in Figure 8.8.

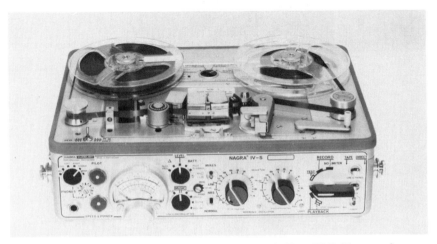

Figure 8.8 A typical professional portable two-track recorder: the Nagra IV-S. (Courtesy of Hayden Laboratories Ltd.)

8.3.4 The multitrack machine

Multitrack machines come in a variety of track configurations and quality levels. The professional multitrack machine tends to be quite massively engineered and is designed to give consistent, reliable performance on a par with the stereo mastering machine. The transport needs to be particularly fine so that consistent performance across the tracks is achieved. A full reel of 2 inch tape is quite heavy, and powerful spooling motors and brakes are required to keep it under control. Apart from the increased number of tracks, multitrack machines are basically the same as their stereo counterparts and manufacturers tend to offer a range of track configurations within a given model type. Alignment of course takes a lot longer, and computer control of this is most welcome when one considers that 24 tracks implies 168 separate adjustments! A typical 24 track tape machine is pictured in Figure 8.9.

A useful feature to have on a multitrack recorder is an automatic repeat function or autolocate. The real-time counter can be programmed so that the machine will repeat a section of the tape over and over again within the specified start and end points to facilitate mixdown rehearsals. Multitrack recorders will be equipped with a number of unique features which are vital during recording sessions. For example, sync replay (see Fact File 8.3), gapless, noiseless punch-in (allowing any track to be dropped into record at any point without introducing a gap or a click) and spot erasure (allowing a track to be erased manually over a very small portion of tape).

Various semi-professional multitrack machines have appeared over the years, making use of decreased track widths in the interests of cheaper transports and lower tape running costs at the expense of quality. Quarter-inch tape has been used for four, eight and even sixteen tracks; 1 inch tape has also been used for eight and

Figure 8.9 A typical professional multitrack recorder: the Saturn 824. (Courtesy of Saturn Research)

FACT FILE 8.3 — **Sync replay**

The overdubbing process used widely in multitrack recording requires that musicians can listen to existing tracks on the tape whilst recording others. If replay was to come from the replay head and the new recording was to be made on to the record head, a recorded delay would arise between old and new material due to the distance between the heads. Sync replay allows the record head to be used as a replay head on the tracks which are not currently recording, thus maintaining synchronisation. The sound quality coming off the record head (called the sync head in this mode) is not always as good as that coming off the replay head, because the gap is larger, but it is adequate for a cue feed. Often separate EQ is provided for sync replay to optimise this. Mixdown should always be performed from the replay head.

Some manufacturers have optimised their head technology such that record and replay heads are exactly the same, and thus there is no difference between true replay and sync replay.

---

sixteen tracks. Some models incorporate domestic-type noise reduction systems such as Dolby C, and a recent machine makes use of the Dolby S system, derived from professional SR (see section 9.2.5).

Cheap multitrack machines will have domestically orientated input and output sockets, and will be unbalanced. Comprehensive alignment facilities may not be provided. Phase errors between the tracks can be high, and the transport will not be first class. Crosstalk between tracks can be poor. The machine may not last as long as its professional counterpart, nor may it stay in alignment for as long.

8.4 Track formats

8.4.1 Mono, two-track and stereo formats

The professional stereo format is known as half track, because each track occupies approximately half of the tape width. Full-track mono recorders record the signal across the whole of the tape width (see Figure 8.10). Domestic open-reel two-track machines conformed to the quarter-track format, in which left and right channels were recorded in the first (upper) and third quarters of the tape's width. The tape could then be turned over so that fresh material could be recorded on the other two quarters of the tape. Twice the recording time could therefore be accommodated. The 'other side' of the tape, as it is conveniently called, is actually the same side but uses a different part of the tape's area. The disadvantages are that reduced track widths mean higher distortion and poorer signal-to-noise ratios; greater possibility of 'drop-out' (momentary loss of signal) due to head-to-tape contact being more critical; and the fact that the editing of a programme will also chop up any material recorded on the other side. Some semi-professional machines are quarter track, and one occasionally finds such a machine in a studio to cater for quarter-track tapes which may sometimes be brought in. Alternatively, some half-track machines have a second replay head in quarter-track form.

Two different, professional two-track formats exist, NAB and DIN, as described in Fact File 8.4.

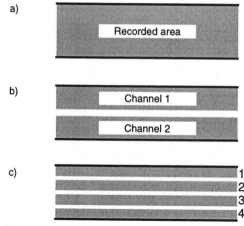

a) Recorded area

b) Channel 1

Channel 2

c) 1
2
3
4

Figure 8.10 Track patterns for quarter-inch recording. (a) Mono, full-track. (b) Stereo, half-track. (c) Four track, or stereo quarter-track (tracks 1 and 3 in one direction and tracks 2 and 4 in the other)

FACT FILE

8.4

NAB and DIN formats

It is important to differentiate between a 'stereo' machine and a 'two-track' machine. With the latter, it is possible to record on the two tracks at separate times if required. Synchronised recordings are also possible if the record head can be switched to perform the replay function instead, thus enabling monitoring of an existing recording on, say, track 1, whilst recording on to track 2. Because entirely separate and unconnected sounds may sometimes need to be recorded on the two tracks, crosstalk between the two needs to be kept to a minimum. The unmodulated band in between the tracks needs to be wider therefore, and so the head guard bands are wider than is the case with the 'stereo' machine. The wide-guard-band format is called the NAB format (which is not necessarily related to NAB equalisation). The narrow-guard-band stereo format is called the DIN format.

This has consequences with respect to compatibility between the two types of machine. A recording made on a DIN machine will occupy more of the guard band than will be the case with the NAB machine. If a recording made on the stereo machine is erased on the two track, the latter's erase head will not completely remove all the signal and traces of it will still be heard on replay. The stereo machine's erase head is full track. Also, if NAB tapes are replayed on DIN heads there will be a marginal increase in noise of 1–2 dB.

8.4.2 Multitrack formats

The professional standard for tape width adheres as far as possible to the scaling derived from the quarter-inch two-track mastering machine. A four-track machine therefore uses half-inch-width tape. Next comes the eight-track which makes use of 1 inch tape. Sixteen track utilises 2 inch tape, and 24 tracks are also accommodated across 2 inches. Comparable quality levels of all the tracks across all the formats are therefore achieved, with no audio degradation as one moves, say, from four track to eight track.

Tracks are normally numbered from 1 at the top of the tape to the highest number at the bottom.

8.5 Magnetic recording levels

It has already been said that the equivalent of electrical current in magnetic terms is magnetic flux, and it is necessary to understand the relationship between electrical levels and magnetic recording levels on tape (a little was said about this in Fact File 7.3). The performance of an analogue tape recorder depends very much on the magnetic level recorded on the tape, since at high levels one encounters distortion and saturation, whilst at low levels there is noise (see Figure 8.11). A window exists, between the noise and the distortion, in which the audio signal must be recorded, and the recording level must be controlled to lie optimally within this region. For this reason the relationship between the electrical input level to the tape machine and the flux level on tape must be established so that the engineer knows what meter indication on a mixer corresponds to what magnetic flux level. Once a relationship has been set up it is possible largely to forget about magnetic flux levels and concentrate on the meters. Fact File 8.5 discusses magnetic flux reference levels.

8.6 What are test tapes for?

A test tape is a reference standard recording containing pre-recorded tones at a guaranteed magnetic flux level. A test tape is the only starting point for aligning a tape machine, since otherwise there is no way of knowing what magnetic level will end up on the tape during recording. During alignment, the test tape is replayed, and a 1 kHz tone at the specified magnetic flux level (say 320 nWb m^{-1}) produces a certain electrical level at the machine's output. The output level would then be adjusted for the desired electrical level, according to the studio's standard (say 0 dBu), to read at a standard meter indication (say PPM 4). It is then absolutely clear that if the output level of the tape machine is 0 dBu then the magnetic level on tape is 320 nWb m^{-1}. After this relationship has been set up it is then possible to *record* a signal on tape at a known magnetic level — for example, a 1 kHz tone at 0 dBu could be fed to the input of the tape machine, and the *input* level adjusted until the output read 0 dBu also. The 1 kHz tone would then be recording at a flux level of 320 nWb m^{-1}.

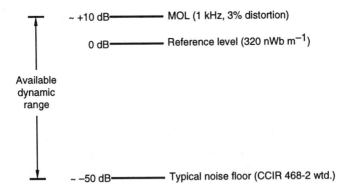

Figure 8.11 The available dynamic range on an analogue tape lies between the noise floor and the MOL. Precise figures depend on tape and tape machine

FACT FILE	Magnetic
8.5	reference levels

Magnetic flux density on tape is measured in nanowebers per meter (nWb m⁻¹), the weber being the unit of magnetic flux. Modern tapes have a number of important specifications, probably the most significant being maximum output level (MOL), HF saturation point and noise level. (These parameters are also discussed in Appendix 1.) The MOL is the flux level at which third-harmonic distortion reaches 3% of the fundamental's level, measured at 1 kHz (or 5% and 315 Hz for cassettes), and can be considered as a sensible peak recording level unless excessive distortion is required for some reason. The MOL for a modern high-quality tape lies at a magnetic level of around 1000 nWb m⁻¹, or even slightly higher in some cases, and thus it is wise to align a tape machine such that this magnetic level corresponds fairly closely to the peak level indication on a mixer's meters.

A common reference level in electrical terms is 0 dBu, which often lines up with PPM 4 or −4 VU on a mixer's meter. This must be aligned to correspond to a recognised *magnetic* reference level on the tape, such as 320 nWb m⁻¹. Peak recording level, in this case, would normally be around +8 dBu if the maximum allowed PPM indication was to be 6, as is conventional. This would in turn correspond to a magnetic recording level of 804 nWb m⁻¹, which is close to the MOL of the tape and would probably result in around 2% distortion.

There are a number of accepted magnetic reference levels in use worldwide, the principal ones being 200, 250 and 320 nWb m⁻¹. There is 4 dB between 200 and 320 nWb m⁻¹, and thus a 320 nWb m⁻¹ test tape should replay 4 dB higher in level on a meter than a 200 nWb m⁻¹ test tape. American test tapes often use 200 nWb m⁻¹ (so-called NAB level), whilst German tapes often use 250 nWb m⁻¹ (sometimes called DIN level). Other European tapes tend to use 320 nWb m⁻¹ (sometimes called IEC level). Test tapes are discussed further in the main text.

There is currently a likelihood in recording studios that analogue tapes are being under-recorded, since the performance characteristics of modern tapes are now good enough to allow higher peak recording levels than before. A studio which aligned PPM 4 to equal 0 dBu, in turn to correspond to only 200 nWb m⁻¹ on tape, would possibly be leaving 4–6 dB of headroom unused on the tape, sacrificing valuable signal-to-noise ratio.

Test tapes also contain tones at other frequencies for such purposes as azimuth alignment of heads and for frequency response calibration of replay EQ (see below). A test tape with the required magnetic reference level should be used, and it should also conform to the correct EQ standard (NAB or CCIR, see section 8.2.2). Tapes are available at all speeds, standards and widths, with most being recorded across the full width of the tape.

8.7 Tape machine alignment

8.7.1 Head inspection and demagnetisation

Heads and tape guides must be periodically inspected for wear. Flats on guides and head surfaces should be looked for; sometimes it is possible to rotate a guide so that a fresh portion contacts the tape. Badly worn guides and heads cause sharp angles to contact the tape which can damage the oxide layer. Heads have been made of several materials. Mu-metal heads have good electromagnetic properties, but are not particularly hard wearing. Ferrite heads wear extremely slowly and their gaps can be machined to tight tolerances. The gap edges can, however, be rather brittle and require careful handling. Permalloy heads last a long time and give a good overall performance, and are often chosen. Head wear is revealed by the presence of a flat area on the surface which contacts the tape. Slight wear does not necessarily indicate that head replacement is required, and if performance is found to be satisfactory during alignment with a test tape then no action need be taken.

Replay-head wear is often signified by exceptionally good high-frequency response, requiring replay EQ to be reduced to the lower limit of its range. This seems odd but is because the replay gap on many designs gets slightly narrower as the head wears down, and is at its narrowest just before it collapses!

Heads should be cleaned regularly using either isopropyl alcohol and a cotton bud, or a freon spray. They should also be demagnetised fairly regularly, since heads can gradually become slightly permanently magnetised, especially on older machines, resulting in increased noise and a type of 'bubbling' modulation noise in the background on recordings. A demagnetiser is a strong AC electromagnet which should be switched on well away from the tape machine, keeping it clear of anything else magnetic or metal. This device will erase a tape if placed near one! Once turned on the demagger should be drawn smoothly and slowly along the tape path (without a tape present), across the guides and heads, and drawn away gently on the far side. Only then should it be turned off.

8.7.2 Replay alignment

Replay alignment should be carried out before record alignment, as explained above. The method for setting replay and record levels has already been covered in the previous section. HF tones for azimuth adjustment normally follow (see Fact File 8.6). The test tape will contain a sequence of tones for replay frequency response alignment, often at 10 or 20 dB below reference level so that tape saturation is avoided at frequency extremes, starting with a 1 kHz reference followed by, say, 31.5 Hz, 63 Hz, 125 Hz, 250 Hz, 500 Hz, 2 kHz, 4 kHz, 8 kHz and 16 kHz. Spoken identification of each section is provided. As the tape runs, the replay equalisation is adjusted so as to achieve the flattest frequency response. Often both LF and HF replay adjustment is provided, sometimes just HF, but normally one should only adjust HF response on replay, since LF can suffer from the head bumps described in Fact File 8.2 and a peak or dip of response may coincide with a frequency on the test tape, leading to potential misalignment. Also full-track test tapes can cause 'fringing' at LF, whereby flux from the guard band leaks on to adjacent tracks. (Although it seems strange, replay LF EQ is normally adjusted during recording, to obtain the flattest record–replay response.)

8.7.3 Record alignment

The frequency response of the machine during recording is considerably affected by bias adjustment, and therefore bias is aligned first before record equalisation. The effects and alignment of bias are described in Fact File 8.7. It is wise to set a roughly correct input level before adjusting bias, by sending a 1 kHz tone at reference level to the tape machine and adjusting the input gain until it replays at the same level.

After bias levels have been set, record azimuth can be adjusted if necessary (see Fact File 8.6) by recording an HF tone and monitoring the now correctly aligned replay output. It may also be necessary to go back and check the 1 kHz record level

FACT FILE 8.6 — Azimuth alignment

Azimuth

Azimuth describes the orientation of the head gap with respect to the tape. The gap should be exactly perpendicular to the edge of the tape otherwise two consequences follow. Firstly, high frequencies are not efficiently recorded or replayed because the head gap becomes effectively wider as far as the tape is concerned, as shown in the diagram (B is wider than A). Secondly, the relative phase between tracks is changed.

The high-frequency tone on a test tape (8, 10, or 16 kHz) can be used with the outputs of both channels combined, adjusting replay azimuth so as to give maximum output level which indicates that both channels are in phase. Alternatively, the two channels can be displayed separately on a double-beam oscilloscope, one wave being positioned above the other on the screen, where it can easily be seen if phase errors are present. Azimuth is adjusted until the two sine waves are in step. It is advisable to begin with a lower-frequency tone than 8 kHz if a large azimuth error is suspected, since there is a danger of ending up with tracks a multiple of 360° out of phase otherwise.

In multitrack machines a process of trial and error is required to find a pair of tracks which most closely represents the best phase alignment between all the tracks. Head manufacturing tolerances result in gaps which are not perfectly aligned on all tracks. Cheap multitrack machines display rather wider phase errors between various tracks than do expensive ones.

Azimuth of the replay head is normally adjusted regularly, especially when replaying tapes made on other machines which may have been recorded with a different azimuth. Record-head azimuth is not modified unless there is reason to believe that it may have changed.

Height

Absolute height of the head should be such that the centre of the face of the head corresponds with the centre of the tape. Height can be adjusted using a test tape that is *not* recorded across the full width of the tape but with two discrete tracks. The correct height gives both equal output level from both channels and minimum crosstalk between them. It is also possible to buy tapes which are only recorded in the guard band, allowing the user to adjust height for minimum breakthrough onto the audio tracks. It can also sometimes be adjusted visually.

Zenith

Zenith is the vertical orientation of the head with respect to the surface of the tape. The head should neither lean forwards towards the tape, nor lean backwards, otherwise uneven wrap of the tape across the surface of the head results causing inconsistent tape-to-head contact and uneven head wear. Zenith is not normally adjusted unless the head has been changed or there is reason to believe that the zenith has changed.

Wrap

Wrap is the centrality of the head gap in the area of tape in contact with the head. The gap should be exactly in the centre of that portion, so that the degree of approach and recede contact of the tape with respect to the gap is exactly equal. Uneven frequency response can be caused if this is not the case. Wrap can be adjusted by painting the head surface with a removable dye and running the tape across it. The tape will remove the dye over the contact area, and adjustments can be made accordingly.

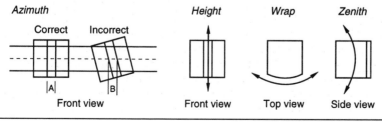

Azimuth	Height	Wrap	Zenith
Correct Incorrect			
|A| |B|			
Front view	Front view	Top view	Side view

if large changes have been made to bias.

Record equalisation can now be aligned. Normally only HF EQ is available on record. A 1 kHz tone is recorded at between 10 and 20 dB below reference level and the meter gain adjusted so that this can be seen easily on replay. Spot frequencies are then recorded to check the machine's frequency response, normally only at the extremes of the range. A 5 kHz tone, followed by tones at 10 kHz and 15 kHz can be recorded and monitored off tape. The HF EQ is adjusted for the flattest possible response. The LF *replay* EQ (see above) can similarly be adjusted, sweeping the oscillator over a range of frequencies from, say, 40 Hz to 150 Hz, and adjusting for the best compromise between the upper and lower limits of the 'head bumps'.

Some machines have a built-in computer which will automatically align it to any tape. The tape is loaded and the command given, and the machine itself runs the tape adjusting bias, level and EQ as it goes. This takes literally seconds. Several settings can be stored in its memory so that a change of tape type can be accompanied simply by telling the machine which type is to be used, and it will automatically set its bias and EQ to the previously stored values. This is of particular value when aligning multitrack machines!

Once the tape machine has been correctly aligned for record and replay, a series of tones should be recorded at the beginning of every tape made on the machine. This allows the replay response of any machine which might subsequently be used for replaying the tape to be adjusted so as to replay the tape with a flat frequency

---

FACT FILE

8.7

Bias adjustment

Bias level affects the performance of the recording process and the correct level of bias is a compromise between output level, distortion, noise level and other factors. The graph below shows a typical tape's performance with increasing bias, and it can be seen that output level increases up to a point, after

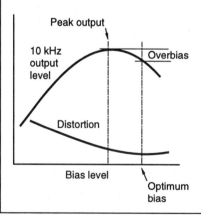

which it falls off. Distortion and noise go down as bias increases, but unfortunately the point of minimum noise and distortion is not quite the same as the point of maximum output level. Typically the optimum compromise between all the factors, offering the best dynamic range, is where the bias level is set just slightly higher than the point giving peak output. In order to set bias, a 10 kHz tone is recorded at, say, 10 dB below reference level, whilst bias is gradually increased from the minimum. The output level from the tape machine gradually rises to a peak and then begins to drop off as bias continues to increase. Optimum bias is set for a number of decibels of fall-off in level after this peak – the so-called 'overbias' amount.

The optimum bias point depends on tape speed and formulation, but is typically around 3 dB of overbias at a speed of 15 ips (38 cm s^{-1}). At 7.5 ips the overbias increases to 6 dB and at 30 ips it is only around 1.5 dB. If bias is adjusted at 1 kHz there is much less change of output level with variation in bias, and thus only between 0.5 and 0.75 dB of overbias is required at 15 ips. This is difficult to read on most meters.

response. The minimum requirement should be a tone at 1 kHz at reference level, followed by tones at HF and LF (say 10 kHz and 63 Hz) at either reference level (if the tape can cope) or at –10 dB. The levels and frequencies of these tones must be marked on the tape box (e.g.: 'Tones @ 1 kHz, 320 nWb m^{-1} (= 0 dB); 10 kHz and 63 Hz @ –10 dB). Designations on the tape box such as '1 kHz @ 0 VU' mean almost nothing, since 0 VU is not a magnetic level. What the engineer means in this case is that he sent a tone from his desk to the tape machine, measuring 0VU on the meters, but this gives no indication of the magnetic level that resulted on the tape. Noted on the box should also be an indication of where peak recording level lies in relation to the 1 kHz reference level (e.g.: 'peak recording level @ 8 dB above 320 nWb m^{-1}'), in order that the replay chain can be set up to accommodate the likely signal peaks. In broadcasting, for example, it is most important to know where the peak signal level will be, since this must be set to peak at PPM 6 on a program meter, corresponding to maximum transmitter modulation.

When this tape comes to be replayed, the engineer will adjust the replay level and EQ controls of the relevant machine, along with replay azimuth, to ensure that the recorded magnetic reference level replays at his or her studio's electrical reference level, and to ensure a flat response. *This is the only way of ensuring that a tape made on one machine replays correctly on another day or on another machine.*

8.8 Mechanical transport functions

Properly, mechanical alignment of the tape transport should be looked at before electrical alignment, because the electromagnetic performance is affected by it, but

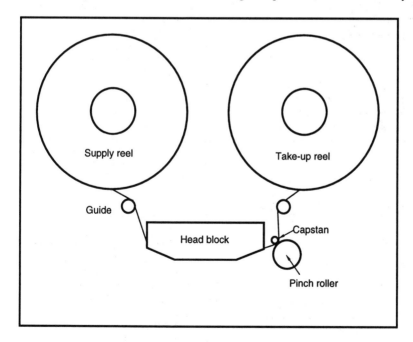

Figure 8.12 Typical layout of mechanical components on the deckplate of an analogue open-reel recorder

the converse is not the case. Mechanical alignment should be required far less frequently than electrical adjustments, and sometimes it also requires rather specialised tools. Because most mechanical alignments are fairly specialised, and because they differ with each tape machine, detailed techniques will not be covered further here. The manual for a machine normally details the necessary procedures. Looking at the diagram in Figure 8.12, it can be seen that the tape unwinds from the reel on the left, passes through various guides on its way to the head block, and then through various further guides and on to the take-up reel on the right. Some tape guides may be loaded with floppy springs which give on the instant of start-up, then slowly swing back in order to control the tension of the tape as the machine starts. The capstan is the shaft of a motor which pokes up through the deck of the machine by a couple of centimetres or so (more of course for multitrack machines with their increased tape widths) and lies fairly close to the tape when the tape is at rest, on the right-hand side of the head block. A large rubber wheel will be located close to the capstan but on the opposite side of the tape. This is called the pinch roller or pinch wheel. The capstan motor rotates at a constant and carefully controlled speed, and its speed of rotation defines the speed at which the tape runs. When record or play is selected the pinch roller rapidly moves towards the capstan, firmly sandwiching the tape in between the two. The rotation of the capstan now controls the speed of tape travel across the heads.

The take-up reel is controlled by a motor which applies a low anticlockwise torque so that the tape is wound on to it. The supply reel on the left is also controlled by a motor, which now applies a low clockwise torque, attempting to drag the tape back in the opposite direction, and this 'back tension' keeps the tape in firm contact with the heads. Different reel sizes require different degrees of back tension for optimum spooling, and a reel size switch will usually be provided although this is sometimes automatic. One or two transports have been designed without pinch rollers, an enlarged diameter capstan on its own providing speed control. The reel motors need to be rather more finely controlled during record and replay so as to avoid tape slippage across the capstan. Even capstanless transports have appeared, the tape speed being governed entirely by the reel motors.

When fast wind or rewind is selected the tape is lifted away from the heads by tape lifters, whilst spooling motors apply an appropriately high torque to the reel which is to take up the tape and a low reverse torque to the supply reel to control back tension. The tape is kept away from the heads so that its rapid movement does not cause excessive heating and wear of the tape heads. Also, very high-level, high-frequency energy is induced into the playback head if the tape is in contact with it which can easily damage speakers, particularly tweeters and HF horns. Nevertheless, a facility for moving the tape into contact with the heads during fast spooling is provided so that a particular point in the tape can be listened for.

Motion sensing and logic control is an important feature of a modern open-reel machine. Because the transport controls are electronically governed on modern machines, one can go straight from, say, rewind to play, leaving the machine itself to store the command and bring the tape safely to a halt before allowing the pinch wheel to approach the capstan. Motion sensing can be implemented by a number of means, often either by sensing the speed of the reel motors using tachometers, or by counting pulses from a roller guide.

The tape counter is usually driven by a rotating roller between the head block and that reel. Slight slippage can be expected, this being cumulative over a complete reel of tape, but remarkably accurate real-time counters are nevertheless to be found.

8.9 The Compact Cassette

8.9.1 Background

The Compact Cassette was invented by Philips, and was launched in 1963. It was originally intended as a convenient low-quality format suitable for office dictation machines and the like. It was envisaged that domestic tape recording would be open-reel, and a boom in this area was predicted. Pre-recorded open-reel tapes were launched. The expected boom never really materialised, however, and the sheer convenience of the cassette medium meant that it began to make inroads into the domestic environment. The format consists of tape one-eighth of an inch wide (3 mm), quarter track, running at a speed of 1.875 ips. Such drastically reduced dimensions and running speed compared with open-reel returned a poor level of audio performance, and if it was to be used for reasonable-quality music reproduction considerable development was needed.

Leaving the standards which had been set for the Compact Cassette unaltered, tape and machine manufacturers worked hard to develop the format, and the level of performance now available from this medium is quite impressive given its humble beginnings. It is worth mentioning that in the mid 1970s Sony introduced a rival called the Elcaset. The cassette housing was larger to accommodate 0.25 inch (6 mm) wide tape and the tape speed was 3.75 ips, promising rather better quality. But the format came too late, the Compact Cassette already being well established particularly since Dolby B noise reduction had been widely exploited since the early 1970s.

8.9.2 Cassette housing and transport

The cassette housing incorporates a felt pressure pad on the opposite side of the tape's recording surface which maintains the tape in intimate contact with the head during record and replay, this being particularly important when dealing with such small dimensions. During record and replay the heads are moved towards the cassette housing, and through appropriate access holes to contact the tape. Originally specified as a two-head format, an erase and a single record/replay head, no provision was made for a third head which would allow record and replay to be carried out by separate dedicated heads optimised for each purpose. In the continuing process of development manufacturers wished to add a third head and some chose to incorporate the record and playback heads in one housing, to be positioned in the normal record/replay-head position. Others chose to use a separate housing locating a third head elsewhere, using a separate access hole in the cassette. No pressure pad is provided for this, and the machine's transport needs to be up to the task of providing optimum back tension of the tape to give good head-to-tape contact. Three heads of course means that off-tape monitoring is possible and 'Double Dolby', 'Double dbx', etc. nolse reduction circuits are employed which encode the incoming signal and simultaneously decode the off-tape signal for monitoring.

Dual-capstan drives have been offered with a capstan and pinch roller placed at each end of the cassette. The capstan which is placed at the end which supplies the tape to the heads is engineered to run at a marginally slower speed than the other one, the latter defining the actual tape speed. Such an arrangement ensures very consistent tensioning of the tape across the heads and the mechanical performance

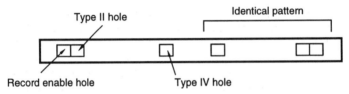

Figure 8.13 Holes in the top edge of a compact cassette may be uncovered to signify different tape types and to prevent recording

of the cassette reels no longer affects the results. Some machines push the pressure pad away from the head so that it does not influence the performance. Auto-reverse has been seen, as has automatic 'preview' whereby the machine senses the silences between recorded items on the tape and then plays the first few seconds of each item before fast winding on to the next.

8.9.3 Tape selection

The different tape formulations were discussed in section 8.1.3. A number of machines automatically select EQ and bias for the tape type which is loaded, sensing special holes in the cassette housing which are provided for this purpose (see Figure 8.13). Internal bias adjustment is usually provided in a cassette machine, but record and replay EQ are usually fixed. If it is desired to bias a machine for a particular tape, a good way to do this in the absence of EQ adjustments is to record pink noise at a level 20 dB below zero on the meters from a test record, or 'white' noise provided by FM tuner interstation noise, increasing the bias from a low level until the sound coming off tape is as near as possible indistinguishable from the sound going on. Too much bias produces a dull sound; too little gives an overbright sound. With a three-head machine this is very easy since the source/tape switch can be flicked too and fro for instant comparison, but two-head machines require regular rewinding and comparison.

It is very important to set the same replay level for the noise source and the off-tape noise, otherwise the ear hears frequency response differences which may simply be a subjective result caused by the different levels. The process should be carried out with noise reduction switched off, as this will exaggerate response errors. After alignment the noise reduction should be switched on and another section of noise recorded. The noise reduction should not introduce significant degradation of the signal off tape. In many cheaper recorders the bias control either affects all tape types equally, or simply works for ferric tapes which have the widest range of requirements.

8.9.4 Other alignments

Cassette test tapes are available, enabling frequency response and azimuth checks. Thorough cleaning and demagnetisation of the machine should be carried out before one is used. Small azimuth adjustments can bring particularly worthwhile improvements in cassette performance, especially when replaying tapes recorded on another machine. Azimuth can simply be adjusted for maximum subjective HF response using the little spring-loaded screw on one side of the record/replay head. Some machines incorporate computer systems similar to those found in certain professional open-reel models which automatically align the machine for a particular type of tape. Settings for several types can be stored in the computer's memory.

Automatic replay azimuth adjustment is also possible. The two channels of the stereo output are filtered, converted into square waves and then fed to a comparator. Phase differences produce an output control voltage and this drives a small low-speed motor which adjusts the azimuth setting of the replay head. When azimuth is correct no control voltage is produced and the azimuth is left alone. The system is continuously active throughout the replay process and it is designed to extract the best performance from pre-recorded musicassettes and recordings made on other machines.

8.9.5 Multitrack cassette recorders

In the late 1970s the Japanese TEAC company introduced a machine called the Portastudio. It was a four-channel multitrack cassette recorder with mixing facilities and multiple inputs built in. The tape ran at twice normal speed, 3.75 ips, and the four tracks were recorded across the full width of the tape. Each track could be recorded on separately, sync facilities were provided, and 'bounce down' could be achieved in the manner of a professional multitrack machine whereby signals recorded on, say, tracks 1, 2 and 3 could be mixed and recorded on to the fourth track, freeing the other three tracks for further use. The final four-track tape could then be mixed down into stereo, these stereo outputs being fed to a conventional cassette recorder (or even open-reel).

One mixer company even offered an eight-track cassette-based system which incorporated a mixing section that offered facilities such as multiband EQ and auxiliary sends.

8.9.6 Cassette duplication

Cassette duplication is an important area of activity, pre-recorded musicassettes being very popular. The vast majority of such cassettes are produced on special machines which run the tape at 16, 32 or 64 times normal speed so that a length of tape for a 20 minute cassette can be duplicated in a few seconds. In one system, used mainly for short runs, banks of cassettes are duplicated in one go by each particular duplicator, and the labelling and packaging is carried out automatically. Master tapes to be duplicated are copied on to a half-inch 'loop-bin master' tape which is fed from a vacuum bin where the tape is stored in loose loops to allow for high-speed repeated reproduction. Tones are recorded on to the tape during duplication which tell the cassette loading machine where the beginning and end of each section is.

A speed 32 times normal for duplication requires record bias frequencies in the megahertz range and record-head gaps need to be rather less than a micron in width for high frequencies to be adequately recorded. Level and frequency response errors will be magnified by Dolby encode/decode processes. Consistent head-to-tape contact is difficult to maintain at such high speeds, high frequency loss normally being the result. Such duplicating equipment therefore needs to be maintained in first-class working order if the final cassette is to stand comparison with an equivalent home recorded example. Many musicassettes are found to be recorded at too low a level, failing to exploit the maximum dynamic range of the medium. Many sound dull, and several actually sound rather better with Dolby switched out during replay. Some manufacturers use higher-quality tape than others, and several chromium dioxide examples are available, but recorded using 120 μs equalisation.

An alternative to high-speed duplication is real-time copying. Instead of the high-speed duplicators a bank of carefully maintained cassette recorders are used which all chug away in real time, programme signal being provided by an open-reel machine or even a digital recorder, linked to a distribution amplifier to feed the cassette machines. Such a setup is ideal for the production of relatively low numbers of cassettes when it may not be economical to prepare the special production tapes necessary for the high-speed duplication process. Sound quality can of course be as good as the medium is capable of.

8.10 The tape cartridge

8.10.1 Background

The eight-track cartridge has been available since the mid 1950s. It consists of an endless loop of tape housed in a plastic cartridge which is capable of being continuously played. Only one hub is housed in the cartridge and no 'rewind' facility is appropriate. Machines can, however, provide 'fast forward'. In the original format four stereo pairs of tracks were recorded across the whole width of the tape in the same direction, and the tape ran at 3.75 ips. The splice in the tape which joined the beginning and end was covered by a piece of metal foil which the replay machine sensed. The replay head would then move down to play the second pair of tracks, and so on downwards as the beginning of the tape was again reached. Pre-recorded cartridges were produced throughout the 1960s and 1970s and replay machines were popular in motor vehicles. The Compact Cassette finally won the day though.

The NAB cartridge or 'cart', however, lives on in professional circles and is much used in the broadcast industry, and also to a lesser extent in the theatre industry. The present format consists of the so-called 'AA'-sized cartridge (several sizes were introduced) with a continuous loop of tape. The standard running speed is 7.5 ips (some machines can be switched to 15 ips) and a single pair of signals are recorded across the whole of the width of the tape in the manner of the half-track open-reel mastering machine. The back of the tape is lubricated so that it slides smoothly from the centre of the pancake in the cartridge and under the layers of tape before reaching the playback head, as shown in Figure 8.14. The tape is driven by a capstan and pinch roller, and back tension is provided by the non-rotating hub around which the tape is wound. Various tape lengths are available, including 10 seconds, 20, 40, 70, 100, ..., up to 10 minutes. The longest examples have a lot of tape wound inside, and back tension is not very consistent.

Record machines have two heads, record and replay, but no erase head. Erasure is carried out either by a bulk eraser or by a special machine which plays the tape at several times normal speed while an erase head is in contact with it. The machine will also locate the splice in the tape so that recording can begin just after it has been passed in order to avoid its detrimental effect. Dedicated machines are normally used for replay, in which the record head and associated circuitry is absent. A commercial machine and cartridge is shown in Figure 8.15. The cart machine is mechanically virtually silent in operation, which enables it to be used by DJs in the broadcast studio for playing jingles and the like without any mechanical noise entering the microphone. Theatres also make use of them where the bangs and clunks associated with open-reel transports are unacceptable. Noise reduction is used in high-quality work.

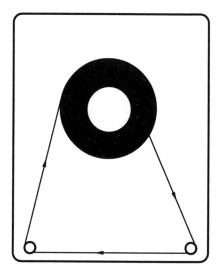

Figure 8.14 The tape in a cartridge is wound in an endless loop

Figure 8.15 A typical broadcast cart machine and cartridge from ITC. (Courtesy of FWO Bauch)

8.10.2 Recording and replay

The 'record' button is pressed. An illumination acknowledges this. When 'play' is pressed the machine runs the tape and automatically records a 1 kHz signal of half a second duration on a special cue track located in between the two stereo audio tracks. This 'primary' cue signal enables the machine to sense when the beginning of the tape is again reached and it will automatically bring the transport to a halt. The audio signal is simultaneously recorded on to the tape in the normal manner. After the programme which is being recorded has ended, the 'fast' or 'fast forward' button is pressed. This time a 150 Hz tone is automatically recorded on the cue track and the machine then drops out of record and the capstan speeds up to three times normal speed to fast-forward the tape. This 150 Hz 'secondary' tone will subsequently be sensed on replay, causing the transport to go into fast forward automatically. When the 1 kHz primary tone is sensed the transport stops moving the tape and it is now lined up at the beginning of the recording.

Replaying the tape, one finds that the machine has lined the tape up at the beginning of the recording with the aid of the 1 kHz primary tone. After the programme has finished the machine senses the 150 Hz secondary tone and goes into fast forward, muting the audio outputs as it does so. After a few seconds the machine stops, indicating that the beginning of the loop has again been reached with its 1 kHz stop tone. The cart can be withdrawn from the machine and labelled.

It is often desirable to record several sound effects on a single cart so that they can be played in fairly rapid succession. The above recording procedure is used but instead of pressing the fast forward button after the first effect has been recorded the stop button is pressed. The record button is then pressed which prepares the machine for recording, and the play button is pressed to record the next item on to the tape. In this way several sound effects can be recorded on to the cart, a separate stop tone being present at the beginning of each item. A fast forward tone can be put on to the tape after the last item is recorded.

Another possibility is the truly endless loop. The record machine has a switch which disables the stop tone so that only the programme material is recorded. This can be faded up over a second or so and the tape timed with a stopwatch. As the beginning of the tape is again approached, off-tape monitoring tells when overlap between the previously recorded signal and the new signal begins. The machine is now stopped, and a continuously recorded loop has been formed without any stop tones. Seawash or wind, for instance, can now form an extended backdrop.

The 1 kHz stop tone needs to be read by the machine during normal replay speed and also three times the replay speed during fast wind when its frequency will now be 3 kHz. The 150 Hz secondary tone needs only be read at normal replay speed, being a cue to put the machine into fast forward. An 8 kHz tertiary cue can also be recorded which can signal another machine to start, or a projector to change slides for instance. FSK (Frequency Shift Keying) information can also be recorded on to the cue track which when decoded by a computer can give information about the programme content of the cart. This can be useful, for instance in broadcast studios, if the cart machine is used for playing commercials which need to be logged.

Recommended further reading

Jorgensen, F. (1988) *The Complete Handbook of Magnetic Recording*. TAB Books
Mallinson, J. C. (1987) *The Foundations of Magnetic Recording*. Academic Press

See also *General further reading* at the end of this book.

Chapter 9

Noise reduction

Noise reduction techniques have been applied to analogue tape machines of all formats, radio microphones, radio transmission and reception, land lines, satellite relays, gramophone records, and even some digital tape machines. The general principles of operation will be outlined, followed by a discussion of particular well-known examples. Detailed descriptions of some individual systems are referred to in the Further reading list at the end of this chapter.

9.1 Why is noise reduction required?

A noise reduction system, used correctly, reduces the level of unwanted signals introduced in a recording–replay or transmission–reception process (see Figure 9.1). Noise such as hiss, hum, and interference may be introduced, as well as, say, print-through in analogue recording, due to imperfections in the storage or transmission process. In communications, a signal sent along a land line may be prone to interference from various sources, and will therefore emerge with some of this interference signal mixed with it. A signal recorded on a cassette machine replays with high-frequency hiss. Unwanted noise already present in a signal before recording or transmitting, though, is very difficult to remove without also removing a part of the wanted signal. One could roll off the treble of a cassette recording to reduce the hiss, but one would also lose the high-frequency information from the sound, causing it to sound muffled and 'woolly'.

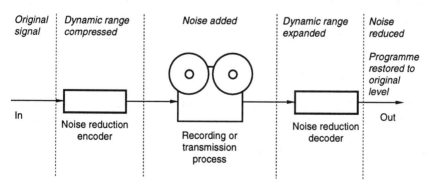

Figure 9.1 Graphical representation of a companding noise reduction process

9.2 Methods of reducing noise

9.2.1 Variable pre-emphasis

Pre-emphasis (see Fact File 9.1) is a very straightforward solution to the problem of noise reduction, but is not a panacea. Many sound sources, including music, have a falling energy content at high frequencies, so lower-level HF signals can be boosted to an extent without too much risk of saturating the tape. But tape tends to saturate more easily at HF than at LF (see the previous chapter), so high levels of distortion and compression would result if too much pre-emphasis were applied at the recording stage. What is needed is a circuit which senses the level of the signal on a continuous basis, controlling the degree of pre-emphasis so as to be non-existent at high signal levels but considerable at low signal levels (see Figure 9.2). This can be achieved by incorporating a filter into a side-chain which passes only high-frequency, low-level signals , adding this component into the un-pre-emphasised signal. On replay, a reciprocal de-emphasis circuit could then be used. The lack of noise reduction at high signal levels does not matter, since high-level signals have a masking effect on low-level noise (see Fact File 2.3).

Such a process may be called a *compansion* process, in other words a process which compresses the dynamic range of a signal during recording and expands it on replay. The variable HF emphasis described above is an example of selective compansion, acting only on a certain band of frequencies. It is most important to notice that the decoding stage is an exact mirror image of the encoding process, and that it is not possible to use one without the other. Recordings not *encoded* by a noise reduction system cannot simply be passed through a decoder to reduce their noise.

FACT FILE 9.1 Pre-emphasis

One approach to the problem of reducing the apparent level of noise could be to pre-condition the incoming signal in some way so as to raise it further above the noise. Hiss is most annoying at high frequencies, so one could boost HF on recording. On replay, HF signals would therefore be reproduced with unnatural emphasis, but if the same region is now attenuated to bring the signal down to its original level any hiss in the same band will also be attenuated by a corresponding amount, and so a degree of noise reduction can been achieved without affecting the overall frequency balance of the signal. This is known as *pre-emphasis* (on record) and *de-emphasis* (on replay), as shown in the diagram.

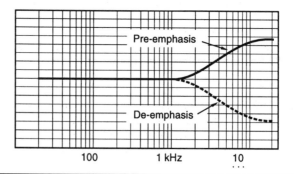

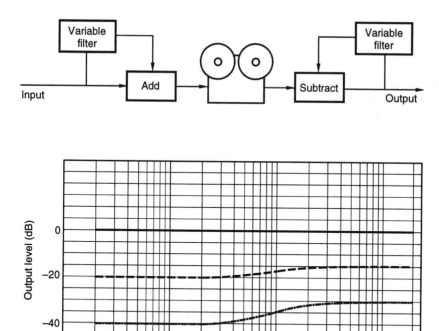

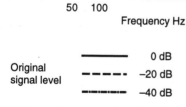

Figure 9.2 A simple complementary noise reduction system could boost high frequencies at low signal levels during encoding, and cut them on decoding (encoding characteristic shown)

Similarly, encoded tapes sound unusual unless properly decoded, normally sounding overbright and with fluctuations in HF level.

9.2.2 Dolby B

The above process is used as the basis for the Dolby B noise reduction system, found in most cassette decks. Specifically, the threshold below which noise reduction comes into play is around 20 dB below a standard magnetic reference level known as 'Dolby level' (200 nWb m^{-1}). The maximum HF boost of the Dolby B system is 10 dB above 8 kHz, and therefore a maximum of 10 dB of noise reduction is provided. A high-quality cassette deck, without noise reduction, using a good ferric tape, will yield a signal-to-noise ratio of about 50 dB ref. Dolby level. When Dolby B noise reduction is switched in, the 10 dB improvement brings this up to 60 dB (which is more adequate for good-quality music and speech recording). The quoted improvement is seen when noise is measured according to the CCIR 468-2 weighting curve (see section A1.5) and will not be so great when measured unweighted.

Dolby B incorporates a sliding band over which pre-emphasis is applied, such that the frequency above which compansion occurs varies according to the nature of the signal. It may slide as low as 400 Hz. This aims to ensure that maximum masking of low-level noise always occurs, and that high-level signals at low frequencies do not result in 'noise pumping' (a phenomenon which arises when a high level signal in one band causes less overall noise reduction, causing the noise in another band to rise temporarily, often not masked by the high-level signal due to the difference in frequency of the signal and the noise).

The Dolby process, being level dependent, requires that the reproduced signal level on decoding is exactly the same with respect to Dolby level as on encoding. This means that a particular cassette machine must be set up internally so that Dolby-encoded tapes recorded on it or other machines will replay into the decoder at the correct electrical level for proper decoding. This is independent of the actual output level of the machine itself, which varies from model to model. If the replay level, for instance, is too low, the decoder applies too much treble cut because the –20 dB threshold level will have moved downwards, causing recorded signal levels above this to be de-emphasised also. Frequency response error will therefore be the result. Similarly, if the frequency response of a cassette machine shows significant errors at HF, these will be exaggerated by the Dolby record/replay process.

A so-called MPX (multiplex) filter is mandatory with Dolby B systems, and removes the 19 kHz pilot tone present in FM stereo radio broadcasts. This is needed because the pilot tone may still be present in the output of an FM tuner, artificially affecting the encoded level of HF signals on a recording from the radio. Since the frequency response of many cassette machines does not extend to 20 kHz the tone would not be reproduced on replay, and thus the decoder would not track the encoded signal correctly, leading to noise pumping and response errors. On some recorders the filter is switchable. On cheaper machines the filter simply rolls off everything above 15 kHz, but on better machines it is a notch at 19 kHz.

9.2.3 Dolby C

Dolby B became widely incorporated into cassette players in the early 1970s, but by the end of the 1970s competition from other companies offering greater levels of noise reduction prompted Dolby to introduce Dolby C, which gives 20 dB of noise reduction. The system acts down to a lower frequency than Dolby B (100 Hz), and incorporates additional circuitry (known as 'anti-saturation') which reduces HF tape squashing when high levels of signal are present. Most of the noise reduction action takes place between 1 kHz and 10 kHz, and less action is taken on frequencies above 10 kHz (where noise is less noticeable) in order to desensitise the system to HF response errors from such factors as azimuth misalignment which would otherwise be exaggerated (this is known as 'spectral skewing'). Dolby C, with its greater compression/expansion ratio compared with Dolby B, will exaggerate tape machine response errors to a correspondingly greater degree, and undecoded Dolby C tapes will sound extremely bright.

9.2.4 Dolby A

Dolby A was introduced in 1965, and is a professional noise reduction system. In essence there is a similarity to the processes described above, but in the Dolby A

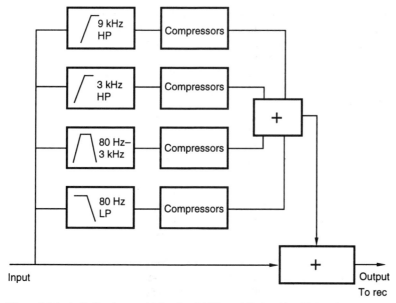

Figure 9.3 In the Dolby A system a low-level 'differential' signal is added to the main signal during encoding. This differential signal is produced in a side-chain which operates independently on four frequency bands. The differential signal is later subtracted during decoding

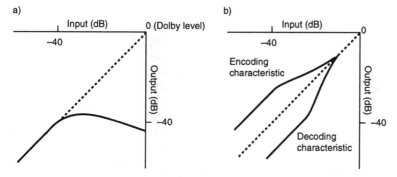

Figure 9.4 (a) Differential signal component produced in a Dolby A side-chain. (b) Input level plotted against output level of Dolby A unit after adding or subtracting differential component

encoder the noise reduction process is divided into four separate frequency bands, as shown in Figure 9.3. A low-level 'differential' component is produced for each band, and the differential side-chain output is then recombined with the main signal. The differential component's contribution to the total signal depends on the input level, having maximum effect below –40 dB ref. Dolby level (see Figures 9.4(a) and (b)).

The band splitting means that each band acts independently, such that a high-level signal in one band does not cause a lessening of noise reduction effort in another low-level band, thus maintaining maximum effectiveness with a wide range of programme material. The two upper bands are high pass and overlap, offering noise reduction of 10 dB up to around 5 kHz, rising to 15 dB at the upper end of the spectrum.

The decoder is the mirror image of the encoder, except that the differential signal produced by the side-chain is now *subtracted* from the main signal, restoring the signal to its original state and reducing the noise introduced between encoding and decoding.

9.2.5 Dolby SR

The late 1980s saw the introduction of Dolby SR – Spectral Recording – which gives greater noise reduction of around 25 dB. It has been successful in helping to prolong the useful life of analogue tape machines, both stereo mastering and multitrack, in the face of the coming of digital tape recorders. Dolby SR differs from Dolby A in that whereas the latter leaves the signal alone until it drops below a certain threshold, the former seeks to maintain full noise reduction (i.e.: maximum signal boost during recording) across the whole frequency spectrum until the incoming signal rises above the threshold level. The band of frequencies where this happens is then subject to appropriately less boost. This is rather like looking at the same process from opposite directions, but the SR system attempts to place a comparably high recording level on the tape across the whole frequency spectrum in order that the dynamic range of the tape is always used optimally.

This is achieved by ten fixed- and sliding-band filters with gentle slopes. The fixed-band filters can vary in gain. The sliding-band filters can be adjusted to cover different frequency ranges. It is therefore a fairly complex multiband system, requiring analysis of the incoming signal to determine its energy at various frequencies. Spectral skewing and anti-saturation are also incorporated (see section 9.2.3). Dolby SR is a particularly inaudible noise reduction system, more tolerant of level mismatches and replay speed changes than previous systems. A simplified 'S'-type version has been introduced for the cassette medium, and is also used on some semi-professional multitrack recorders.

9.2.6 dbx

dbx is another commonly encountered system. It offers around 30 dB of noise reduction and differs from the various Dolby systems as follows. dbx globally compresses the incoming signal across the whole of the frequency spectrum, and in addition gives pre-emphasis at high frequencies (treble boost). It is not level dependent, and seeks to compress an incoming signal with, say, a 90 dB dynamic range into one with a 60 dB dynamic range which will now fit into the dynamic range capabilities of the analogue tape recorder. On replay, a reciprocal amount of expansion is applied together with treble de-emphasis.

Owing to the two factors of high compansion ratios and treble pre- and de-emphasis, frequency response errors can be considerably exaggerated. Therefore, dbx type 1 is offered which may be used with professional equipment and type 2 is to be used with domestic equipment such as cassette decks where the noise reduction at high frequencies is relaxed somewhat so as not to exaggerate response errors unduly. The degree of compression/expansion is fixed, that is it does not depend on the level of the incoming signal. There is also no division of noise reduction between frequency bands. These factors sometimes produce audible modulation of background hiss with critical programme material such as wide dynamic range classical music, and audible 'pumping' noises can sometimes be heard. The system does, however, offer impressive levels of noise reduction, particularly welcome with the cassette medium, and does not require accurate level alignment

9.2.7 telcom c4

The ANT telcom c4 noise reduction system arrived somewhat later than did Dolby and dbx, in 1978. Capitalising on the experience gained by those two systems, the telcom c4 offers a maximum noise reduction of around 30 dB, is level dependent like Dolby, and also splits the frequency spectrum up into four bands which are then treated separately. The makers claim that the c4 system is less affected by record/replay-level errors than is Dolby A. The system works well in operation, and side-effects are minimal.

There is another system offered by the company, called 'hi-com', which is a cheaper, simpler version intended for home studio setups and domestic cassette decks.

9.3 Line-up of noise reduction systems

In order to ensure unity gain through the system on recording and replay, with correct tracking of a Dolby decoder, it is important to align the noise reduction signal chain. Many methods are recommended, some more rigorous than others, but in a normal studio operation for everyday alignment, the following process should be satisfactory. It should be done after the tape machine has been aligned (this having been done with the NR unit bypassed).

For a Dolby A encoder, a 1 kHz tone should be generated from the mixer at +4 dBu (usually PPM 5), and fed to the input of the NR unit. The unit should be in 'NR out' mode, and set to 'record'. The input level of the NR unit should normally be adjusted so that this tone reads on the 'NAB' level mark on the meter (see Figure 9.5). The output of the unit should then be adjusted until its electrical level is also +4 dBu. (If the tape machine has meters then the level can be read here, provided that these meters are reliable and the line-up is known.)

It is customary to record a passage of 'Dolby tone' (in the case of Dolby A) or Dolby Noise (in the case of Dolby SR) at the beginning of a Dolby-encoded tape, along with the other line-up tones (see section 8.7.3). During record line-up, the Dolby tone is generated by the Dolby unit itself, and consists of a frequency-modulated 700 Hz tone at the Dolby's internal line-up reference level, which is easily recognised and distinguished from other line-up tones which may be present on a tape. Once the output level of the record Dolby has been set then the Dolby tone button on the relevant unit should be pressed, and the tone recorded at the start of the tape.

To align the replay Dolby (set to 'NR out', 'replay' mode), the recorded Dolby tone should be replayed and the input level adjusted so that the tone reads at the NAB mark on the internal meter. The output level should then be adjusted for +4 dBu, or

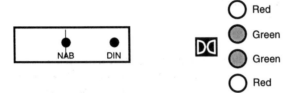

Figure 9.5 Dolby level is indicated on Dolby units using either a mechanical meter (shown left), or using red and green LEDs (shown right). The meter is normally aligned to the '18.5 NAB' mark or set such that the two green LEDs are on together

so that the mixer's meter reads PPM 5 when switched to monitor the tape machine replay.

For operation, the record and replay units should be switched to 'NR in'.

Dolby SR uses pink noise instead of Dolby tone, to distinguish tapes recorded with this system, and it is useful because it allows for line-up of the replay Dolby in cases where accurate level metering is not available. Since level misalignment will result in response errors the effects will be audible on a band of pink noise. A facility is provided for automatic switching between internally generated pink noise and off-tape noise, allowing the user to adjust replay-level alignment until there appears to be no audible difference between the spectra of the two. In normal circumstances Dolby SR systems should be aligned in a similar way to Dolby A, except that a noise band is recorded on the tape instead of a tone. Most systems use LED meters to indicate the correct level, having four LEDs as shown in Figure 9.5.

9.4 Operational considerations

A word may be said about varispeed. It is not uncommon for the replay speed of a tape to need to be adjusted slightly to alter pitch, or total playing time. In creative work massive amounts of speed change are sometimes employed. The pitch change means that Dolby decoding will be inaccurate since the frequency bands will not now correspond to those during the recording process, and Dolby mistracking will result.

Professional noise reduction systems are available as single-channel units, stereo packages, and conveniently grouped multiples of 8, 16 and 24 for multitrack work. They generally fit into standard 19 inch (48 cm) racks. Certain models are designed to fit straight into multitrack recorders so that the complete recorder plus noise reduction combination is conveniently housed in one unit.

Each noise reduction channel is manually switchable between encode for record and decode for replay, and in addition a special input is usually provided which accepts a remote DC signalling voltage, which will switch the unit into encode. Removal of the DC causes the unit to revert to decode ready for replay. Professional tape machines can usually provide this DC requirement, linking it to the record status of each track. Those tracks which are switched to record will now automatically switch the appropriate noise reduction channels to encode ready for recording. The system enables the selection of correct noise reduction status to be left to the recorder itself which is a very convenient feature particularly when a large number of channels are in use.

9.5 Single-ended noise reduction

9.5.1 General systems

Several companies offer so-called 'single-ended' noise reduction systems, and these are intended to 'clean up' an existing noisy recording or signal. They operate by sensing the level of the incoming signal, and as the level falls below a certain threshold the circuit begins to roll off the treble progressively, thereby reducing the level of hiss. The wanted signal, being low in level, in theory suffers less from this treble reduction than would a high-level signal due to the change in response of the ear with level (see Fact File 2.2). High-level signals are left unprocessed. The system is in fact rather similar to the Dolby B decoding process, but of course the proper reciprocal Dolby B encoding is absent. The input level controls of such systems must be carefully adjusted so as to bring in the effect of the treble roll-off at the appropriate threshold for the particular signal being processed so that a suitable compromise can be achieved between degree of hiss reduction and degree of treble loss during quieter passages. Such single-ended systems should be judiciously used – they are not intended to be left permanently in circuit – and value judgements must always be made as to whether the processed signal is in fact an improvement over the unprocessed one.

If a single-ended system is to be used on a stereo programme, units which are capable of being electronically 'ganged' must be employed so that exactly the same degree of treble cut is applied to each channel; otherwise varying frequency balance between channels will cause stereo images to wander.

9.5.2 Noise gates

The noise gate can be looked upon as another single-ended noise reduction system. It operates as follows. A threshold control is provided which can be adjusted such that the output of the unit is muted (the gate is 'closed') when the signal level falls below the threshold. During periods when signal level is very low (possibly consisting of tape or guitar amplifier noise only) or absent the unit shuts down. A very fast attack time is employed so that the sudden appearance of signal opens up the output without audible clipping of the initial transient. The time lapse before the gate closes, after the signal has dropped below the chosen threshold level, can also be varied. The close threshold is engineered to be lower than the open threshold (known as *hysteresis*) so that a signal level which is on the borderline does not confuse the unit as to whether it should be open or closed, which would cause 'gate flapping'.

Such units are useful when, for instance, a noisy electric guitar setup is being recorded. During passages when the guitarist is not playing the output shuts down so that the noise is removed from the mix. They are sometimes also used in a similar manner during multitrack mixdown where they mute outputs of the tape machine during the times when the tape is unmodulated, thus removing the noise contribution from those tracks.

The noise gate is frequently heard in action during noisy satellite link broadcasts and long-distance telephone-line operation. An impressive silence reigns when no-one is talking, but when speech begins a burst of noise abruptly appears and accompanies the speaker until he or she stops talking. This can sometimes be disconcerting for the speaker at the other end of the line because he or she gets the impression that the line has been cut off when the noise abruptly disappears.

Noise gates can also be used as effects in themselves, and the 'gated snare drum' is a common effect on pop records. The snare drum is given a heavy degree of gated reverb, and a high threshold level is set on the gate so that around half a second or so after the drum is hit the heavy 'foggy' reverb is abruptly cut off. Drum machines can mimic this effect, as can some effects processors.

9.5.3 Digital noise extraction

Extremely sophisticated single-ended computer-based noise reduction systems have been developed. They are currently expensive, and are offered as a service by a limited number of facilities. A given noisy recording will normally have a short period somewhere in which only the noise is present without any programme, for instance the run-in groove of an old 78 rpm shellac disc recording provides a sample of that record's characteristic noise. This noise is analysed by a computer and can subsequently be recognised as an unwanted constituent of the signal, and then extracted electronically from it. Sudden discontinuities in the programme caused by scratches and the like can be recognised as such and removed. The gap is filled by new material which is made to be similar to that which exists either side of the gap. Not all of these processes are currently 'real time', and it may take several times longer than the progamme's duration for the process to be carried out, but as the speed of digital signal processing increases more operations become possible in real time.

Recommended further reading

Dolby, R. (1967) An audio noise reduction system. *J. Audio Eng. Soc.*, vol. 15, pp. 383–388

Dolby, R. (1970) A noise reduction system for consumer tape applications. Presented at the *39th AES Convention. J. Audio Eng. Soc. (Abstracts)*, vol. 18, p. 704

Dolby, R. (1983) A 20 dB audio noise reduction system for consumer applications. *J. Audio Eng. Soc.*, vol. 31, pp. 98–113

Dolby, R. (1986) The spectral recording process. Presented at the *81st AES Convention*. Preprint 2413 (C-6). Audio Engineering Society

See also *General further reading* at the end of this book.

Chapter 10

Digital recording

10.1 Digital and analogue recording contrasted

In analogue recording, as described in the previous chapters, sound is recorded by converting continuous variations in sound pressure into continuous variations in electrical voltage, using a microphone. This varying voltage is then converted into a varying pattern of magnetisation on a tape, or, alternatively, into a pattern of light and dark areas on an optical-film sound track, or a groove of varying deviation on an LP.

Because the physical characteristics of analogue recordings relate closely to the sound waveform, replaying them is a relatively simple matter. Variations in the recorded signal can be converted directly into variations in sound pressure using a suitable collection of transducers and amplifiers. The problem, though, is that the replay system is unable to tell the difference between *wanted* signals and *unwanted* signals. Unwanted signals might be distortions, noise and other forms of interference introduced by the recording process. For example, a record player cannot distinguish between the stylus movement it experiences because of a scratch on a record (unwanted) and that caused by a loud transient in the music (wanted). Imperfections in the recording medium are reproduced as clicks, crackles and other noises.

Digital recording, on the other hand, converts the electrical waveform from a microphone into a series of binary numbers, each of which represents the amplitude of the signal at a unique point in time, recording these numbers in a coded form which allows the system to detect whether the replayed signal is correct or not. A reproducing device is then able to distinguish between the wanted and the unwanted signals introduced above, and is thus able to reject all but the wanted original information in most cases. Digital audio is much more tolerant of a poor recording channel than analogue audio, and distortions and imperfections in the storage or transmission process need not affect the sound quality of the signal provided that they remain within the design limits of the system. These issues are given further coverage in Fact File 10.1.

Digital audio has made it possible for sound engineers to take advantage of developments in the computer industry, and this is particularly beneficial because the size of that industry results in mass production (and therefore cost savings) on a scale not possible for audio products alone. Today it is common for sound to be recorded, processed and edited on relatively low cost desktop computer equipment, and this is a trend likely to continue.

FACT FILE 10.1 — Analogue and digital information

Analogue information is made up of a continuum of values, which at any instant may have any value between the limits of the system. For example, a rotating knob may have one of an infinite number of positions – it is therefore an analogue controller (see the diagram).

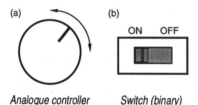

(a) Analogue controller   (b) Switch (binary)

A simple switch, on the other hand, can be considered as a digital controller, since it has only two positions – off or on. It cannot take any value in between. The brightness of light which we perceive with our eyes is analogue information, and as the sun goes down the brightness falls gradually and smoothly, whereas a household light without a dimmer may be either on or off – its state is binary (that is it has only two possible states).

Electrically, analogue information may be represented as a varying voltage or current. If a rotary knob is connected to a variable resistor and a voltage supply, its position will affect the output voltage as shown here. The switch may be connected to a similar voltage supply, and in this case the output voltage can only be either zero volts or +V. In other words the electrical information which results is binary.

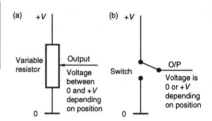

The high (+V) state could be said to correspond to a binary one, and the low state to binary zero (although in many real cases it is actually the other way around).

Binary information is inherently more resilient to noise and interference than analogue information, as shown below. If noise is added to an analogue signal then it becomes very difficult to tell at any later stage in the signal chain what is the wanted signal and what is the unwanted noise, since there is no means of distinguishing between the two. If noise is added to a digital signal it is possible to extract the important information at a later stage, since only two states matter – the one and zero states. By comparing the signal amplitude with a fixed decision point it is possible for a receiver to decide that everything above the decision point is 'high', whilst everything below it is 'low'. For any noise to influence the state of a digital signal it must be at least large enough in amplitude to cause a high level to be interpreted as 'low', or vice versa.

The timing of digital signals may also be corrected to some extent, using a similar method, which gives digital signals another advantage over analogue. If the timing of bits in a digital message becomes unstable, such as after having been passed over a long cable, resulting in timing 'jitter', the signal may be reclocked at a stable rate ensuring that the timing stability of the information is restored.

Analogue signal plus noise

Binary signal plus noise

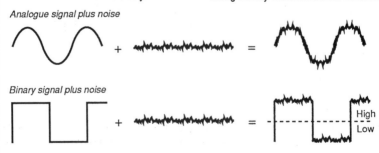

This chapter contains an introduction to the main principles of digital recording, described in as non-technical a manner as possible. Further reading recommendations at the end of this chapter are given for those who want to study the subject in more depth.

10.2 The digital audio signal chain

Figure 10.1 shows the signal chain involved in any digital recording or broadcasting system. First the analogue audio signal (a time-varying electrical voltage) is passed through an analogue-to-digital (A/D) convertor where it is transformed from a continuously varying voltage into a series of 'samples', which are 'snapshots' of the analogue signal taken many thousand times per second. Each sample is represented by a number. This series of samples is coded into a form which makes it suitable for recording or broadcasting (a process known as coding or channel coding), and the signal is then recorded or transmitted. Upon replay or reception the signal is decoded and subjected to error correction, and it is this latter process which works out what damage has been done to the signal since it was coded. Any errors in timing or value of the samples are corrected if possible, and the result is fed to a digital-to-analogue (D/A) convertor which turns the numerical data back into a time-continuous analogue audio signal.

In the following sections each of the main processes involved in this chain will be explained, followed by a discussion of the implementation of this technology in real audio systems. Only the most straightforward principles are described here, since much of the theoretical matter involved in digital audio is above the intended level of this book.

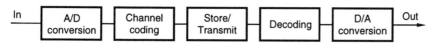

Figure 10.1 Block diagram of the typical digital recording or broadcasting signal chain

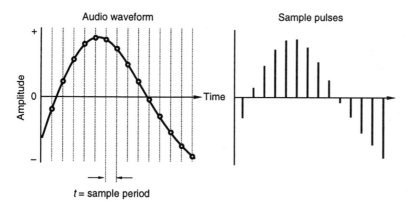

Figure 10.2 An arbitrary audio signal is sampled at regular intervals of time *t* to create short sample pulses whose amplitudes represent the instantaneous amplitude of the audio signal at each point in time

10.3 Analogue-to-digital (A/D) conversion

10.3.1 Sampling

The sampling process employed in an A/D convertor involves the measurement or 'sampling' of the amplitude of the audio waveform at regular intervals in time (see Figure 10.2). From this diagram it will be clear that the sample pulses represent the instantaneous amplitudes of the signal at each point in time. The samples can be considered as like instantaneous 'still frames' or 'snapshots' of the audio signal which together and in sequence form a representation of the continuous waveform, rather as the still frames which make up a movie film give the impression of a continuously moving picture when played in quick succession.

In order to represent the fine detail of the signal it is necessary to take a large number of these samples per second, and the mathematical sampling theorem proposed by Shannon indicates that at least two samples must be taken per audio cycle if the necessary information about the signal is to be conveyed. In other words, the sampling frequency must be at least twice as high as the highest frequency of audio to be handled by the system (this is often called the Nyquist criterion). Another way of visualising this issue is presented in Fact File 10.2, which considers sampling as a modulation process.

10.3.2 Filtering and aliasing

A simple example in Figure 10.3 shows that if too few samples are taken per cycle of the audio signal then the samples may be interpreted as representing a wave other

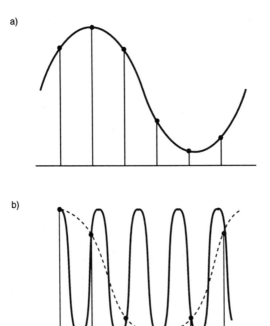

a)

b)

Figure 10.3 (a) Wave sampled at an adequate rate. (b) Wave sampled at an inadequate rate leading to aliasing

FACT FILE

10.2

Sampling – frequency domain

One way of visualising the sampling process is to consider it in terms of modulation. The continuous audio waveform is used to modulate a regular chain of pulses. The frequency of these pulses is the sampling frequency (f_s). Before modulation all these pulses have the same amplitude (height), but after modulation the amplitude of the pulses is modified according to the instantaneous amplitude of the audio signal at that point in time. This process is called pulse amplitude modulation (PAM).

The frequency spectrum of the modulated signal is as shown in the diagram. The shape of the spectrum is not significant – it is just an example of a complex audio signal such as music. It will be seen that in addition to the 'baseband' audio signal (the original spectrum before sampling) there are now a number of additional spectra, each centred on multiples of the sampling frequency. Sidebands have been produced either side of the sampling frequency and its multiples, as a result of the amplitude modulation, and these extend above and below the sampling frequency and its multiples to the extent of the base bandwidth. In other words these sidebands are pairs of mirror images of the audio band.

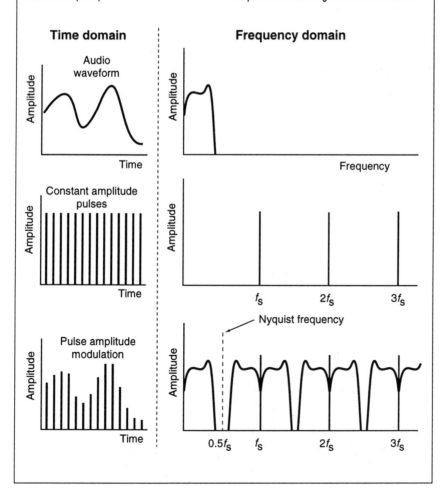

than that originally sampled, and this is one way of understanding the phenomenon known as aliasing. An 'alias' is an unwanted by-product of the original signal. It arises when the signal is reconstructed as an analogue signal during D/A conversion. If the sample values in Figure 10.3(b) were presented to a D/A convertor the lower frequency (dotted) wave could be reconstructed, which is not correct. There is no information in the samples to indicate the rapid changes in the wave that take place in between.

To understand the issue in more detail one must consider the signals in the frequency domain. It is relatively easy to see why the sampling frequency must be at least twice the highest audio frequency from Figure 10.4. Here an extension of the audio signal's bandwidth (baseband) above the Nyquist frequency (half the sampling frequency) results in the lower sideband of the first spectral repetition overlapping the upper end of the baseband, resulting in an alias region where they overlap. Two further examples are shown to illustrate the point – the first in which a sine wave tone is sampled, having a low enough frequency for the sampled sidebands to lie above the audio frequency range, and the second in which a much higher frequency tone (above the Nyquist frequency) causes a sampled sideband to fall well within the baseband. This would be audible as an alias of the original tone.

The aliasing phenomenon can also be witnessed visually in the case of the well-known 'spoked-wheel' effect on films, since moving pictures are also an example of a sampled signal. In film, still pictures (image samples) are normally taken at a rate of 24 per second. If a rotating wheel with a marker on it is filmed it will appear to move round in a forward direction as long as the rate of rotation is much slower

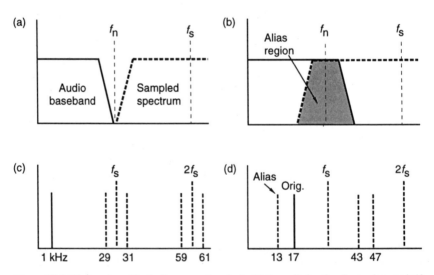

Figure 10.4 Aliasing viewed in the frequency domain. In (a) the audio baseband extends up to half the sampling frequency (the Nyquist frequency f_n) and no aliasing occurs. In (b) the audio baseband extends above the Nyquist frequency and consequently overlaps the lower sideband of the first spectral repetition, giving rise to aliased components in the shaded region. In (c) a tone at 1 kHz is sampled at a sampling frequency of 30 kHz, creating sidebands at 29 and 31 kHz (and at 59 and 61 kHz, etc.). These are well above the normal audio frequency range, and will not be audible. In (d) a tone at 17 kHz is sampled at 30 kHz, putting the first lower sideband at 13 kHz – well within the normal audio range. The 13 kHz sideband is said to be an alias of the original wave

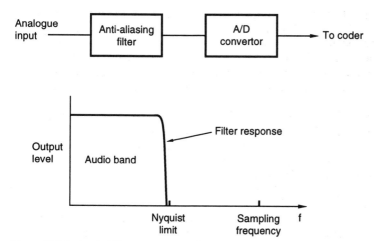

Figure 10.5 In simple A/D convertors an analogue anti-aliasing filter is used prior to conversion, which removes input signals with a frequency above the Nyquist limit

than the rate of the still photographs, but as its rotation rate increases it will appear to slow down, stop, and then appear to start moving backwards. The virtual impression of backwards motion gets faster as the rate of rotation of the wheel gets faster, and this backwards motion is the aliased result of sampling at too low a rate. Clearly the wheel is not really rotating backwards, it just appears to be – it is an incorrect reconstruction of the original information, resulting from sampling the picture at too slow a rate.

If audio signals are allowed to alias in digital recording one hears the audible equivalent of the backwards-rotating wheel – that is, sound components in the audible spectrum which were not there in the first place, moving downwards in frequency as the original frequency of the signal increases. In basic convertors, therefore, it is necessary to filter the baseband audio signal before the sampling process, as shown in Figure 10.5, so as to remove any components having a frequency higher than half the sampling frequency.

In real systems, and because filters are not perfect, the sampling frequency is made slightly higher than twice the highest audio frequency to be represented, allowing for the filter to roll off slightly more gently. Choice of sampling frequency is discussed further in Fact File 10.3.

10.3.3 Quantisation

Once the audio signal has been sampled, the samples must be converted into a series of numbers in a process known as quantisation. The result of sampling the sound signal is that a train of pulses of varying amplitude is produced, and the quantisation process assigns a numerical value to each of these samples according to its amplitude. It is rather like taking a tape measure to each of the sample pulses and measuring its height, above or below the 0 volt line (see Figure 10.6). Each sample is assigned a value from a range of fixed possibilities, and in the example a ruler scale from 1 to 10 is used for both positive and negative ranges. Each sample must be given one of these integer values, and no fractions or decimal places are allowed;

FACT FILE

10.3

Sampling rates

The choice of sampling rate determines the maximum audio bandwidth available. The table here is an attempt to summarise the variety of sampling frequencies in existence and their applications. A wide range of rates is used in desktop PCs and sound cards, and only the most popular are shown here. Arguments have been presented for the standardisation of higher sampling rates such as 88.2 and 96 kHz, quoting evidence from sources claiming that information above 20 kHz is important for sound quality. It is certainly true that the ear's frequency response does not cut off completely at 20 kHz, but there is very little properly supported evidence that listeners can reliably distinguish between signals containing higher frequencies and those which do not. Doubling the sampling frequency leads to a doubling in the overall data rate of a digital audio system, and a consequent halving in storage time.

Low sampling frequencies such as those below 30 kHz are sometimes encountered in PC workstations for lower quality sound applications such as the storage of speech samples, the generation of internal sound effects and so forth.

Commonly encountered sampling frequencies

$f_s(kHz)$	Application
8	Used in telephony. Poor audio quality. CCITT G711 standard. IMA RP rate*
~11.025	One quarter of the CD sampling rate, used in the sound hardware of some desktop computers, particularly the Apple Macintosh, for low quality applications. IMA RP rate*
16	Used in some telephony applications. G722 data reduction
18.9	CD-ROM/XA and CD-I standard for low–moderate quality audio using ADPCM to extend playing time
~22.05	Half the CD rate is 22.05 kHz. The original Apple Macintosh audio sampling rate was 22254.5454..... IMA RP rate*
32	Used in some broadcast systems, e.g. NICAM 3, NICAM 728, DAT long play mode
37.8	CD-ROM/XA and CD-I sampling rate for intermediate quality audio using ADPCM
44.056	A slight modification of the 44.1 rate used in some older equipment to align digital audio with the NTSC television frame rate of 29.97 frames per second. Occasionally still encountered in the USA
44.1	CD sampling rate. Used widely for professional audio recording in many formats. Some professionally modified DAT machines will operate at this rate from analogue inputs. IMA RP rate*
47.952	Occasionally encountered when 48 kHz equipment is used in NTSC video operations. To be avoided
48	'Professional' rate, as specified in AES5-1984, and encountered mainly in digital video recorder sound tracks. Many DAT machines will only sample at this rate through analogue inputs
88.2 and 96	Twice the 44.1 and 48 k standard rates. Found in some audiophile equipment, such as certain non-standard DAT machines.

*IMA RP rates were selected in the International Multimedia Association Recommended Practice for Enhancing Digital Audio Compatibility in Multimedia Systems, Rev. 3.00, Oct. 1992, for sound file transfer between workstations

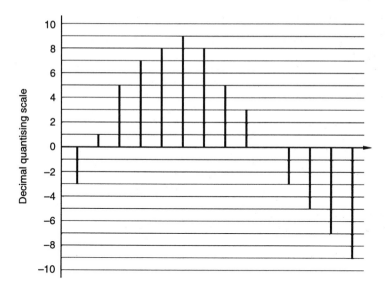

Resulting output data sequence = –3, 1, 5, 7, 8, 9, 8, 5, 3, 0, –3, –5, –7, –9

Figure 10.6 Sample pulses must be quantised to give them numerical values (decimal scale shown). Each sample is forced to the closest quantising level

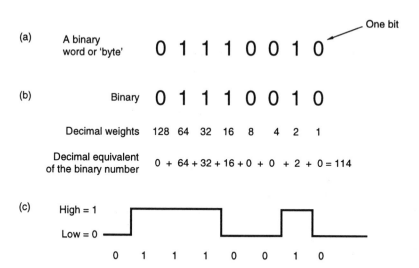

Figure 10.7 (a) A binary number (word or 'byte') consists of a number of bits. (b) Each bit represents a power of two. (c) Binary numbers can be represented electrically in pulse-code modulation (PCM) by a string of high and low voltages

thus each sample is rounded to the nearest whole number during quantisation. The resulting sequence of numbers is shown in the figure.

In a digital audio system, the number system used is *binary* rather than decimal. This has great advantages, since it allows numbers to be represented in two-state form (on/off, high/low, true/false, voltage/no voltage, etc.), and it allows audio information to be stored in the same way as computer data is stored. In the decimal number system, each digit of a number represents a power of ten. In a binary system each digit represents a power of two, as shown in Figure 10.7. Each *bi*nary dig*it* is called a bit. It is possible to calculate the decimal equivalent of a binary integer (whole number) by using the method shown. A number made up of more than one bit is called a binary 'word', and an 8 bit word is called a 'byte' (from 'by eight'). Four bits is called a 'nibble'. The more bits there are in a word the larger the number of states it can represent, with eight bits allowing 256 (2^8) states and sixteen bits allowing 65 536 (2^{16}). The bit with the lowest weight (2^0) is called the least significant bit or LSB, and that with the greatest weight is called the most significant bit or MSB. The term kilobyte or kbyte is used to mean 1024 or 2^{10} bytes, and the term megabyte or Mbyte represents 1024 kbytes.

A binary number (word) with 3 bits is capable of representing 2^3 (8) different values, and an example is shown in Figure 10.8 of a quantising scale using 3 bit binary instead of decimal values. The numbers have been arranged so that binary zero is at 0 volts (as explained in Fact File 10.4).

Because it is rare for samples to lie precisely at the amplitudes represented by each quantising step, there will usually be a slight difference between the quantised

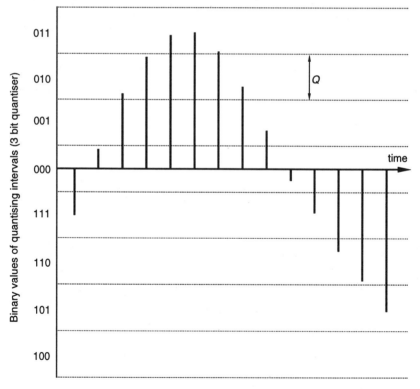

Figure 10.8 Each sample is mapped to the closest quantising interval Q, and given the binary value assigned to that interval. (Example of a 3 bit quantiser shown, with binary zero at 0 volts). On D/A conversion each binary value is assumed to represent the voltage at the mid point of the quantising interval

FACT FILE 10.4 Representing negative values

Negative sample values (those that lie under the 0 volts line) must be designated as such, since a digital audio system may need to perform mathematical operations on the data in order to perform such functions as adding two signals together. A number system known as two's complement is normally used for this purpose, which represents negative binary numbers by making the leftmost bit a one. The leftmost bit is called the most significant bit or MSB, since it represents the greatest power of two, and for positive values the MSB is a zero.

Negative values are worked out by taking the positive equivalent, inverting all the bits and adding a one. Thus to get the 4 bit binary equivalent of decimal minus five (-5_{10}) in binary two's complement form:

$$5_{10} = 0101_2$$
$$-5_{10} = 1010 + 0001 = 1011_2$$

Two's complement numbers have the advantage that the MSB represents the sign (1 = negative, 0 = positive), and that arithmetic may be performed between positive and negative numbers giving the correct result:

e.g. (in decimal):   5
                     +(−3)
                      =2

or (in binary):   0101
                   + 1101
                   = 0010

The carry bit which may result from adding the two MSBs is ignored.

An example was shown in Figure 10.8 of a 3 bit quantising scale using two's complement numbering, and it will be seen that the binary value changes from all zeros to all ones as it crosses the 0 volts line. Further, the maximum positive value is 011, and the maximum negative value is 100, so the values wrap around from max. positive to max. negative.

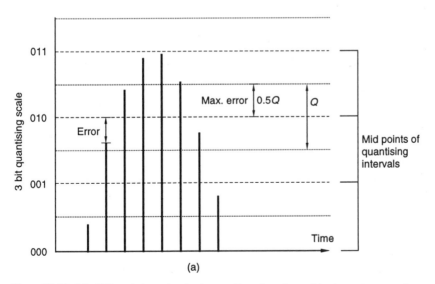

(a)

Figure 10.9 In (a) a 3 bit scale is used and only a small number of quantising intervals covers the analogue voltage range, making the maximum quantising error quite large. The second sample in this picture will be assigned the value 010, for example, the corresponding voltage of which is somewhat higher than that of the sample.

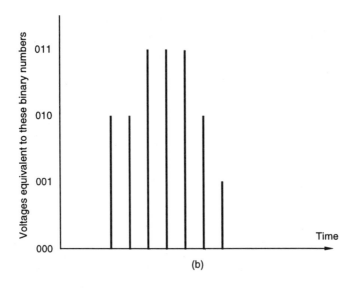

(b)

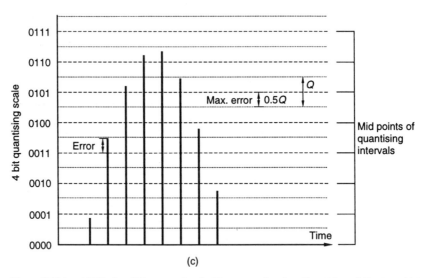

(c)

Figure 10.9 (contd.) During D/A conversion the binary sample values from (a) would be turned into pulses with the amplitudes shown in (b), where many samples have been forced to the same level owing to quantising. In (c) the 4 bit scale means that a larger number of intervals is used to cover the same range and the quantising error is reduced. (Expanded positive range only shown for clarity)

sample and its original amplitude. This is called the quantising error. The maximum value of the quantising error is plus or minus half of one quantising interval, since if a sample's amplitude is more than half of a quantising interval above a step it is rounded to the next one. If more steps are used to cover the same quantising range, the potential quantising error is smaller, since the individual quantising intervals are smaller (see Figure 10.9).

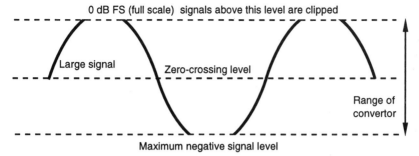

Figure 10.10 Signals exceeding peak level in a digital system are hard-clipped, since no more digits are available to represent the sample value

Four bit quantisation is quite coarse, and leads to a large quantising error, since only sixteen steps are available. A system using 4 bit quantisation would sound very distorted because the difference between the quantised signal and the true shape of the signal would be considerable. What, though, is the effect of the error which arises in quantisation? It is really a form of distortion, since it changes the shape of the audio signal very slightly, depending on the magnitude of the error. The error signal which results is called quantising noise or quantising distortion, and provided that the quantised audio signal is of suitably high level and has a semi-random nature (as most real signals do), the error will manifest itself as low level noise. The addition of an intentional low level noise signal to the audio signal prior to conversion, known as 'dither', helps to randomise the effect of the error and make it more noise like, particularly at low signal levels.

The dynamic range of a digital audio system is limited at high signal levels by the point at which the range of the convertor has been 'used up' (in other words, when there are no more bits available to represent a higher level signal). At this point the waveform will be hard clipped (see Figure 10.10) and will become very distorted. This point will normally be set to occur at a certain electrical input voltage, such as +24 dBu in some professional systems. (The effect is very different from that encountered in analogue tape recorders which tend to produce gradually more distortion as the recording level increases. Digital recorders remain relatively undistorted as the recording level rises until the overload point is reached, at which point very bad distortion occurs.) At low signal levels the dynamic range is limited by quantisation noise. (Fact File 10.5 addresses the choice of quantising resolution in digital audio systems.)

The quantised output of an A/D convertor is sometimes carried on a collection of wires, each carrying 1 bit of the binary word which represents the audio signal, so a 16 bit convertor would have sixteen single-bit outputs. When each bit of the audio sample is carried on a separate wire, the signal is said to be in a *parallel* format. If the data is transmitted down a single channel, one bit after the other, the data is said to be in *serial* format. The process of converting an analogue signal into serial binary data is called pulse-code modulation (PCM).

10.3.4 Oversampling in A/D conversion

Oversampling involves sampling audio at a higher frequency than strictly necessary to satisfy the Nyquist criterion (in other words, much higher than twice the highest

FACT FILE
10.5

Quantising resolution

The number of bits per sample dictates the signal-to-noise ratio or dynamic range of a digital audio system. The theoretical dynamic range of a linear PCM system is approximately 6 dB per bit. Thus a 16 bit system is theoretically capable of around 96 dB dynamic range. Unfortunately the situation in practice is nothing like so straightforward, since many systems are capable of reproducing signals below the theoretical limit due partly to modulation of the dither signal, and oversampling convertors with noise shaping exhibit noise characteristics which may appear subjectively to be lower. Additionally, many so-called 16 bit convertors do not have full 16 bit linearity, and thus the resulting noise floor may be higher than predicted in theory. The table is a summary of the applications for different sample resolutions.

For many years now 16 bit linear PCM has been the norm for high quality audio applications. This is the CD standard and is capable of offering good dynamic range of over 90 dB. For most purposes this is adequate, but it is often the case that for professional recording purposes one needs a certain amount of 'headroom' – in other words some unused dynamic range above the normal peak recording level which can be used in unforeseen circumstances such as when a signal overshoots its expected level. This can be particularly necessary in live recording situations where one is never quite sure what is going to happen with recording levels. 20 and 24 bit recording formats are therefore becoming increasingly popular because of their extra dynamic range

At the lower quality end, some PC sound cards and internal sound generators operate at resolutions as low as 4 bits. 8 bit resolution is quite common in desktop computers, and this proves adequate for moderate quality sound through the PC's internal loudspeakers. It gives a dynamic range of nearly 50 dB undithered.

Linear quantising resolution

Bits per sample	Approx dyn. range with dither	Application
8	44 dB	Low–moderate quality for older PC internal sound generation. Some multimedia applications. Usually in the form of unsigned binary numbers
12	68 dB	Older Akai samplers, e.g. S900
14	80 dB	Original EIAJ format PCM adaptors, such as Sony PCM-100
16	92 dB	CD standard. DAT standard. Most widely used high quality resolution for consumer media and many professional recorders. Many multimedia PCs. Two's complement (signed) binary numbers
20	116 dB	High quality professional audio recording and mastering applications. Good convertors available
24	140 dB	Maximum resolution of most new prof. recording systems, also of AES/EBU digital interface. Dynamic range would exceed psychoacoustic requirements. Hard to convert at this resolution

audio frequency). The sampling rate required just to satisfy the Nyquist criterion is often known as the Nyquist rate. Using a sampling rate well above the audio range (often tens or hundreds of times the Nyquist rate) means that the spectral repetitions resulting from PAM are a long way from the upper end of the audio band (see Figure

10.11). Steep analogue filters are therefore not required, resulting in a better sound quality. The analogue anti-aliasing filter used in conventional convertors is replaced by a digital decimation filter which is capable of a more precise frequency and phase response.

Oversampling also makes it possible to introduce 'noise shaping' into the conversion process. Noise shaping is a means by which noise within the most audible parts of the audio frequency range (see Chapter 2) is reduced at the expense of increased noise at other frequencies, using a process which 'shapes' the spectral energy of the quantising noise so that most of it lies above the most sensitive regions of the human hearing range.

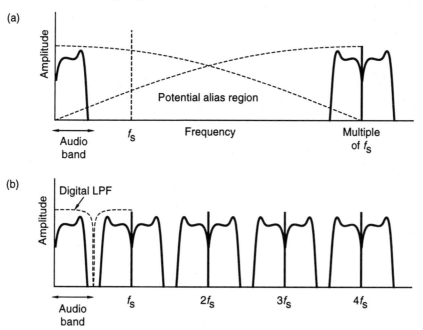

Figure 10.11 (a) Oversampling in A/D conversion initially creates spectral repetitions that lie a long way from the top of the audio baseband. The dotted line shows the theoretical extension of the baseband and the potential for aliasing, but the audio signal only occupies the bottom part of this band. (b) Decimation and digital low pass filtering limits the baseband to half the sampling frequency, thereby eliminating any aliasing effects, and creates a conventional collection of spectral repetitions at multiples of the sampling frequency

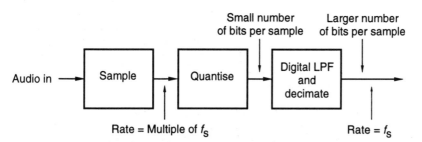

Figure 10.12 Block diagram of oversampling A/D conversion process

Although oversampling A/D convertors often use sampling rates of up to 128 times the standard Nyquist rates of 44.1 or 48 kHz, the actual rate at the digital output of the convertor may be no more than the Nyquist rate. Samples acquired at the high rate are quantised to only a few bits resolution and then digitally filtered to reduce the sampling rate back to the Nyquist rate, as shown in Figure 10.12. The digital low-pass filter limits the bandwidth of the signal to the Nyquist frequency in order to avoid aliasing, and this is coupled with 'decimation'. Decimation reduces the sampling rate by dropping samples from the oversampled stream. A result of the low-pass filtering operation is to increase the word length of the samples very considerably. This is not simply an arbitrary extension of the wordlength, but an accurate calculation of the correct value of each sample, based on the values of surrounding samples. The sample resolution can then be shortened as necessary to produce the desired word length.

Recently a system called 'Direct Stream Digital' has been proposed which involves oversampling *without* necessarily using decimation to reduce the sampling rate before recording. This is intended for the storage of audio signals with a bandwidth much wider than the conventional 20 kHz, in order that very high frequency information may be preserved. The use of a very high sampling rate means that recordings occupy more storage space than those made with conventional PCM, but proponents argue that information above 20 kHz may be subjectively important and that one should attempt to retain as much information as possible in archive recordings.

10.4 Digital-to-analogue (D/A) conversion

10.4.1 A basic D/A convertor

In order to reproduce digital recordings it is necessary to convert the binary data back into analogue sound signals. The basic D/A conversion process is shown in Figure 10.13. Audio sample words are converted back into a staircase-like chain of voltage levels corresponding to the quantised sample values. This is achieved in simple convertors by using the states of bits to turn current sources on or off, making up the required pulse amplitude by the combination of outputs of each of these sources. This staircase is then 'resampled' to reduce the width of the pulses before

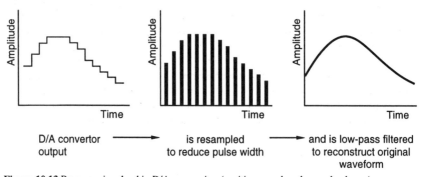

Figure 10.13 Processes involved in D/A conversion (positive sample values only shown)

they are passed through a low-pass reconstruction filter, whose cut-off frequency is half the sampling frequency. The effect of the reconstruction filter is to join up the sample points to make a smooth waveform. Resampling is necessary because otherwise the averaging effect of the filter would result in a reduction in the amplitude of high-frequency audio signals (the so-called 'aperture effect'). Aperture effect may be reduced by limiting the width of the sample pulses to perhaps one-eighth of the sample period. Equalisation may be required to correct for aperture effect.

10.4.2 Oversampling in D/A conversion

Oversampling may be used in D/A conversion, as well as in A/D conversion. In the D/A case new samples must be calculated and inserted at appropriate intervals in between the Nyquist rate samples, in order that conversion can be performed at a higher sampling rate. These samples are then converted back to analog at the higher rate, thereby avoiding the need for steep analog reconstruction filters. Noise shaping may also be introduced at the D/A stage, depending on the design of the convertor, to reduce the subjective level of the noise.

A number of advanced D/A convertor designs exist which involve oversampling at a high rate, creating samples with only a few bits of resolution. The extreme version of this approach involves very high rate conversion of single bit samples (so-called 'bit stream conversion'), with noise shaping to optimise the noise spectrum of the signal. The theory of these convertors is outside the scope of this book.

10.5 Introduction to digital audio signal processing

Just as processing operations like equalisation, fading and compression can be performed on analogue sound signals, so they can on digital sound signals. Indeed it is often possible to achieve certain operations in the digital domain with fewer side-effects such as phase distortion. It is also possible to perform operations in the digital domain which are either very difficult or impossible in the analogue domain. High quality, authentic-sounding artificial reverberation is one such example, in which the reflection characteristics of different halls and rooms can be accurately simulated. Digital signal processing (DSP) involves the high speed manipulation of the binary data representing audio samples. This is carried out by computing devices capable of performing fast mathematical operations on data, often many millions per second. It may involve changing the values and timing order of samples, and it may involve the combining of two or more streams of audio data. DSP can affect the sound quality of digital audio in that it can add noise or distortion, although one must assume that the aim of good design is to minimise any such degradation in quality.

In the sections which follow an introduction will be given to some of the main applications of DSP in audio products such as mixers, effects devices and workstations without delving into the mathematical principles involved. In some cases the description is an over-simplification of the process, but the aim has been to illustrate concepts not to tackle the detailed design considerations involved.

10.5.1 Gain changing (level control)

It is relatively easy to change the level of an audio signal in the digital domain. It is

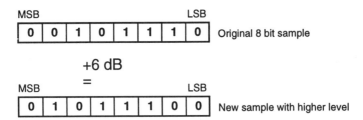

Figure 10.14 The gain of a sample may be changed by 6 dB simply by shifting all the bits one step to the left or right

most easy to shift its gain by 6 dB since this involves shifting the whole sample word either one step to the left or right (see Figure 10.14). Effectively the original value has been multiplied or divided by a factor of two. More precise gain control is obtained by multiplying the audio sample value by some other factor representing the increase or decrease in gain. The number of bits in the multiplication factor determines the accuracy of gain adjustment. The result of multiplying two binary numbers together is to create a new sample word which may have many more bits than the original, and it is common to find that digital mixers have internal structures capable of handling 32 bit words, even though their inputs and outputs may handle only 20.

The values used for multiplication in a digital gain control may be derived from any user control such as a fader, rotary knob or on-screen representation, or they may be derived from stored values in an automation system. A simple 'old-fashioned' way of deriving a digital value from an 'analogue' fader is to connect the fader to a fixed voltage supply and connect the fader wiper to an A/D convertor, although it is quite common now to find controls capable of providing a direct binary output relating to their position. The 'law' of the fader (the way in which its gain is related to its physical position) can be determined by creating a suitable look-up table of values in memory which are then used as multiplication factors corresponding to each physical fader position.

10.5.2 Crossfading

Crossfading is employed widely in digital audio workstations at points where one section of sound is to be joined to another (edit points). It avoids the abrupt change of waveform which might otherwise result in an audible click, and allows one sound to take over smoothly from the other. (This is discussed further later.)

The process is illustrated conceptually in Figure 10.15. It involves two signals each undergoing an automated fade (binary multiplication), one downwards and the other upwards, followed by an addition of time-coincident samples of the two signals. By controlling the rates and coefficients involved in the fades one can create different styles of crossfade for different purposes.

10.5.3 Mixing

Mixing is the summation of independent data streams representing the different audio channels. Time coincident samples from each input channel are summed to

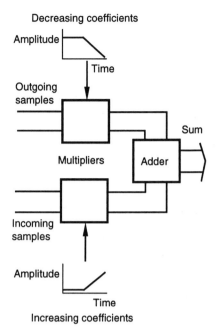

Figure 10.15 Conceptual block diagram of the crossfading process, showing two audio signals multiplied by changing coefficients, after which they are added together

produce a single output channel sample. Clearly it is possible to have many mix 'buses' by having a number of separate summing operations for different output channels. The result of summing a lot of signals may be to increase the overall level considerably, and the architecture of the mixer must allow enough headroom for this possibility. In the same way as an analogue mixer, the gain structure within a digital mixer must be such that there is an appropriate dynamic range window for the signals at each point in the chain, also allowing for operations such as equalisation which change the signal level.

10.5.4 Digital filters and equalisation

Digital filtering is something of a 'catch-all' term, and is often used to describe DSP operations which do not at first sight appear to be filtering. A digital filter is essentially a process which involves the time delay, multiplication and recombination of audio samples in all sorts of configurations, from the simplest to the most complex. Using digital filters one can create low and high-pass filters, peaking and shelving filters, echo and reverberation effects, and even adaptive filters which adjust their characteristics to affect different parts of the signal.

To understand the basic principle of digital filters it helps to think about how one might emulate a certain analogue filtering process digitally. Filter responses can be modelled in two main ways – one by looking at their frequency domain response and the other by looking at their time domain response. (There is another approach involving the so-called z-plane transform, but this is not covered here.). The frequency domain response shows how the amplitude of the filter's output varies with frequency, whereas the time domain response is usually represented in terms of an impulse response (see Figure 10.16). An impulse response shows how the

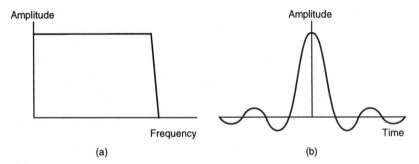

Figure 10.16 Examples of (a) the frequency response of a simple filter, and (b) the equivalent time domain impulse response

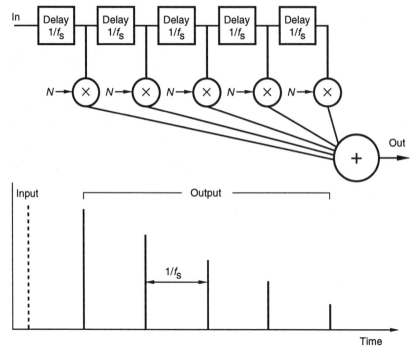

Figure 10.17 A simple FIR filter (transversal filter). N = multiplication coefficient for each tap. Response shown below indicates successive outputs samples multipled by decreasing coefficients

filter's output responds to stimulation at the input by a single short impulse. Every frequency response has a corresponding impulse response because the two are directly related. If you change the way a filter responds in time you also change the way it responds in frequency. A mathematical process known as the Fourier transform is often used as a means of transforming a time domain response into its equivalent frequency domain response (see Chapter 1). They are simply two ways of looking at the same thing.

Digital audio is time discrete because it is sampled. Each sample represents the amplitude of the sound wave at a certain point in time. It is therefore normal to create certain filtering characteristics digitally by operating on the audio samples in the time domain. In fact if it was desired to emulate a certain analogue filter character-istic digitally one would need only to measure its impulse response and model this in the digital domain. The digital version would then have the same frequency response as the analogue version, and one can even envisage the possibility for favourite analogue filters to be recreated for the digital workstation. The question, though, is how to create a particular impulse response characteristic digitally, and how to combine this with the audio data.

As mentioned earlier, all digital filters involve delay, multiplication and recom-bination of audio samples, and it is the arrangement of these elements that gives a filter its impulse response. A simple filter model is the finite impulse response (FIR) filter, or transversal filter, shown in Figure 10.17. As can be seen, this filter consists of a tapped delay line with each tap being multiplied by a certain coefficient before being summed with the outputs of the other taps. Each delay stage is normally a one sample period delay. An impulse arriving at the input would result in a number of separate versions of the impulse being summed at the output, each with a different amplitude. It is called a finite impulse response filter because a single impulse at the input results in a finite output sequence determined by the number of taps. The more taps there are the more intricate the filter's response can be made, although a simple low pass filter only requires a few taps.

The other main type is the infinite impulse response (IIR) filter which is also known as a recursive filter, because there is a degree of feedback between the output and the input (see Figure 10.18). The response of such a filter to a single impulse is an infinite output sequence, because of the feedback. IIR filters are often used in audio equipment because they involve fewer elements for most variable equalisers

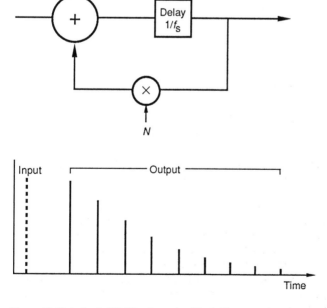

Figure 10.18 A simple IIR filter (recursive filter). The output impulses continue indefinitely but become very small. N in this case is about 0.8. A similar response to the previous FIR filter is achieved but with fewer stages

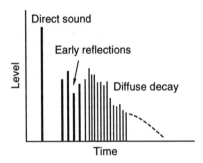

Figure 10.19 The impulse response of a typical reflective room

than equivalent FIR filters, and they are useful in effects devices. They are unfortunately not phase linear, though, whereas FIR filters can be made phase linear.

10.5.5 Digital reverberation and other effects

It can probably be seen that the IIR filter described in the previous section forms the basis for certain digital effects, such as reverberation. The impulse response of a typical room looks something like Figure 10.19, that is an initial direct arrival of sound from the source, followed by a series of early reflections, followed by a diffuse 'tail' of densely packed reflections decaying gradually to almost nothing. Using a number of IIR filters, perhaps together with a few FIR filters, one could create a suitable pattern of delayed and attenuated versions of the original impulse to simulate the decay pattern of a room. By modifying the delays and amplitudes of the early reflections and the nature of the diffuse tail one could simulate different rooms.

The design of convincing reverberation algorithms is a skilled task, and the difference between crude approaches and good ones is very noticeable. Some audio workstations offer limited reverberation effects built into the basic software package, but these often sound rather poor because of the limited DSP power available and the crude algorithms involved. More convincing reverberation processors are available which exist either as stand-alone devices or as optional plug-in cards to the workstation, having more DSP capacity and tailor-made software.

Other simple effects can be introduced without much DSP capacity, such as double-tracking and phasing/flanging effects. These often only involve very simple delaying and recombination processes. Pitch shifting can also be implemented digitally, and this involves processes similar to sample rate conversion, as described below. High quality pitch shifting requires quite considerable horsepower because of the number of calculations required.

10.5.6 Dynamics processing

Digital dynamics processing involves gain control which depends on the instantaneous level of the audio signal. A simple block diagram of such a device is shown in Figure 10.20. A side chain produces coefficients corresponding to the instantaneous gain change required, which are then used to multiply the delayed audio samples. First the r.m.s. level of the signal must be determined, after which it needs

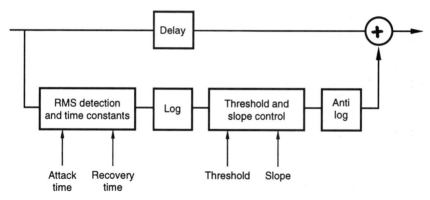

Figure 10.20 A simple digital dynamics processing operation

to be converted to a logarithmic value in order to determine the level change in decibels. Only samples above a certain threshold level will be affected, so a constant factor must be added to the values obtained, after which they are multiplied by a factor to represent the compression slope. The coefficient values are then antilogged to produce linear coefficients by which the audio samples can be multiplied.

10.6 Low bit rate audio coding

Considerable advances have been made in recent years in the field of audio bit rate reduction or low bit rate coding. Conventional PCM audio has a high data rate, and there are many applications for which it would be an advantage to have a lower data rate without much (or any) loss of sound quality. 16 bit linear PCM at a sampling rate of 44.1 kHz ('CD quality digital audio') results in a data rate of:

$$16 \times 44100 = 705\ 600 \text{ bits per second per channel (about 700 kbit s}^{-1})$$

For multimedia applications, broadcasting, communications and some consumer purposes the data rate may be reduced to a fraction of that shown above with almost no effect on the perceived sound quality. Simple techniques for reducing the data rate, such as reducing the sampling rate or number of bits per sample would have a very noticeable effect on sound quality, so most modern low bit rate coding works by exploiting the phenomenon of auditory masking (see Fact File 2.3) to 'hide' the increased noise resulting from bit rate reduction in parts of the audio spectrum where it will hopefully be inaudible. The following is a very brief overview of how one approach works, based on the technology involved in the MPEG (Moving Pictures Expert Group) standards.

As shown in Figure 10.21, the incoming digital audio signal is filtered into a number of narrow frequency bands. Parallel to this a computer model of the human hearing process analyses a short portion of the audio signal (a few milliseconds). This analysis is used to determine what parts of the audio spectrum will be masked, and to what degree, during that short time period. In bands where there is a strong signal, quantising noise can be allowed to rise considerably without it being heard, because one signal is very efficient at masking another lower level signal in the

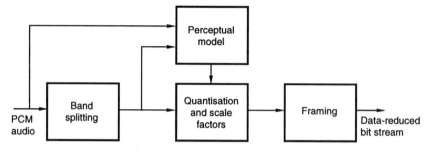

Figure 10.21 Generalised block diagram of a psychoacoustic low bit rate coder

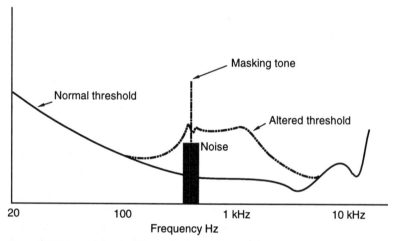

Fgure 10.22 Quantising noise lying under the masking threshold will normally be inaudible

same band as itself (see Figure 10.22). Provided that the noise is kept below the masking threshold in each band it should be inaudible.

Blocks of audio samples in each narrow band are scaled (low level signals are amplified so that they use more of the most significant bits of the range), and the scaled samples are then reduced in resolution (requantised) by reducing the number of bits available to represent each sample – a process which results in increased quantising noise. The output of the hearing model is used to control the requantising process so that the sound quality remains as high as possible for a given bit rate. The greatest number of bits is allocated to frequency bands where noise would be most audible, and the fewest to those bands where the noise would be effectively masked by the signal. Control information is sent along with the blocks of bit-rate-reduced samples to allow them to be reconstructed at the correct level and resolution upon decoding.

The above process is repeated every few milliseconds, so that the masking model is constantly being updated to take account of changes in the audio signal. Carefully implemented, such a process can result in a reduction of the data rate to anything from about one quarter to less than one tenth of the original data rate. A decoder uses the control information transmitted with the bit-rate-reduced samples to restore the

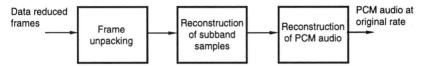

Figure 10.23 Generalised block diagram of an MPEG audio decoder

Table 10.1 MPEG-Audio layers

Layer	Complexity	Min. delay	Bit rate range	Target
1	Low	19 ms	32 – 448 kbit/s	192 kbit/s
2	Moderate	35 ms	32 – 384 kbit/s*	128 kbit/s
3	High	59 ms	32 – 320 kbit/s	64 kbit/s

* In Layer 2, bit rates of 224 kbit/s and above are for stereo modes only

samples to their correct level, and can determine how many bits were allocated to each frequency band by the encoder, reconstructing linear PCM samples and then recombining the frequency bands to form a single output (see Figure 10.23). A decoder can be much less complex, and therefore cheaper, than an encoder, because it does not need to contain the auditory model processing.

A standard known as MPEG, published by the International Standards Organisation (ISO 11172-3), defines a number of 'layers' of complexity for low bit rate audio coders as shown in Table 10.1.

Each of the layers can be operated at any of the bit rates within the ranges shown (although some of the higher rates are intended for stereo modes), and the user must make appropriate decisions about what sound quality is appropriate for each application. The lower the data rate, the lower the sound quality that will be obtained. At high data rates the encoding–decoding process has been judged by many to be audibly 'transparent' – in other words they cannot detect that the coded and decoded signal is different from the original input.

Layer 2 has been chosen as the standard for Digital Audio Broadcasting (DAB) in Europe and some other parts of the world. It allows a large number of digital stereo radio channels to be transmitted in less radio frequency spectrum space than conventional VHF FM stereo channels, with higher sound quality and lower interference.

There are a number of types of low bit rate coding used in audio systems, working on similar principles, and used for applications such as consumer disk and tape systems (DCC, MiniDisc, DVD) and digital cinema sound (Dolby SR-D, Sony SDDS, DTS).

10.7 Digital tape recording

10.7.1 The future for dedicated tape formats

In the sections that follow, the basic features of dedicated tape recording formats are outlined. Before the explosion in the use of computer workstations and mass storage for digital audio, tape recorders were the main means of recording and editing.

Today there are still a considerable number of dedicated digital tape recording formats in existence, although they are slowly being superseded by computer-based products using removable disks or other mass storage media. Even though tape storage will continue to be used in the future (tape is much cheaper than most disks), it is more likely to be formatted in a computer file system manner than as a dedicated audio recording format, using tape drives connected by standard data interfaces to a host processor.

Tape's main advantages are its cheapness and familiarity. One can easily remove a tape from the recorder at the end of a session and walk away with it. Also, a dedicated tape format can easily be interchanged between recorders, provided that another machine operating to the same standard can be found. Disks, on the other hand, come in a very wide variety of sizes and formats, and even if the disk fits a particular drive it may not be possible to access the audio files thereon, owing to the multiplicity of levels at which compatibility must exist between systems before interchange can take place. This issue is discussed in much greater detail in *The Audio Workstation Handbook* by Francis Rumsey, as detailed at the end of the chapter.

Tape is relatively slow to access, because it is a linear storage medium. One must spool through the tape to reach the desired point. That said, the absence of a formal filing structure on dedicated tape formats makes it much more straightforward to use for people used to conventional recording, since an operator can drop directly in to record mode anywhere along the tape's length without having to worry about file names and so forth. It is clear to the operator exactly where one recording stops and the next one starts, whereas there is a greater degree of abstraction with computer filing structures.

10.7.2 Background to digital tape recording

When commercial digital audio recording systems were first introduced in the 1970s and early 1980s it was necessary to employ recorders with sufficient bandwidth for the high data rates involved (a machine capable of handling bandwidths of a few megahertz was required). Conventional audio tape recorders were out of the question because their bandwidths extended only up to around 35 kHz at best. Video recording, however, required high bandwidths and recording densities for a high-quality picture to be adequately represented, and so existing video tape recorders (VTRs) were drafted in as suitable storage devices. PCM adaptors converted digital audio data into a waveform which resembled a television waveform, suitable for recording on to a VTR. The Denon company of Japan developed such a system in partnership with the NHK broadcasting organisation (Japan's equivalent to the UK's BBC) and they released the world's first PCM recording on to LP in 1971. In the early 1980s, devices such as Sony's PCM-F1 (pictured in Figure 10.24) became available at modest prices, allowing 16 bit, 44.1 kHz digital audio to be recorded on to a consumer VTR, resulting in wide-spread proliferation of stereo digital recording.

Dedicated open-reel digital recorders using stationary heads were also developed (see Fact File 10.6). High-density tape formulations were then manufactured for digital use, and this, combined with new channel codes (see below), improvements in error correction and better head design, led to the use of a relatively low number of tracks per channel, or even single-track recording of a given digital signal, combined with playing speeds of 15 or 30 ips.

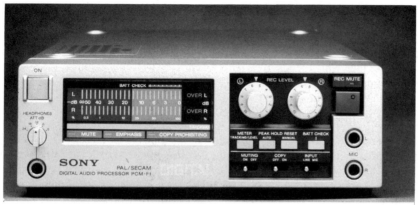

Figure 10.24 An early PCM adaptor: the Sony PCM-F1

FACT FILE

10.6

Rotary and stationary heads

There are two fundamental mechanisms for the recording of digital audio on tape, one which uses a relatively low linear tape speed and a quickly-rotating head, and one which uses a fast linear tape speed and a stationary head. In the rotary-head system the head either describes tracks almost perpendicular to the direction of tape travel, or it describes tracks which are almost in the same plane as the tape travel. The former is known as transverse scanning and the latter is known as helical scanning. Transverse scanning uses more tape when compared with helical scanning. It is not common for digital tape recording to use the transverse scanning method. The reason for using a rotary head is to achieve a high head-to-tape speed, since it is this which governs the available bandwidth. Rotary-head recordings can not easily be splice edited because of the track pattern, but they can be electronically edited using at least two machines.

Stationary heads allow the design of tape machines which are very similar in many respects to analogue transports. With stationary-head recording it is possible to record a number of narrow tracks in parallel across the width of the tape. Tape speed can be traded off against the number of parallel tracks used for each audio channel, since the required data rate can be made up by a combination of recordings made on separate tracks. This approach is used in the DASH format, where the tape speed may be 30 ips (76 cm s^{-1}) using one track per channel, 15 ips using two tracks per channel, or 7.5 ips using four tracks per channel.

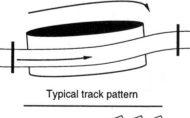

Typical track pattern

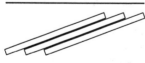

Head is stationary

Tape travel

Track pattern

Dedicated rotary-head systems not using video tape recorders, have also been developed recently – the RDAT format being the most well known.

Digital recording tape is thinner (27.5 microns) than that used for analogue recordings; long playing times can be accommodated on a reel, but also thin tape contacts the machine's heads more intimately than does standard 50 micron thickness tape which tends to be stiffer. Intimate contact is essential for reliable recording and replay of such a densely packed and high-bandwidth signal.

10.7.3 Channel coding for dedicated tape formats

Since 'raw' binary data is normally unsuitable for recording directly on to tape, a 'channel code' is used which matches the data to the characteristics of the recording system, uses storage space efficiently, and makes the data easy to recover on replay. A wide range of channel codes exists, each with characteristics designed for a specific purpose. The channel code converts a pattern of binary data into a different pattern of transitions in the recording or transmission medium. It is another stage of modulation, in effect. Thus the pattern of bumps in the optical surface of a CD bears little resemblance to the original audio data, and the pattern of magnetic flux transitions on a DAT cassette would be similarly different. Given the correct code book, one could work out what audio data was represented by a given pattern from either of these systems.

Many channel codes are designed for a low DC content (in other words, the data is coded so as to spend, on average, half of the time in one state and half in the other) in cases where signals must be coupled by transformers (see section 13.1), and others may be designed for narrow bandwidth or a limited high-frequency content. Certain codes are designed specifically for very high-density recording, and may have a low clock content with the possibility for long runs in one binary state or the other without a transition. Channel coding involves the incorporation of the data to be recorded with a clock signal, such that there is a sufficient clock content to allow the data and clock to be recovered on replay (see Fact File 10.7). Channel codes vary as to their robustness in the face of distortion, noise and timing errors in the recording channel.

Some examples of channel codes used in audio systems are shown in Figure 10.25. FM is the simplest, being an example of binary frequency modulation. It is otherwise known as 'bi-phase mark', one of the Manchester codes, and is the channel code used by SMPTE/EBU timecode (see Chapter 16). MFM and Miller-squared are more efficient in terms of recording density. MFM is more efficient than FM because it eliminates the transitions between successive ones, only leaving them between successive zeros. Miller^2 eliminates the DC content present in MFM by removing the transition for the last one in an even number of successive ones.

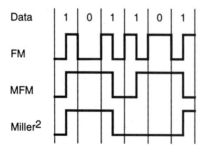

Figure 10.25 Examples of three channel codes used in digital recording. Miller^2 is the most efficient of those shown since it involves the least number of transitions for the given data sequence

FACT FILE
10.7
Data recovery

Channel-coded data must be decoded on replay, but first the audio data must be separated from the clock information which was combined with it before recording. This process is known as data and sync separation (see diagram).

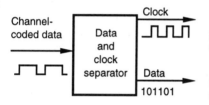

It is normal to use a phase-locked loop for the purpose of regenerating the clock signal from the replayed data (see diagram), this being based around a voltage-controlled oscillator (VCO) which runs at some multiple of the off-tape clock frequency. A phase comparator compares the relative phases of the divided VCO output and the clock data off tape, producing a voltage proportional to the error which controls the frequency of the VCO. With suitable damping, the phase-locked

oscillator will 'flywheel' over short losses or irregularities of the off-tape clock.

Recorded data is usually interspersed with synchronising patterns in order to give the PLL in the data separator a regular reference in the absence of regular clock data from the encoded audio signal, since many channel codes have long runs without a transition.

Even if the off-tape data and clock have timing irregularities, such as might manifest themselves as 'wow' and 'flutter' in analogue reproducers (see section A1.6), these can be removed in digital systems. The erratic data

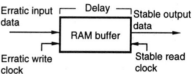

(from tape or disk, for example) is written into a short-term solid-state memory (RAM) and read out again a fraction of a second later under control of a crystal clock (which has an exceptionally stable frequency), as shown in the diagram. Provided that the average rate of input to the buffer is the same as the average rate of output, and the buffer is of sufficient size to soak up short-term irregularities in timing, the buffer will not overflow or become empty.

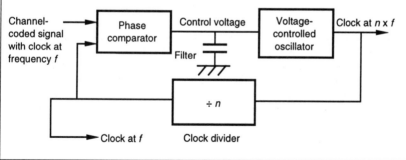

Group codes, such as that used in the Compact Disc and R-DAT, involve the coding of patterns of bits from the original audio data into new codes with more suitable characteristics, using a look-up table or 'code book' to keep track of the relationship between recorded and original codes. This has clear parallels with coding as used in intelligence operations, in which the recipient of a message requires the code book to be able to understand the message. CD uses a method known as 8-to-14 modulation, in which 16 bit audio sample words are each split into two 8 bit words, after which a code book is used to generate a new 14 bit word for each of the 256 possible combinations of 8 bits. Since there are many more words possible with 14 bits than with 8, it is possible to choose those which have appropriate characteristics for the CD recording channel. In this case, it is those words which have no more than eleven consecutive bits in the same state, and no less than three. This limits the bandwidth of the recorded data, and makes it suitable for the optical pickup process, whilst retaining the necessary clock content.

10.7.4 Error correction

There are two stages to the error correction process used in digital tape recording systems. Firstly the error must be detected, and then it must be corrected. If it cannot be corrected then it must be concealed. In order for the error to be detected it is necessary to build in certain protection mechanisms.

Two principal types of error exist: the burst error and the random error. Burst errors result in the loss of many successive samples and may be due to major momentary signal loss, such as might occur at a tape drop-out or at an instant of impulsive interference such as an electrical spike induced in a cable or piece of dirt on the surface of a CD. Burst error correction capability is usually quoted as the number of consecutive samples which may be corrected perfectly. Random errors result in the loss of single samples in randomly located positions, and are more likely to be the result of noise or poor signal quality. Random error rates are normally quoted as an average rate, for example: 1 in 10^6. Error correction systems must be able to cope with the occurrence of both burst and random errors in close proximity.

Audio data is normally interleaved before recording, which means that the order of samples is shuffled (as shown conceptually in Figure 10.26). Samples which had been adjacent in real time are now separated from each other on the tape. The benefit of this is that a burst error which destroys consecutive samples on tape will result in a collection of single-sample errors in between good samples when the data is de-interleaved, allowing for the error to be concealed. A common process, associated with interleaving, is the separation of odd and even samples by a delay. The greater the interleave delay, the longer the burst error that can be handled. A common example of this is found in the DASH tape format (an open-reel digital recording format), and involves delaying odd samples so that they are separated from even

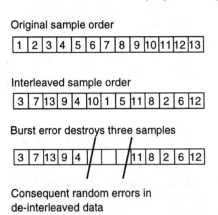

Original sample order

| 1 | 2 | 3 | 4 | 5 | 6 | 7 | 8 | 9 | 10 | 11 | 12 | 13 |

Interleaved sample order

| 3 | 7 | 13 | 9 | 4 | 10 | 1 | 5 | 11 | 8 | 2 | 6 | 12 |

Burst error destroys three samples

| 3 | 7 | 13 | 9 | 4 | | | | 11 | 8 | 2 | 6 | 12 |

Consequent random errors in de-interleaved data

| | 2 | 3 | 4 | | 6 | 7 | 8 | 9 | | 11 | 12 | 13 |

Figure 10.26 Interleaving is used in digital recording and broadcasting systems to rearrange the original order of samples. This can have the effect of converting burst errors into random errors when the samples are de-interleaved

FACT FILE 10.8 — Error handling

True correction
Up to a certain random error rate or burst error duration an error correction system will be able to reconstitute erroneous samples perfectly. Such corrected samples are indistinguishable from the originals, and sound quality will not be affected. Such errors are often signalled by green lights showing 'CRC' failure or 'Parity' failure.

Interpolation
When the error rate exceeds the limits for perfect correction, an error correction system

Original sample amplitudes

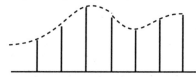

Sample missing due to error

Mathematically-interpolated value inserted

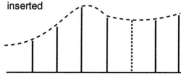

may move to a process involving interpolation between good samples to arrive at a value for a missing sample (as shown in the diagram). The interpolated value is the mathematical average of the foregoing and succeeding samples, which may or may not be correct. This process is also known as concealment or averaging, and the audible effect is not unpleasant, although it will result in a temporary reduction in audio bandwidth. Interpolation is usually signalled by an orange indicator to show that the error condition is fairly serious. In most cases the duration of such concealment is very short, but prolonged bouts of concealment should be viewed warily, since sound quality will be affected. This will usually point to a problem such as dirty heads or a misaligned transport, and action should be taken (see text).

Hold
In extreme cases, where even interpolation is impossible (when there are not two good samples either side of the bad one), a system may 'hold'. In other words, it will repeat the last correct sample value. The audible effect of this will not be marked in isolated cases, but is still a severe condition. Most systems will not hold for more than a few samples before muting. Hold is normally indicated by a red light.

Mute
When an error correction system is completely overwhelmed it will usually effect a mute on the audio output of the system. The duration of this mute may be varied by the user in some systems. The alternative to muting is to hear the output, regardless of the error. Depending on the severity of the error, it may sound like a small 'spit', click, or even a more severe break-up of the sound. In some cases this may be preferable to muting.

samples by 2448 samples, as well as reordering groups of odd and even samples within themselves.

Redundant data is also added before recording. Redundancy, in simple terms, involves the recording of data in more than one form or place. A simple example of the use of redundancy is found in the twin-DASH format, in which all audio data is recorded twice. On a second pair of tracks (handling the duplicated data), the odd–even sequence of data is reversed to become even–odd. Firstly, this results in double protection against errors, and secondly it allows for perfect correction at a splice, since two burst errors will be produced by the splice, one in each set of tracks. Because of the reversed odd–even order in the second set of tracks, uncorrupted odd data can be used from one set of tracks, and uncorrupted even data from the other set, obviating the need for interpolation (see Fact File 10.8).

Cyclic redundancy check (CRC) codes, calculated from the original data and recorded along with that data, are used in many systems to detect the presence and position of errors on replay. Complex mathematical procedures are also used to form codewords from audio data which allow for both burst and random errors to be corrected perfectly up to a given limit. Reed–Solomon encoding is another powerful system which is used to protect digital recordings against errors, but it is beyond the scope of this book to cover these codes in detail.

10.7.5 Overview of digital tape formats

There have been a considerable number of commercial recording formats over the last 20 years, and only a brief summary will be given here of the most common at the time of writing. The list is not intended to be exhaustive. Open reel two track formats are not covered here because there are used very rarely today. Although they were designed because of a perceived need for tape-cut editing on digital machines, they were not adopted widely by the market.

Sony's PCM-1610 and PCM-1630 adaptors dominated the CD-mastering market for a number of years, although by today's standards they use a fairly basic recording format and rely on 60 Hz/525 line U-matic cassette VTRs (as shown in Figure 10.27). The system operates at a sampling rate of 44.1 kHz and uses 16 bit quantisation, being designed specifically for the making of tapes to be turned into CDs. Recordings made in this format may be electronically edited using the Sony DAE-3000 editing system, and the playing time of tapes runs up to 75 minutes using a tape specially developed for digital audio use.

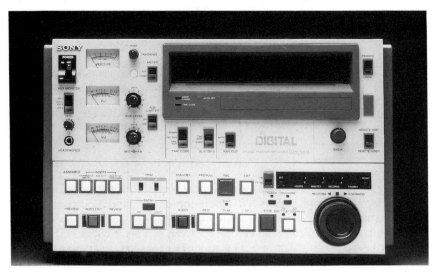

Figure 10.27 Sony DMR-4000 digital master recorder (Courtesy of Sony Broadcast and Professional Europe)

Figure 10.28 Sony PCM7030 professional DAT machine. (Courtesy of Sony Broadcast and Professional Europe)

The R-DAT or DAT format is a small stereo, rotary-head, cassette-based format offering a range of sampling rates and recording times, including the professional rates of 44.1 and 48 kHz. Originally, consumer machines operated at 48 kHz to avoid the possibility for digital copying of CDs, but professional versions are available which will record at either 44.1 or 48 kHz. Modern consumer machines will record at 44.1 kHz, but usually only via the digital inputs. DAT is a 16 bit format. There is a fairly large number of machines available, including truly professional designs offering editing facilities, external sync and IEC-standard timecode. Various non-standard modifications have been introduced in recent years, including a 96 kHz sampling rate DAT machine for audiophile use, and adaptors which will store 20 bit audio on such a high sampling rate machine (but sacrificing the high sampling rate for more bits).

The IEC timecode standard for R-DAT was devised in 1990 as a common format for the insertion of timecode (see Chapter 16) into the subcode area of the helically scanned track. In practice it will allow for SMPTE/EBU timecode of any frame rate to be converted into the internal DAT 'running-time' code, and then converted back into any SMPTE/EBU frame rate on replay. This will involve the installation of an intelligent timecode reader/generator into the DAT machine.

A typical professional R-DAT machine is pictured in Figure 10.28. The format has become exceptionally popular with professionals owing to its low cost, high performance, portability and convenience.

The Nagra-D recorder (see Figure 10.29) was designed as a digital replacement for the world-famous Nagra analogue recorders, and as such is intended for professional use in field recording and studios. The format was designed to have considerable commonality with the audio format used in D1- and D2-format digital VTRs, having rotary heads, although it uses open reels for operational convenience. Allowing for 20–24 bits of audio resolution, the Nagra-D format is appropriate for use with high resolution convertors. The error correction and recording density used in this format are designed to make recordings exceptionally robust, and recording time may be up to 6 hours on a 7 inch (18 cm) reel, in two-track mode. The format is also designed for operation in a four-track mode at twice the stereo tape speed, such that in stereo the tape travels at 4.75 cm s^{-1}, and in four track at 9.525 cm s^{-1}.

The DASH (Digital Audio Stationary Head) format consists of a whole family of open-reel stationary-head recording formats from two tracks up to 48 tracks.

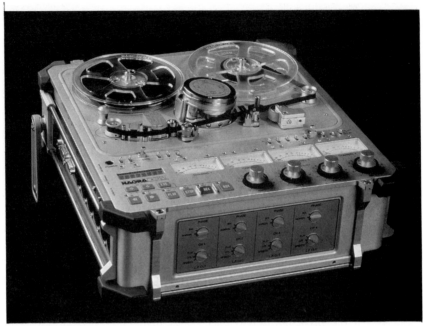

Figure 10.29 Nagra-D open reel digital tape recorder. (Courtesy of Sound PR)

Figure 10.30 An open-reel digital multitrack recorder: the Sony PCM-3348. (Courtesy of Sony Broadcast and Professional Europe)

DASH-format machines may operate at 44.1 kHz or 48 kHz rates (and sometimes optionally at 44.056 kHz), and they allow varispeed ±12.5%. They are designed to allow gapless punch-in and punch-out, splice editing, electronic editing and easy synchronisation. Multitrack DASH machines have gained wide acceptance in

Figure 10.31 A modular digital multitrack machine, Sony PCM-800. (Courtesy of Sony Broadcast and Professional Europe)

studios, but the stereo machines have not. An example of a machine is shown in Figure 10.30. Recent developments have resulted in DASH multitracks capable of storing 24 bit audio instead of the original 16 bits.

The ProDigi (PD) format was introduced by Mitsubishi in the mid 1980s, and a number of PD multitrack machines are to be found in studios around the world. The most common version of the format is in the 32 channel multitrack machines by Mitsubishi and Otari, but the format is now no longer marketed by these companies.

In recent years there has been a move towards the introduction of budget modular multitrack formats which are beginning to take over from the more expensive open reel formats. Most of these are based on 8 track cassettes using rotary head transports borrowed from consumer video technology. The most widely used at the time of writing are the DA-88 format (based on Hi-8 cassettes) and the ADAT format (based on VHS cassettes). These offer most of the features of open reel machines and a number of them may be synchronised to expand the channel capacity. An example is shown in Figure 10.31.

10.7.6 Performance and alignment

Crosstalk between the tracks of a digital recorder is virtually absent, and this frees the engineer from the need to allocate tracks carefully. For example, with an analogue multitrack machine one tends to record vocals on tracks which are physically far away on the tape from, say, drum tracks, since crosstalk from the latter can easily be audible. Unused tracks will often be used to form an extended guard band to separate, say, an electric guitar track or timecode track from a vocal. None of this is necessary with a digital multitrack recorder. Other recording artefacts commonly encountered with analogue recordings, such as wow and flutter, are also absent on digital machines.

Digital tape machines require at least as much care and maintenance as do their analogue counterparts. In fact, the analogue machine can be monitored for slight deterioration in performance rather more easily than can the digital. Slight frequency response fall-off, for example, due to, say, head wear or azimuth misalignment, can quickly be checked for and spotted using a test tape on an analogue machine. Misalignment of a digital machine, however, will cause two types of problem, neither of which is easily spotted unless careful checks are regularly carried out. Firstly, a machine's transport may be misaligned such that although tapes recorded on it will replay satisfactorily on the same machine, another correctly aligned machine may well not be able to obtain sufficient data from the tape to reconstitute the signal adequately. Error correction systems will be working too hard and drop-outs will occur. Secondly, digital recording tends to work without noticeable performance deterioration until things get so bad that drop-outs occur. There is little warning of catastrophic drop-outs, although the regularity of errors is an excellent means of telling how close to the edge a system is in terms of random errors. If a recorder has error status indication, these can be used to tell the prevailing state of the tape or the machine's alignment. Different degrees of seriousness exist, as was discussed in Fact File 10.8. If the machine is generating an almost constant string of CRC errors or interpolations, then this is an indication that either the tape or the machine is badly out of alignment or worn. It is therefore extremely important to check alignment often and to clean the machine's heads using whatever means is recommended. Alignment may require specialised equipment which dealers will usually possess.

10.7.7 Editing digital tape recordings

Razor blade cut-and-splice editing is possible on open-reel digital formats, and the analogue cue tracks are monitored during these operations. This is necessary because the digital tracks are generally only capable of being replayed at speeds that are no more than about 10% away from normal replay speed. The analogue tracks are of low quality – rather lower than a dedicated analogue machine – but usually adequate as a cue.

A 90° butt joint is used for the editing of digital tape. The discontinuity in the data stream caused by the splice would cause complete momentary drop-out of the digital signal if no further action were taken, so circuits are incorporated which sense the splice and perform an electronic crossfade from one side of the splice to the other, with error concealment to minimise the audibility of the splice. It is normally advised that a 0.5 mm gap is left at the splice so that its presence will easily be detected by the crossfade circuitry. The thin tape can easily be damaged during the cut-and-splice edit procedure and a safety copy would be most advisable. Electronic editing is far more desirable, and is the usual method.

Electronic editing normally requires the use of two machines plus a control unit, as shown in the example in Figure 10.32. A technique is employed whereby a finished master tape is assembled from source takes on player machines. This is a relatively slow process, as it involves real time copying of audio from one machine to another, and modifications to the finished master are difficult. The digital editor can often store several seconds of programme in its memory and this can be replayed at normal speed or under the control of a search knob which enables very slow to-and-fro searches to be performed in the manner of rock and roll editing on an analogue machine. Edits may be rehearsed prior to execution. When satisfactory

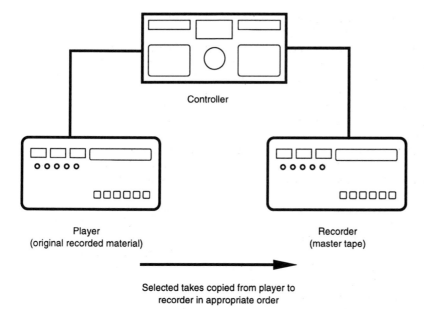

Controller

Player
(original recorded material)

Recorder
(master tape)

Selected takes copied from player to
recorder in appropriate order

Figure 10.32 In electronic tape-copy editing selected takes are copied in sequence from player to
recorder with appropriate crossfades at joins

edit points have been determined the two machines are synchronised using timecode,
and the record machine is switched to drop in the new section of the recording from
the replay machine at the chosen moment. Here a crossfade is introduced between
old and new material to smooth the join. The original source tape is left unaltered.

10.8 Workstations and recording on mass storage media

Once audio is in a digital form it can be handled by a computer like any other data.
The only real difference is that audio requires a high sustained data rate and large
amounts of storage compared with more basic data such as text. The following is a
very brief introduction to some of the technology associated with computer-based
audio workstations. Much more detail will be found in *The Audio Workstation
Handbook* as detailed in the Further Reading list. It should be noted that the
terminology for such recording systems has not yet reached a point of universal
agreement. It has become quite common to refer to 'non-linear recording', meaning
recording on media where the data storage is not linear (as opposed to tape, which
is a linear medium). This should not be confused with non-linear coding of audio,
which is a technique used in A/D conversion and data reduction. Other terms used
for recording on disk-based mass storage media include 'random access' and 'direct
access' recording, also 'tapeless recording'.

10.8.1 Storage requirements of digital audio

Table 10.2 shows the data rates required to support a single channel of digital audio
at various resolutions. Media to be used as primary storage would need to be able

Table 10.2 Data rates and capacities required for linear PCM

Sampling rate kHz	Resolution bits	Bit rate kbit/s	Capacity/min. Mbytes/min	Capacity/hour Mbytes/hour
96	16	1536	11.0	659
48	20	960	6.9	412
48	16	768	5.5	330
44.1	16	706	5.0	303
44.1	8	353	2.5	151

to sustain data transfer at a number of times these rates to be useful for multimedia workstations. The table also shows the number of megabytes of storage required per minute of audio, showing that the capacity needed for audio purposes is considerably greater than that required for text or simple graphics applications. Storage requirements increase pro rata with the number of audio channels to be handled.

Storage systems may use removable media but many have fixed media. It is advantageous to have removable media for audio purposes because it allows different jobs to be kept on different media and exchanged at will, but unfortunately the highest performance is still obtainable from storage systems with fixed media. Although the performance of removable media drives is improving all the time, fixed media drives have so far retained their advantage.

10.8.2 Disk drives

Disk drives are probably the most common form of mass storage. They have the advantage of being random-access systems – in other words any data can be accessed at random and with only a short delay. This may be contrasted with tape drives which only allow linear access – that is by winding through the tape until the desired data is reached, resulting in a considerable delay. Disk drives come in all shapes and sizes from the commonly encountered floppy disk at the bottom end to high performance hard drives at the top end. The means by which data are stored is usually either magnetic or optical, but some use a combination of the two, as described below. There exist both removable and fixed media disk drives, but in almost all cases the fixed media drives have a higher performance than removable media drives. This is because the design tolerances can be made much finer when the drive does not have to cope with removable media, allowing higher data storage densities to be achieved. Although removable disk media can appear to be expensive compared with tape media, the cost must be weighed against the benefits of random access and the possibility that some removable disks can be used for primary storage whereas a tape can not.

The general structure of a disk drive is shown in Figure 10.33. It consists of a motor connected to a drive mechanism which causes one or more disk surfaces to rotate at anything from a few hundred to many thousands of revolutions per minute. This rotation may either remain constant or may stop and start, and it may either be at a constant rate or a variable rate, depending on the drive. One or more heads are mounted on a positioning mechanism which can move the head across the surface of the disk to access particular points, under the control of hardware and software called a disk controller. The heads read data from and write data to the disk surface by whatever means the drive employs. It should be noted that certain disk types are read-only, some are write-once–read-many (WORM), and some are fully erasable and rewritable.

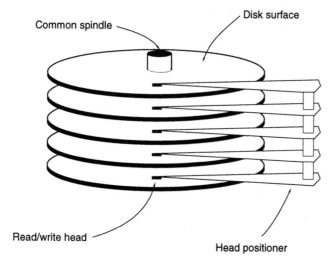

Common spindle

Disk surface

Read/write head

Head positioner

Figure 10.33 General mechanical structure of a disk drive

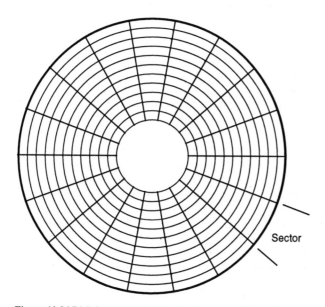

Sector

Figure 10.34 Disk formatting divides the storage area into tracks and sectors

The disk surface is normally divided up into tracks and sectors, not physically but by means of 'soft' formatting (see Figure 10.34). Formatting places logical markers which indicate block boundaries, amongst other processes. On most hard disks the tracks are arranged as a series of concentric rings, but with some optical disks there is a continuous spiral track.

Disk drives look after their own channel coding, error detection and correction so there is no need for system designers to devise dedicated audio processes for disk-based recording systems. The formatted capacity of a disk drive is all available for the storage of 'raw' audio data, with no additional overhead required for redundancy and error checking codes. 'Bad blocks' are mapped out during the formatting of a disk, and not used for data storage. If a disk drive detects an error when reading a block of data it will attempt to read it again. If this fails then an error is normally generated and the file cannot be accessed, requiring the user to resort to one of the many file recovery packages on the market. Disk-based audio systems do not resort to error interpolation or sample hold operations, unlike tape recorders. Replay is normally either correct or not possible.

10.8.3 Recording audio on to disks

The discontinuous 'bursty' nature of recording on to disk drives requires the use of a buffer RAM (Random Access Memory) during replay, which accepts this interrupted data stream and stores it for a short time before releasing it as a continuous stream. It performs the opposite function during recording, as shown in Figure 10.35. Several things cause a delay in the retrieval of information: the time it takes for the head positioner to move across a disk, the time it takes for the required data in a particular track to come around to the pickup head, and the transfer of the data from the disk via the buffer RAM to the outside world, as shown in Figure 10.36. Total delay, or data access time, is in practice several milliseconds. The instantaneous rate at which the system can accept or give out data is called the transfer rate and varies with the storage device.

Sound is stored in named data files on the disk, the files consisting of a number of blocks of data stored either separately or together. A directory stored on the disk keeps track of where the blocks of each file are stored so that they can be retrieved in correct sequence. Each file normally corresponds to a single recording of a single channel of audio, although some stereo file formats exist.

Multiple channels are handled by accessing multiple files from the disk in a time-shared manner, with synchronisation between the tracks being performed subsequently in RAM. The storage capacity of a disk can be divided between channels in whatever proportion is appropriate, and it is not necessary to pre-allocate storage space to particular audio channels. For example, a 360 Mbyte disk will store about 60 minutes of mono audio at professional rates. This could be subdivided to give 30 minutes of stereo, 15 minutes of four track, etc, or the proportions could be shared unequally. A feature of the disk system is that unused storage capacity is not necessarily 'wasted' as can be the case with a tape system. During recording of a

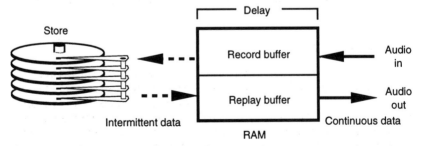

Figure 10.35 Short-term memory buffering converts continuous audio data into bursts for recording onto disks, and vice versa on replay

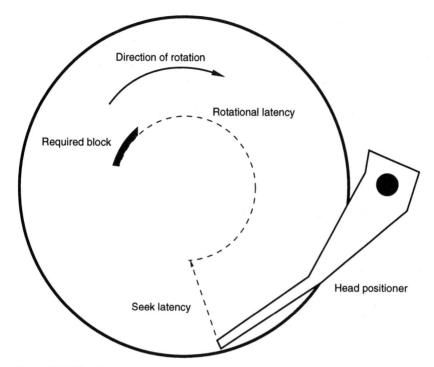

Figure 10.36 The delays involved in reading a block of data from a disk

multitrack tape there will often be sections on each track with no information recorded, but that space cannot be allocated elsewhere. On a disk these gaps do not occupy storage space and can be used for additional space on other channels at other times.

The number of audio channels which can be recorded or replayed simultaneously depends on the performance of the storage device and the host computer. Slow systems may only be capable of handling a few channels whereas faster systems with multiple disk drives may be capable of expansion up to a virtually unlimited number of channels. Manufacturers are tending to make their systems modular, allowing for expansion of storage and other audio processing facilities as means allow, with all modules communicating over a high speed data bus, as shown in Figure 10.37.

10.8.4 Non-linear editing

Speed and flexibility of editing is probably one of the greatest benefits obtained from non-linear recording. With non-linear editing the editor may preview a number of possible masters in their entirety before deciding which should be the final one. Even after this, it is a simple matter to modify the edit list to update the master. Edits may also be previewed and experimented with in order to determine the most appropriate location and processing – an operation which is less easy with other forms of editing.

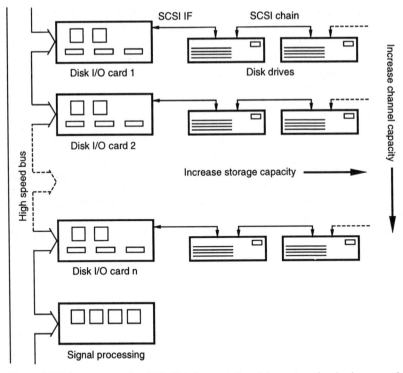

Figure 10.37 Arrangement of multiple disks in a typical modular system, showing how a number of disks can be attached to a single SCSI chain to increase storage capacity, and how additional disk I/O cards can be added to increase data throughput for additional audio channels

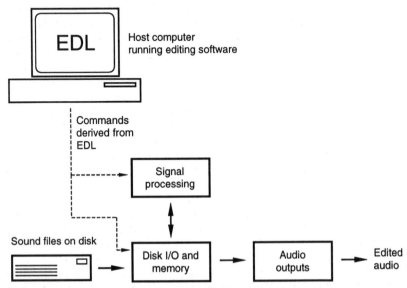

Figure 10.38 Instructions from an edit decision list (EDL) are used to control the replay of sound file segments from disk, which may be subjected to further processing (also under EDL control) before arriving at the audio outputs

The majority of music editing is done today using digital audio workstations, indeed these are now taking over from dedicated audio editing systems because of the speed with which takes may be compared, crossfades modified, and adjustments made to equalisation and levels, all in the digital domain. Non-linear editing has also come to feature very widely in post-production for video and film, because it has a lot in common with film post-production techniques involving a number of independent mono sound reels.

Non-linear editing is truly non-destructive in that the edited master only exists as a series of instructions to replay certain parts of certain sound files at certain times, with certain signal processing overlaid, as shown in Figure 10.38. The original sound files remain intact at all times, and a single sound file can be used as many times as desired in different locations and on different tracks without the need for copying the actual audio data. Editing may involve the simple joining of sections, or it may involve more complex operations such as long crossfades between one album track and the next, or gain offsets between one section and another. The beauty of non-linear editing is that all these things are possible without in any way affecting the original source material.

10.8.5 Digital audio workstations

There are fundamentally two types of audio workstation: dedicated systems or systems which use additional hardware and software installed in a standard desktop computer. Most systems conform to one or other of these models, with a tendency for dedicated systems to be more expensive than desktop computer-based systems (although this is by no means always the case).

It was most common in the early days of hard disk audio systems for manufacturers to develop dedicated systems with fairly high prices. This was mainly because mass produced desktop computers were insufficiently equipped for the purpose, and because large capacity mass storage media were less widely available than they are now, having a variety of different interfaces and requiring proprietary file storage strategies. It was also because the size of the market was relatively small to begin with, and considerable R&D investment had to be recouped. There are considerable advantages to dedicated systems, and they are very popular with professional facilities. Rather than a mouse and a QWERTY keyboard, the user controls the system using an interface designed specifically for the purpose, with perhaps more ergonomically appropriate devices. An example of such a system is pictured in Figure 10.39. It has a touch screen and dedicated controls for many functions, as well as rotary and slider controls for continuously variable functions. It is also becoming common for cheaper dedicated editing systems to be provided with an interface to a host computer so that more comprehensive display and control facilities can be provided.

In recent years desktop multimedia computers have been introduced with built-in basic AV (audio-video) facilities, providing limited capabilities for editing and sound manipulation. The quality of the built-in convertors in desktop computers is necessarily limited by price, but they are capable of 16 bit, 44.1 kHz audio operation in many cases. Better audio quality is achieved by using third party hardware.

Many desktop computers lack the processing power to handle digital audio and video directly, but by adding third-party hardware and software it is possible to turn a desktop computer into an AV workstation, capable of storing audio for an almost unlimited number of tracks with digital video alongside. An audio signal processing

Figure 10.39 A dedicated digital audio workstation. (Courtesy of Digital Audio Research)

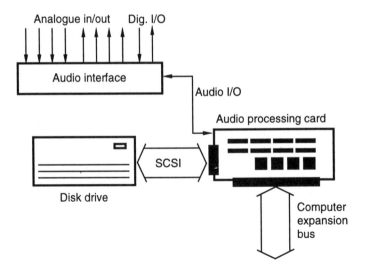

Figure 10.40 Typical system layout of a desktop computer-based audio workstation

card is normally installed in an expansion slot of the computer, as shown in Figure 10.40. The card would be used to handle all sound editing and post-processing operations, using one or more DSP chips, with the host computer acting mainly as a user interface. The audio card would normally be connected to an audio interface, perhaps containing a number of A/D and D/A convertors, digital audio interfaces, probably a SMPTE/EBU timecode interface, and in some cases a MIDI interface. A SCSI interface (the most common high-speed peripheral interface for mass storage media) to one or more disk drives is often provided on the audio expansion card in order to optimise audio file transfer operations, although some basic systems use the computer's own SCSI bus for this purpose.

10.9 Digital audio interfaces

It is often necessary to interconnect audio equipment so that digital audio data can be transferred between them without converting back to analog form. This preserves sound quality, and is normally achieved using one of the standard point-to-point digital interfaces described below. These are different from computer data networks in that they are designed purely for the purpose of carrying audio data in real time, and cannot be used for general purpose file transfer applications.

10.8.1 Computer networks and digital audio interfaces compared

Real-time digital audio interfaces are the digital audio equivalent of signal cables, down which digital audio signals for one or more channels are carried in real time from one point to another, possibly with some auxiliary information attached. A real-time audio interface uses a data format dedicated specifically to audio purposes, unlike a computer data network which is not really concerned with what is carried in its data packets. A recording transferred over a digital interface to a second machine may be copied 'perfectly' or cloned, and this process takes place in real time, requiring the operator to put the receiving device into record mode such that it simply stores the incoming stream of audio data. The auxiliary information may or may not be recorded (usually most of it is not).

In contrast, a computer network typically operates asynchronously and data is transferred in the form of packets. One would expect a number of devices to be interconnected using a single network, and for an addressing structure to be used such that data might be transferred from a certain source to a certain destination. Bus arbitration is used to determine the existence of network traffic, and to avoid conflicts in bus usage. The 'direction' of data flow is determined simply by the source and destination addresses. The format of data on the network is not necessarily audio specific (although protocols optimised for audio transfer may be used), and one network may carry text data, graphics data and E-mail, all in separate packets, for example.

10.8.2 Interface standards

Professional digital audio systems, and some consumer systems, have digital interfaces conforming to one of the standard protocols and allow for a number of channels of digital audio data to be transferred between devices with no loss of sound quality. Any number of generations of digital copies may be made without affecting the sound quality of the latest generation, provided that errors have been fully corrected. The digital outputs of a recording device are taken from a point in the signal chain after error correction, which results in the copy being error corrected. Thus the copy does not suffer from any errors which existed in the master, provided that those errors were correctable.

There are a number of types of digital interface, some of which are international standards and others of which are manufacturer-specific. They all carry digital audio with at least 16 bit resolution, and will operate at the standard sampling rates of 44.1 and 48 kHz, as well as at 32 kHz if necessary, with a degree of latitude for varispeed. Most of the interface standards are one- or two-channel only, but one of them, known commonly as MADI, is a multi-channel interface.

Manufacturer-specific interfaces may also be found on equipment from other manufacturers if they have been considered necessary for the purposes of signal interchange, especially on devices manufactured prior to the standardisation and proliferation of the AES/EBU interface (see Fact File 10.9). The interfaces vary as to how many physical interconnections are required, since some require one link per channel plus a synchronisation signal, whilst others carry all the audio information plus synchronisation information over one cable. The reader is referred to *The Digital Interface Handbook* by Rumsey and Watkinson, as well as to the standards themselves, if a greater understanding of the intricacies of digital audio interfaces is required.

The consumer interface (historically related to SPDIF – the Sony/Philips digital interface) is very similar to the professional AES/EBU interface, but uses unbal-

---

FACT FILE 10.9   AES/EBU interface

The AES/EBU interface allows for two channels of digital audio to be transferred serially over one balanced interface. The interface allows audio to be transferred over distances up to 100 m, but longer distances may be covered using combinations of appropriate cabling, equalisation and termination. Standard XLR-3 connectors are used, often labelled DI (for digital in) and DO (for digital out).

One frame of data is made up of two sub-frames (see the diagram), and each sub-frame begins with one of three synchronising patterns to identify the sample as either A or B channel, or to mark the start of a new channel status block. Additional data is carried within the subframe in the form of 4 bits of auxiliary data (which may either be used for additional audio resolution or for other purposes such as low-quality speech),

a validity bit (V), a user bit (U), a channel status bit (C) and a parity bit (P), making 32 bits per subframe and 64 bits per frame.

One frame is transmitted in the time period of one audio sample, and thus the data rate varies with the sampling rate. Channel status bits are aggregated at the receiver to form a 24 byte word every 192 frames, and each bit of this word has a specific function relating to interface operation. Examples of bit usage in this word are the signalling of sampling rate and pre-emphasis, as well as the carrying of a sample address 'timecode' and labelling of source and destination. Bit 1 of the first byte signifies whether the interface is operating according to the professional (set to 1) or consumer (set to 0) specification.

Bi-phase mark coding, the same channel code as used for SMPTE/EBU timecode, is used in order to ensure that the data is self-clocking, of limited bandwidth, DC free, and polarity independent. The interface has to accommodate a wide range of cable types, and a nominal 110 ohms characteristic impedance is recommended.

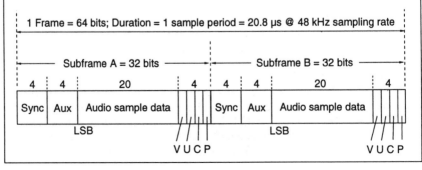

anced electrical interconnection over a coaxial cable having a characteristic impedance of 75 ohms. It can be found on many items of semi-professional or consumer digital audio equipment, such as CD players and DAT machines. It usually terminates in an RCA phono connector, although some hi-fi equipment makes use of optical fibre interconnects carrying the same data. Format convertors are available for converting consumer format signals to the professional format, and vice versa, and for converting between electrical and optical formats.

Recommended further reading

Pohlmann, K. (1995) *Principles of Digital Audio*. McGraw Hill
Pohlmann, K. (1989) *The Compact Disc*. Oxford University Press
Rumsey, F. (1996) *The Audio Workstation Handbook*. Focal Press
Rumsey, F. and Watkinson, J. (1995) *The Digital Interface Handbook*. Focal Press
Watkinson, J. (1994) *An Introduction to Digital Audio*. Focal Press
Watkinson, J. (1994) *The Art of Digital Audio*. Focal Press

Chapter 11

Record players

In this digital age, the gramophone record, with its hundred-odd-year-old principle of a sharp point rattling around mechanically in a groove, is something of an anachronism. Nevertheless, at its best it achieves superb results and of course there are many existing analogue records which still offer valuable material, a lot of which is still not adequately represented in the CD catalogue. A wealth of useful sound effects are to be found on this medium.

11.1 Principles

11.1.1 Cutting and stamping disks

It is necessary go back to Alan D. Blumlein's stereo patent of 1931 to find the origins of the techniques used to record two audio channels in a single record groove. Blumlein described a heated cutter stylus to which was attached two systems of coils and magnets, driven by the two channels of a stereo signal. A groove would thereby be cut into the acetate blank, in which each channel is represented by movements of the cutting stylus in a direction 45° to the vertical, as shown in Figure 11.1. Figure 11.2 shows plan views of three simple types of groove which result from the action of the cutting stylus.

After cutting, stampers are produced from which the records are made, the latter being a much harder vinyl co-polymer version of the original acetate, suitable for playback. The replay stylus transmits the groove information via a cantilever to a

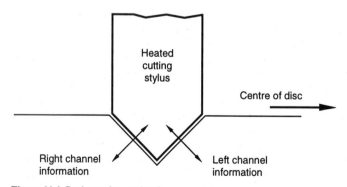

Figure 11.1 Cutting stylus motion for stereo disk recording

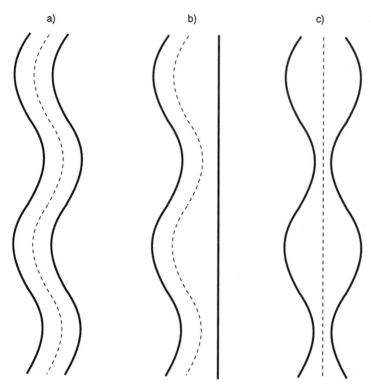

Figure 11.2 Groove patterns. (a) Two channels recorded in phase. (b) One channel only recorded. (c) Two channels recorded out of phase

system of coils and magnets in the cartridge body which in turn generate the electrical output ready for amplification and equalisation. Looking again at the groove, a signal on the left wall causes the replay stylus to move in the appropriate direction 45° to the vertical. A signal on the right wall causes stylus movement in the other 45° direction. A mono signal moves the stylus from side to side in a direction parallel to the surface of the record. Two signals of equal amplitude but of opposite phase cause up-and-down movements of the stylus, perpendicular to the record surface, and this kind of groove is difficult for the stylus to track securely. Another way of looking at the stylus motion is that sideways movements represent the mono sum signal (M) of the two stereo channels, whereas up-and-down movements represent the stereo 'difference' (S) information (see Fact File 4.5).

Disk-cutting engineers are always on the lookout for out-of-phase information on the tape from which the record is to be cut. Sometimes they have to resort to processing the tape to attenuate such signal components in order that the final record will be capable of being played by the majority of domestic record players. Another type of signal which is difficult for a stylus to track is low-frequency information if it is present mainly on one channel, and therefore on one wall of the groove. These large, relatively slow groove wall modulations are best avoided or at least carefully attenuated, and the usual method is to blend the two stereo channels to an extent at low frequencies so that a relatively easy-to-track lateral groove results at these low frequencies.

11.1.2 Pickup mechanics

The replay stylus motion should describe an arc offset from the vertical by 20°, as shown in Figure 11.3. This will be achieved if the arm height at the pivot is adjusted such that the arm tube is parallel to the surface of the record when the stylus is resting in the groove. The stylus tip should have a cone angle of 55°, as shown in Figure 11.4. The point is rounded such that the tip makes no contact with the bottom of the groove. Stylus geometry is discussed further in Fact File 11.1.

The arm geometry is arranged so that a line drawn through the cartridge body, front to back, forms a tangent to the record groove at a point where the stylus rests in the groove, at two points across the surface of the record: the outer groove and the inner position just before the lead-out groove begins. Figure 11.5 illustrates this. Note that the arm tube is bent to achieve the correct geometry. Alternatively, the arm tube can be straight with the cartridge headshell set at an offset angle which achieves the same result. The arc drawn between the two stylus positions shows the horizontal path of the stylus as it plays the record. Due to the fact that the arm has a fixed pivot, it is not possible for the stylus to be exactly tangential to the groove throughout its entire travel across the record's surface, but setting up the arm to meet this ideal at the two positions shown gives a good compromise, and a correctly designed and installed arm can give less than ±1° tracking error throughout the whole of the playing surface of the disk.

Alignment protractors are available which facilitate the correct setting up of the arm. These take the form of a rectangular piece of card with a hole towards one end which fits over the centre spindle of the turntable when it is stationary. It has a series of parallel lines marked on it (tangential to the record grooves) and two points corresponding to the outer and inner groove extremes. The stylus is lowered on to these two points in turn and the cartridge and arm are set up so that a line drawn through the cartridge from front to back is parallel to the lines on the protractor.

FACT FILE 11.1 Stylus profile

Two basic cross-sectional shapes exist for a replay stylus – conical and elliptical, as shown in the diagram. The elliptical profile can be seen to have a smaller contact area with the wall of the groove, and this means that for a given tracking weight (the downforce exerted by the arm on to the record surface) the elliptical profile exerts more force per unit area than does the conical tip. To compensate, elliptical styli have a specified tracking force which is less than that for a conical tip. The smaller contact area of the elliptical tip enables it to track the small, high-frequency components of the signal in the groove walls, which have short wavelengths, more faithfully. This is particularly advantageous towards the end of the side of the record where the groove length per revolution is shorter and therefore the recorded wavelength is shorter for a given frequency. Virtually all high-quality styli have an elliptical profile or esoteric variation of it, although there are still one or two high-quality conical designs around. The cutting stylus is, however, always conical.

Conical   Elliptical

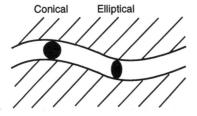

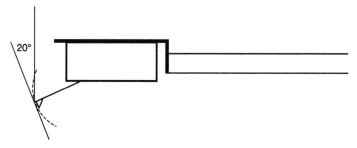

Figure 11.3 Stylus vertical tracking geometry

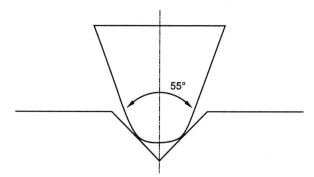

Figure 11.4 Stylus cone angle

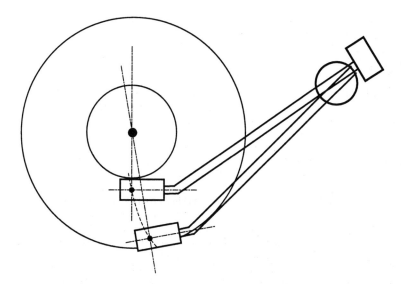

Figure 11.5 Ideal lateral tracking is achieved when a line through the head-shell forms a tangent to the groove

The original cutting stylus is driven across the acetate in a straight line towards the centre, using a carriage which does not have a single pivot point like the replay arm, and it can therefore be exactly tangential to the groove all the way across the disk. The cutting lathe is massively engineered to provide an inert, stable platform. There are some designs of record player which mimic this action so that truly zero tracking error is achieved on replay. The engineering difficulties involved in implementing such a technique are probably not justified since a well-designed and well-set-up arm can achieve excellent results using just a single conventional pivot.

A consequence of the pivoted arm is that a side thrust is exerted on the stylus during play which tends to cause it to skate across the surface of the record. This is a simple consequence of the necessary stylus overhang in achieving low tracking error from an arm which is pivoted at one end. Consider Figure 11.6. Initially it can be considered that the record is not rotating and the stylus simply rests in the groove. Consider now what happens when the record rotates in its clockwise direction. The stylus has an immediate tendency to drag across the surface of the record in the arrowed direction towards the pivot rather than along the record groove. The net effect is that the stylus feels a force in a direction towards the centre of the record causing it to bear harder on the inner wall of the groove than on the outer wall. One stereo channel will therefore be tracked more securely than the other, and uneven wear of the groove and stylus will also result. To overcome this, a system of bias compensation or 'anti-skating' is employed at the pivot end of the arm which is arranged so that a small outward force is exerted on the arm to counteract its natural inward tendency. This can be implemented in a variety of ways, including a system of magnets; or a small weight and thread led over a pulley which is contrived so as to pull the arm outwards away from the centre of the record; or a system of very light

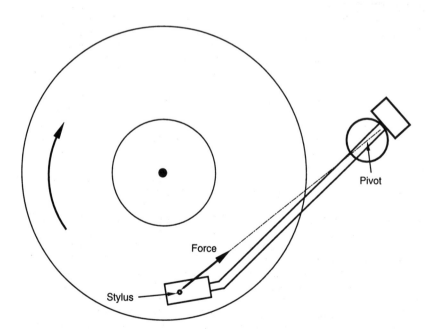

Figure 11.6 The rotation of the disk can create a force which pulls the arm towards the centre of the disk

springs. The degree of force which is needed for this bias compensation varies with different stylus tracking forces but it is in the order of one-tenth of that value (see Fact File 11.2).

The cartridge is mounted into the headshell using two bolts, and the standard spacing of the bolts is half an inch. It must be mounted so that the sides of the cartridge (if they are parallel, which is by no means always the case due to cosmetic factors) form a tangent to the outer and inner grooves, as has been discussed. If the cartridge sides do not provide a usable line of sight for this, perhaps the front or back of the cartridge provides one.

Although every cartridge will fit into every headshell apart from one or two special types, one must be aware of certain specifications of both the arm and the cartridge in order to determine whether the two are compatible. In order for the stylus to move about in the groove, the cantilever must be mounted in a suitable suspension system so that it can move to and fro with respect to the stationary cartridge body. This suspension has compliance or springiness and is traditionally specified in (cm/dyne) $\times$ 10^{-6} , abbreviated to cu ('compliance units'). This is a measure of how many centimetres (in practice, fractions of a centimetre!) the stylus will deflect when a force of 1 dyne is exerted on it. A low-compliance cartridge will have a compliance of, say, 8 cu. Highest compliances reach as much as 45 cu. Generally, values of 10–30 cu are encountered, the value being given in the maker's specification.

FACT FILE 11.2 Tracking weight

The required weight varies from cartridge to cartridge. A small range of values will be quoted by the manufacturer such as '1 gram $\pm$ 0.25 grams' or '1–2 grams' and the exact force must be determined by experiment in conjunction with a test record. Firstly, the arm and cartridge must be exactly balanced out so that the arm floats in free air without the stylus moving either down towards the record surface or upwards away from it, i.e.: zero tracking force. This is generally achieved by moving the counterweight on the end of the arm opposite to the cartridge either closer to or further away from the pivot until an exact balance point is found. The counterweight is usually moved by rotating it along a thread about the arm, or alternatively a separate secondary weight is moved. This should be carried out with bias compensation off.

When a balance has been achieved, a tracking weight should be set to a value in the middle of the cartridge manufacturer's values.

Either the arm itself will have a calibrated tracking force scale, or a separate stylus balance must be used. The bias compensation should then be set at the appropriate value, which again will have either a scaling on the arm itself or an indication in the setting up instructions. A good way to set the bias initially is to lower the stylus towards the play-in groove of a rotating record such that the stylus initially lands mid-way between these widely spaced grooves on an unused part of the surface of the record. Too much bias will cause the arm to move outwards before dropping into the groove. Too little bias will cause the arm to move towards the centre of the record before dropping into the groove. Just the right amount will leave the arm stationary until the relative movement of the groove itself eventually engages the stylus. From there, the optimum tracking and bias forces can then be determined using the test record according to the instructions given. In general, a higher tracking force gives more secure tracking but increases record wear. Too light a tracking force, though, will cause mistracking and damage to the record grooves.

11.2 RIAA equalisation

The record groove is an analogue of the sound waves generated by the original sources, and this in itself caused early pioneers serious problems. In early electrical cutting equipment the cutter stylus *velocity* remained roughly constant with frequency, for a constant input voltage (corresponding to a falling amplitude response with frequency) except at extreme LF where it became of more constant *amplitude*. Thus, unequalised, low frequencies would cause stylus movements of considerably greater excursion per cycle for a given stylus velocity than at high frequencies. It would be difficult for a pickup stylus and its suspension system inside the cartridge body to handle these relatively large movements, and additionally low frequencies would take up relatively more playing surface or 'land' on the record curtailing the maximum playing time. Low-frequency attenuation was therefore used during cutting to restrict stylus excursions.

In modern record cutting, a standard known as RIAA equalisation has been adopted which dictates a recorded *velocity* response, no matter what the characteristics of the individual cutting head. Electrical equalisation is used to ensure that the recorded velocity corresponds to the curve shown in Figure 11.7(a). A magnetic replay cartridge will have an output voltage proportional to stylus *velocity* (its unequalised output would rise with frequency for a constant amplitude groove) and thus its output must be electrically equalised according to the curve shown in Figure 11.7(b) in order to obtain a flat voltage–frequency response.

The treble pre- and de-emphasis of the RIAA replay curve have the effect of reducing HF surface noise. An additional recommendation is very low 20 Hz bass cut on replay (time constant 7960 µs) to filter out subsonic rumble and non-programme-related LF disturbance. The cartridge needs to be plugged into an input designed for this specific purpose; the circuitry will perform the above discussed replay equalisation as well as amplification.

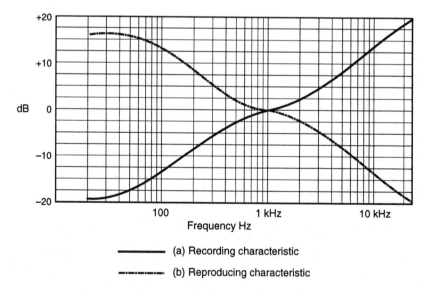

(a) Recording characteristic

(b) Reproducing characteristic

Figure 11.7 RIAA recording and reproducing characteristics

11.3 Cartridge types

The vast majority of cartridges in use are of the moving-magnet type, meaning that the cantilever has small powerful magnets attached which are in close proximity to the output coils. When the stylus moves the cantilever to and fro, the moving magnets induce current in the coils to generate the output. The DC resistance of the coils tends to be several hundred ohms, and the inductance several hundred millihenries (mH). The output impedance is therefore 'medium', and rises with frequency due to the inductance. The electrical output level depends upon the velocity with which the stylus moves, and thus for a groove cut with constant deviation the output of the cartridge would rise with frequency at 6 dB per octave. The velocity of the stylus movements relative to an unmodulated groove is conveniently measured in cm s^{-1}, and typical output levels of moving magnet cartridges are in the order of 1 mV cm^{-1} s^{-1}.

The average music programme produces cartridge outputs of several millivolts, and an upper limit of 40 or 50 mV will occasionally be encountered at mid frequencies. Due to the RIAA recording curve, the output will be less at low frequencies but not necessarily all that much more at high frequencies owing to the falling power content of music with rising frequency. A standard input impedance of an RIAA input of 47 k has been adopted, and around 40 dB of gain ($\times$ 100) is needed at mid frequencies to bring the signal up to line level.

Another type of cartridge which is much less often encountered but has a strong presence in high-quality audio circles is the moving-coil cartridge. Here, the cantilever is attached to the coils rather than the magnets, the latter being stationary inside the cartridge body. These cartridges tend to give much lower outputs than their moving-magnet counterparts because of the need to keep coil mass low by using a small number of turns, and they have a very low output impedance (a few ohms up to around a hundred) and negligible inductance. They require 20–30 dB

FACT FILE   Crosstalk

11.3

Crosstalk between the two stereo channels exists partly due to the fact that the two signals are present in the same record groove and partly due to the close proximity of the electrical components in the cartridge body itself. A poorly set up arm and cartridge will cause the stylus to sit incorrectly in the groove which also degrades crosstalk performance. At mid frequencies cartridges generally produce crosstalk figures of 25–30 dB, meaning that a signal present on one channel only will also appear, 25–30 dB down, on the other channel. At very low frequencies crosstalk worsens a little to around 10–20 dB due to excitation of the cartridge/arm resonance which tends to couple the two channels together. At the highest audio frequencies crosstalk can reach a quite poor 10 dB due to the stylus/ vinyl resonance which again tends to couple the two channels together. Figures worse than these should not, however, be accepted, and an additional check to see that left-to-right crosstalk is substantially the same as right-to-left crosstalk is advised, otherwise incorrect internal alignment of the cartridge's components is indicated.

Despite such relatively poor crosstalk figures, very good stereo images are possible from this medium, since the level difference required between two channels for a sound to appear to be fully to one side of the image is only around 18 dB anyway.

more gain than moving magnets do, and this is often provided by a separate head amplifier or step-up transformer, although many high-quality hi-fi amplifiers provide a moving-coil input facility. The impedance of such inputs is around 100 ohms or so.

Several 'high-output' moving-coil cartridges also exist which give an output level of around 10 dB lower than moving magnets. Such models can usually be used directly into ordinary cartridge inputs without the need for additional head amplification.

From time to time several other operating principles have been used, but the only one likely to be encountered now is the ceramic cartridge which is sometimes employed at the cheapest end of the market. A special ceramic material is employed which can be made to give an electrical output when pressure is exerted on it. Such a device gives a much higher output than the moving-magnet type, and also its output is roughly proportional to displacement (or groove amplitudue), rather than velocity, such that the obtained frequency response tends to compensate for the RIAA recording characteristic. Owing to its modest quality of reproduction anyway, this is deemed sufficiently good to obviate the need for a special cartridge RIAA input stage, and an ordinary relatively low-gain amplifier is therefore used.

Crosstalk may arise in a cartridge, and this is discussed in Fact File 11.3.

11.4 Connecting leads

Owing to the inductive nature of the output impedance of a moving-magnet cartridge, it is sensitive to the capacitance present in the connecting leads and also that present in the amplifier input itself. This total capacitance appears effectively in parallel with the cartridge output, and thus forms a resonant circuit with the cartridge's inductance. It is the high-frequency performance of the cartridge which is affected by this mechanism, and the total capacitance must be adjusted so as to give the best performance. Too little capacitance causes a frequency response which tends to droop several decibels above about 5 kHz, sharply rising again to a 2–3 dB peak with respect to the mid band at around 18–20 kHz, which is the result of the resonant frequency of the stylus/record interface, the exact value depending upon the stylus tip mass. Adding some more capacitance lifts the 5–10 kHz trough and also curtails the tip mass resonant peak to smooth out the frequency response. Too much capacitance causes attenuation of the highest frequencies giving a dull sound.

The optimum value of capacitance will be different for different models of cartridge, but is always in the picofarad (pF) range. The wiring connecting the arm to the amplifier will exhibit a capacitance of typically 100 pF per metre, to which must be added the amplifier input capacitance which will generally be a little below 100 pF. Manufacturers' specifications should give the exact value. The total parallel capacitance due to these two sources seen by the cartridge then will generally be around 200 pF. This is too low a value for many moving-magnet cartridges, and extra capacitance must be added. If the cartridge manufacturer does not specify a capacitance requirement, it must be determined by experiment. A pink noise recording on a test record is very good for this. With a little practice, it becomes quite easy to hear whether the output of a cartridge is drooping at the higher frequencies accompanied by a peak at the highest audible frequencies. A kind of 'hole' can be heard in the upper frequency spectrum. Adding more capacitance fills in this hole giving a smooth, continuous rushing sound on pink noise. Too much causes attenuation of the highest frequencies, and the rushing sound is more distant and less 'exciting'.

Around 300–400 pF total is the usual range of capacitance to be tried. Adding capacitance is most conveniently carried out using special in-line plugs containing small capacitors for this purpose. Assume that 200 pF is present already, and try an extra 100 pF, then 200 pF. Alternatively, small polystyrene capacitors can be purchased and soldered between signal and earth wires of the leads inside the plugs or sockets. Sometimes solder tags are present in the base of the record player itself which are convenient. NEVER solder anything to the cartridge pins. The cartridge can very easily be damaged by doing this.

The length of wiring between the turntable and the amplifier should not be greater than a metre or so, due to the fact that long leads will give excessive capacitance values as well as being more susceptible to interference. Note that equalising the cartridge by ear means that you are actually equalising the complete cartridge/amplifier/speaker combination. Subsequent changes of any of these three should be accompanied by re-equalisation of the cartridge.

Moving-coil cartridges have a very low output impedance and negligible inductance, and their frequency response is therefore not affected by capacitance. Some workers have, however, found that the subjective sound quality can be improved by adding quite high values of parallel capacitance to certain models, up to the nanofarad range (1 nanofarad, 1 nF, equals 1000 pF). This can only be determined by experimentation.

11.5 Arm design

The principal specifications of an arm are its length, effective mass and the bearing frictions, both vertical and horizontal. The standard length is 9 inches (23 cm) or thereabouts. One or two 12 inch (30 cm) models are also available which can yield lower values of tracking error simply because of their increased length, but they exhibit greater effective mass. The effective mass is the inertial mass felt by the stylus, due to the arm and its counterweight. It is not therefore simply the weight of the arm as a whole. Values of effective arm mass range from around 6 g to 25 g. Intermediate values generally encompass high-quality examples which have been designed to give a good compromise between reasonably low effective mass and adequate rigidity. The value of an arm's effective mass is given by the manufacturer, and its significance will be dealt with later. A typical example of a pickup arm is shown in Figure 11.8.

The pivot of the arm can be implemented in several ways, including ball races, knife-edge-type bearings, a 'unipivot' consisting of a single vertical pointed rod like a nail on which the arm seats, or the arm can even hang by a thread as in one esoteric model. Whatever, it is the bearing friction felt by the stylus which really matters, combined with consistent, stable performance. Horizontal bearing friction should be below 80 mg (milligrams) i.e.: less than 80 mg of force is required to move the arm continuously. Vertical bearing friction is generally lower than this, and should be below 30 mg, otherwise the stylus will have to work hard to carry the arm across the record surface with it and up and down over warps. Also, frictions of greater than these values make a nonsense of bias compensation.

The effective mass, coupled with the cartridge's suspension compliance, to-gether form a resonant system, the frequency of which must be contrived such that

Figure 11.8 The SME Series V pickup arm. (Courtesy of SME Ltd.)

it is low enough not to fall within the audio band but high enough to avoid coinciding with record warp frequencies and other LF disturbances which would continually excite the resonance causing insecure tracking and even groove jumping. Occasionally, large, slow excursions of the cones of the speaker woofers can be observed when the record is being played, which is the result of non-programme-related, very low-frequency output which results from an ill-matched arm/cartridge combination.

A value of 10–12 Hz is suitable, and a simple formula exists which enables the frequency to be calculated for a given combination of arm and cartridge:

$$f = 1000/(2\pi\sqrt{(MC)})$$

where f = resonant frequency in hertz, M = effective mass of the arm + mass of the cartridge + mass of hardware (nuts, bolts, washers) in grams, C = compliance of the cartridge in compliance units.

For example, consider a cartridge weighing 6 g, having a compliance of 25 cu; and an arm of effective mass 20 g, additional hardware a further 1 g. The resonant frequency will therefore be 6.2 Hz. This value is below the optimum, and such a combination could give an unsuitable performance due to this resonance being excited by mechanical vibrations such as people walking across the floor, record warps, and vibrations emanating from the turntable main bearing. Additionally, the 'soft' compliance of the cartridge will have difficulty in coping with the high effective mass of the arm, and the stylus will be continually changing its position in the groove somewhat as the arm's high inertia tends to flex the cartridge's suspension and dominate its performance.

If the same cartridge in an arm having an effective mass of 8 g is considered, then f = 8.4 Hz. This is quite close to the ideal, and would be acceptable. It illustrates well the need for low-mass arms when high compliances are encountered. The resonance tends to be high Q, and this sharp resonance underlines the need to get the frequency into the optimum range. Several arms provide damping in the form of a paddle, attached to the arm, which moves in a viscous fluid, or some alternative arrangement. This tends to reduce the amplitude of the resonance somewhat, which helps to stabilise the performance. Damping cannot, however, be used to overcome the effects of a non-optimum resonant frequency, which must still be carefully chosen.

11.6 Turntable designs

11.6.1 Drive mechanisms

The two principal specifications of the turntable itself are rumble, and wow and flutter. Rumble is now a thing of the past, and should not be encountered even in a fairly modestly priced record player. If present, the main sources are likely to be the main bearing at the centre of the platter being poorly designed or manufactured so that movement is allowed, which can transmit low-frequency vibrations to the record and thence to the stylus, and low-frequency motor noise caused by the drive motor being inadequately decoupled from the rotating platter.

One popular method of driving the platter was to employ an idler wheel which had a diameter of around 3 cm and which was positioned between the inside of the outer rim of the platter and the motor drive shaft, the latter having a shank with several diameters to give various gear ratios to provide different playback speeds, as shown in Figure 11.9. This technique was very widely employed at the cheap end of the market, but the relative lack of decoupling of the motor from the platter due to the mutual contact provided by the rubber idler wheel generally returned poor rumble performance. However, one or two high-quality designs existed which demonstrated that careful engineering could yield good results.

The two main methods of rotating the platter are now belt drive and direct drive. Belt drive models employ a fairly large, flat 'elastic band' which runs around a pulley on the motor's drive shaft and also either around a fairly large diameter subplatter on which the platter rests, or sometimes even around the platter itself. Different-sized pullies on the motor's shaft provide the different playing speeds. The compliant nature of the rubber belt decouples the motor from the platter, and very low rumble performance is achieved. The platter is generally quite substantially built to give a good flywheel effect which ensures good short-term speed stability.

The direct drive turntable features a relatively large slow-speed motor, the drive shaft of which forms the centre spindle of the turntable itself, and it thus drives the platter directly. Speed change is implemented electronically. The earliest models employed extremely lightweight platters so that the drive motor had a light load to accelerate and could rapidly attain the correct rotational speed. Unfortunately the relative lack of flywheel effect gave poor short-term speed stability and even the drag of the stylus during heavily modulated sections of the record groove had a small but audible effect upon stability.

Servo systems were used in an attempt to control the speed, and although this was successful when a continuous, steady sine wave was being reproduced from a test

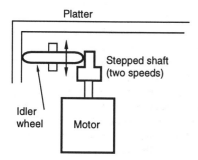

Figure 11.9 Idler wheel turntable drive

FACT FILE

11.4

Wow and flutter

Wow (see section A1.6) can be caused by the record hole not being exactly central, or by deficiencies in the motor/belt performance in belt drive turntables. Flutter has generally been encountered in direct drive turntables which have inadequately massive platters and therefore suffer from poor flywheel action. A substantial platter is a good way of helping to avoid these short-term speed variations due to the good momentum characteristics of the platter itself. Some designs go even further by concentrating much of the metal from which the platter is made around the perimeter to increase the flywheel effect for a given total platter mass. A contaminated drive belt is sometimes the culprit, and a very careful cleaning of the belt and the pullies and subplatter assembly, around which the belt runs, with mildly soapy water followed by thorough drying is the remedy here. Do not handle the belt with the fingers, otherwise skin grease will contaminate the belt causing it to slip.

---

record, a real music programme caused stylus drag to change continually. A servo system can only act after a speed change has been detected, which is already too late. A continuous 'hunting' mechanism therefore resulted, causing a less than smooth audio performance. Later models have been more substantially engineered, and give good performances; the servo systems need now only ensure longer-term speed stability. The fact that the motor is coupled to the platter calls for high-quality engineering if rumble is to be avoided.

11.6.2 Hum induction

Motors radiate hum fields which can be induced into the cartridge if the latter is itself poorly screened, and this manifests itself as low-frequency hum reproduced through the speakers. The bass boost requirement of the RIAA EQ curve magnifies the problem. The hum often increases as the cartridge moves towards the centre of the record, closer to the motor. If this occurs it is usually fair to blame the cartridge for its lack of adequate screening, although sometimes it may be found that the cartridge is incorrectly wired. The connecting pins on the back of the cartridge must be correctly identified and the appropriate colour-coded arm wires should be attached to each via the push-on terminals. Colour coding is as follows: left signal – white, left earth – black, right signal – red, right earth – green.

Additionally, the turntable system or the arm itself will have a separate earthing wire which should be connected to the amplifier at a special terminal provided for the purpose. This earths the metal parts of the turntable and/or arm and greatly attenuates any hum and interference which may be initially encountered. If the cartridge still produces LF continuous hum, an earth loop may be present due to both the record player and the amplifier being earthed at the mains. The solution is to disconnect the record player earth wire from the mains plug, but one must then extend this wire and connect it to a metal part of the amplifier for safety. This should rarely, if ever, be necessary due to the manufacturers' awareness of these problems. If the cartridge still hums, and the hum level varies with the cartridge's relative position between its rest position and over the inner grooves when the platter is rotating, the cartridge is not adequately screened and is not compatible with that particular turntable.

11.6.3 Turntable mounting

Due to the fact that the stylus rests on a fairly light, large-diameter surface during play, namely the record, the record player is susceptible to acoustic feedback, this being caused by the sound from the speakers provoking vibrations in the record-playing system which the stylus picks up to reproduce again through the speakers, colouring the sound. In extreme cases, if the stylus is lowered on to a stationary record and a high replay volume level is set, a continuous drone can be induced due to the fairly rapid build up of this positive feedback. Structure-borne feedback, which produces the same result, can also be present due to the platform the record player is sitting on being vibrated. For these reasons the speaker must never be placed on the same platform or shelving system as the turntable. A degree of feedback must always be present when loudspeakers are in the same room as the record player.

Several high-quality turntables employ a suspended subchassis whereby the arm and platter are mounted on a separate platform inside the plinth, this platform being suspended on a system of springs which give a degree of isolation from the environment. Good resistance from both acoustic feedback and mechanical shock can be achieved, and maximum benefit is attained if the resonant frequency of the suspension system is below the frequency of both programme material and mechanical vibrations such as disturbances of the turntable and the like. Therefore, subchassis resonances of around 4 or 5 hertz are called for, a suspension system being able to isolate the subchassis from any vibrations lying above its own resonant frequency. It is absolutely essential that this resonant frequency is at least several hertz below the arm/cartridge resonance. Such turntables therefore have a bouncy, jelly-like feel to them which requires careful handling when the stylus is being lowered into the groove and when records are being changed. This is an ergonomic compromise which comes with the superior audio performance of well-designed examples of the suspended subchassis system. The turntable mat on which the record rests should be devoid of ridges or cosmetic patterns so that maximum surface area makes contact with the record, this helping to damp vibrations set up in the record itself due to acoustic feedback and 'needle-talk' – the sound produced mechanically by the stylus itself as it traces the groove.

11.6.4 Professional turntables

The professional turntable, such as the one pictured in Figure 11.10, will, however, tend to avoid the above features because a firm, positive feel is essential so that records can be quickly taken off and replaced, the stylus being quickly lowered into the run-in grooves ready for instant play. A delicate bouncy suspension must be forgone if this is to be achieved. A broadcast turntable will need to be capable of being 'back-cued', that is the stylus is lowered on to the stationary record and the platter is actually rotated backwards until the beginning of the track on the record is reached so that instant play can be set up. This means of course that the cartridge suspension system must be specially designed to withstand such treatment; an ordinary hi-fi cartridge would easily be damaged. Additionally, the turntable must reach playing speed almost instantly so that one does not hear the music speeding up to the proper pitch. A substantially engineered belt drive turntable takes several seconds to attain the proper speed which is unacceptable to a DJ. Professional players therefore work on the direct drive principle, a high-torque motor accelerat-

Figure 11.10 A professional turntable: the EMT 948. (Courtesy of FWO Bauch Ltd.)

ing a medium-mass platter up to playing speed in a fraction of a second. Some players mute the audio outputs until speed stability has been attained so that there is no possibility of music speeding up to pitch being broadcast. Acoustic and structure-borne feedback is generally kept at bay by employing mass-damping – engineering the turntable system to be heavy so that its inertia alone tends to resist vibration. Some hi-fi record players also employ this technique.

Because of the rough handling a professional player is subjected to, the cartridges often have a low, 'stiff' compliance together with a substantially built arm. Tracking forces of greater than 2 grams are common. They often contain their own electronics so that the outputs are at line level, and balanced. Some incorporate a sprung suspension system but the need to avoid difficult handling means that the resonant frequency of such suspensions is at least 50 Hz or more, this being virtually useless since the turntable is just where it is needed to provide resistance from low frequencies from speakers, people walking across a wooden floor, and accidental disturbance of the platform the turntable is sitting on. Professional systems therefore achieve their environmental isolation simply from inertia due to their relatively high mass, as has been said.

The arm headshell will be detachable so that if the stylus is damaged a replacement cartridge and headshell can quickly be fitted so that the player suffers no downtime. Styluses are not generally replaceable as they are with virtually all domestic moving-magnet cartridges because of the back-cueing capability. The detachable headshell represents a compromise in the rigidity of the arm, but its convenience is important professionally.

The various compromises present in the professional record player compared with its high-quality domestic counterpart do not in practice lead to greatly inferior sound quality, mainly due to the fact that professional models are very carefully designed and engineered (and are very expensive) so as not to allow the theoretical shortcomings to intrude upon the sonic performance more than is absolutely necessary.

11.7 Laser pickups

The idea of reading a record groove with a laser beam rather than a stylus has been mooted for quite some time, and in 1990 a player using such a technique finally appeared. It is a very attractive proposition for record libraries due to the fact that record wear becomes a thing of the past. However, a laser beam does not push particles of dust aside as does a stylus, and the commercial system needs to be fed with disks which are almost surgically clean, otherwise signal drop-outs occur. Two entirely separate laser beams are in fact used, one reading the information from each wall of the groove. Error concealment circuitry is built in, which suppresses the effects of scratches. Towards the centre of the record, short wavelengths occupy a proportionately smaller area of the groove than at the perimeter of the disk, and the width of the laser beam means that difficulties in reading these high-amplitude HF signals occur. The frequency response of the player therefore droops by around 10 dB if the highest frequencies towards the end of the side are of a high amplitude.

CD-type features are offered such as pause, track repeat and track search, which are very useful. The player does not suffer from traditional record player ills such as LF arm/cartridge resonance, rumble, and wow and flutter. Costing not much under £10 000, the player obviously has a limited appeal, the professional user being the main potential customer.

Recommended further reading

AES (1981) *Disk Recording – An Anthology*, Vols 1 and 2. Audio Engineering Society
BS 7063. British Standards Office
Earl, J. (1973) *Pickups and Loudspeakers*. Fountain Press
Roys, H. E. (1978) ed. *Disk Recording and Reproduction*. Dowden, Hutchinson and Ross

See also *General further reading* at the end of this book.

Chapter 12

Power amplifiers

Power amplifiers are uneventful devices. They are usually big and heavy, take up a lot of rack space, and feature very little (or sometimes nothing) beyond input and output sockets. Because one tends to ignore them, it is all the more important that they are chosen and used with due care. Coming in a variety of shapes, sizes and 'generations', they are all required to do the ostensibly simple job of providing voltage amplification – converting line levels of up to a volt or so into several tens of volts, with output currents in the ampere range to develop the necessary power across the loudspeaker terminals. Given these few requirements, it is perhaps surprising how many designs there are on the market.

12.1 Domestic power amplifiers

The domestic power amplifier, at its best, is designed for maximum fidelity in the true sense of that word, and this will usually mean that other considerations such as long-term overload protection and complete stability into any type of speaker load are not always given the type of priority which is essential in the professional field. A professional power amp may well be asked to drive a pair of 6 ohm speakers in parallel on the other end of 30 metres of cable, at near to maximum output level for hours on end if used in a rock PA rig. This demands large power supplies and heavy transformers, with plenty of heat sink area (the black fins usually found on the outer casing) to keep it from overheating. Cooling fans are frequently employed which will often run at different speeds depending on the temperature of the amplifier.

The domestic amplifier is unlikely to be operated at high output levels for a significant length of time, and the power supplies are often therefore designed to deliver high currents for short periods to take care of short, loud passages. A power supply big enough to supply high currents for lengthy periods is probably wasted in a domestic amplifier. Also, the thermal inertia of the transformer and the heat sinks means that unacceptable rises in temperature are unlikely. Although there are one or two domestic speakers which are notoriously difficult to drive due to various combinations of low impedance, low efficiency (leading to high power demand), and wide phase swings (current and voltage being out of step with each other due to crossover components and driver behaviour in a particular speaker enclosure), the majority of domestic hi-fi speakers are a comfortable load for an amplifier, and usually the speaker leads will be less than 10 metres in length.

It is unlikely that the amplifier will be driven into a short-circuit due to faulty

Class A

The output stage draws a constant high current from the power supply regardless of whether there is an audio signal present or not. Low-current class A stages are used widely in audio circuits. The steady bias current as it is known is employed because transistors are non-linear devices, particularly when operated at very low currents. A steady current is therefore passed through them which biases them into the area of their working range at which they are most linear. The constant bias current makes class A amplification inefficient due to heat generation, but there is the advantage that the output transistors are at a constant steady temperature. Class A is capable of very high sound quality, and several highly specified up-market domestic class A power amplifiers exist.

Class B

No current flows through the output transistors when no audio signal is present. The driving signal itself biases the transistors into conduction to drive the speakers. The technique is therefore extremely efficient because the current drawn from the power supply is entirely dependent upon the level of drive signal. Class B is therefore particularly attractive in battery-operated equipment. The disadvantage is that at low signal levels the output transistors operate in a non-linear region. It is usual for pairs (or multiples) of transistors to provide the output current of a power amplifier. Each of the pair handles opposite halves of the output waveform (positive and negative with respect to zero) and therefore as the output swings through zero from positive to negative and vice versa the signal suffers so-called 'crossover distortion'. The result is relatively low sound quality, but class B can be used in applications which do not require high sound quality such as telephone systems, hand-held security transceivers, paging systems and the like.

Class A–B

In this design a relatively low constant bias current flows through the output transistors to give a low-power class A amplifier. As the input drive signal is increased, the output transistors are biased into appropriately higher-current conduction in order to deliver higher power to the speakers. This part of the operation is the class B part, i.e.: it depends on input drive signal level. But the low-level class A component keeps the transistors biased into a linear part of their operating range so that crossover distortion is largely avoided. The majority of high-quality amplifiers operate on this principle.

Other classes

Class C drives a narrow band of frequencies into a resonant load, and is appropriate to radio-frequency (RF) work where an amplifier is required to drive a single frequency into an appropriately tuned aerial. Class D is 'pulse width modulation' in which an ultrasonic frequency, modulated by the audio signal, is used to drive the output transistors. A low-pass filter is employed after the output stage. This technique has been revived in one or two designs in the late 1980s. Classes E and F were concerned with increasing efficiency, and currently no commercial models conform to these particular categories. Class G incorporates several different voltage rails which progressively come into action as the drive signal voltage is increased. This technique can give very good efficiency because for much of the time only the lower-voltage, low-current supplies are in operation. Such designs can be rather smaller than their conventional class A–B counterparts of comparable output power rating.

Since the early 1980s the MOSFET (Metal Oxide Semiconductor Field-Effect Transistor) has been widely employed for the output stages of power amplifiers. MOSFET techniques claim lower distortion, better thermal tracking (i.e.: good linearity over a wide range of operating temperatures), simpler output stage design, and greater tolerance of adverse loudspeaker loads without the need for elaborate protection circuitry.

speaker lines for any length of time (silence gives an immediate warning), which is not the case with a professional amplifier which may well be one of many, driving a whole array of speakers. A short-circuit developing soon after a show has begun may cause the amplifier to be driven hard into this condition for the whole evening. Protection circuitry needs to be incorporated into the design to allow the professional amplifier to cope with this without overheating or catastrophically failing which can affect other amplifiers in the same part of the rig.

Several 'classes' of amplifier design have appeared over the years, these being labels identifying the type of output stage topology employed to drive the speaker. These are outlined in Fact File 12.1.

12.2 Professional amplifier facilities

The most straightforward power amplifiers have input sockets and output terminals, and nothing else. Single-channel models are frequently encountered, and in the professional field these are often desirable because if one channel of a stereo power amplifier develops a fault then the other channel also has to be shut down, thus losing a perfectly good circuit. The single-channel power amplifier is thus a good idea when multi-speaker arrays are in use such as in rock PA systems and theatre sound.

Other facilities found on power amplifiers include input level controls, output level meters, overload indicators, thermal shutdown (the mains feed is automatically disconnected if the amplifier rises above a certain temperature), earth-lift facility to circumvent earth loops, and 'bridging' switch. This last facility, applicable to a stereo power amplifier, is a facility sometimes provided whereby the two channels of the amp can be bridged together to form a single-channel higher-powered one, the speaker(s) now being connected across the two positive output terminals with the negative terminals left unused. Only one of the input sockets is now used to drive it.

Cooling fans are often incorporated into an amplifier design. Such a force-cooled design can be physically smaller than its convection-cooled counterpart, but fans tend to be noisy. Anything other than a genuinely silent fan is unacceptable in a studio or broadcast control room, or indeed in theatre work, and such models will need to be housed in a separate well-ventilated room. Ventilation of course needs to be a consideration with all power amplifiers.

12.3 Specifications

Power amplifier specifications include sensitivity, maximum output power into a given load, power bandwidth, frequency response, slew rate, distortion, crosstalk between channels, signal-to-noise ratio, input impedance, output impedance, damping factor, phase response, and DC offset. Quite surprising differences in sound quality can be heard between certain models, and steady-state measurements do not, unfortunately, always tell a user what he or she can expect to hear.

12.3.1 Sensitivity

Sensitivity is a measurement of how much voltage input is required to produce the amplifier's maximum rated output. For example, a model may be specified

'150 watts into 8 ohms, input sensitivity 775 mV = 0 dBu'. This means that an input voltage of 775 mV will cause the amplifier to deliver 150 watts into an 8 ohm load. Speakers exhibit impedances which vary considerably with frequency, so this is always a nominal specification when real speakers are being driven. Consideration of sensitivity is important because the equipment which is to drive the amp must not be allowed to deliver a greater voltage to the amplifier than its specification states, otherwise the amplifier will be overloaded causing 'clipping' of the output waveform (a squaring-off of the tops and bottoms of the waveform resulting in severe distortion). This manifests itself as a 'breaking-up' of the sound on musical peaks, and will often quickly damage tweeters and high-frequency horns.

Many amplifiers have input level controls so that if, for instance, the peak output level of the mixer which drives the amplifier is normally say 'PPM 6' – about 2 volts – then the amp's input levels can be turned down to prevent overload. In the given example, 2 volts is 8 dB higher than 775 mV (PPM 4 = 0 dBu) and so the input level control should be reduced by 8 dB to allow for this. If a dB calibration is not provided on the level control, and many are not particularly accurate anyway, a reasonable guide is that, compared with its maximum position of about '5 o'clock', reducing the level to about 2 o'clock will reduce the sensitivity by about 10 dB, or by a factor of three. In this position, the power amplifier with an input sensitivity of 775 mV will now require 0.775 × 3, or about 2 volts, to develop its full output.

If input level controls are not provided, one can build a simple resistive attenuator which reduces the voltage being fed to the amplifier's input. Two examples are shown in Figure 12.1. It is best to place such attenuators close to the power amp input in order to keep signal levels high while they are travelling down the connecting leads. In both cases the 3k3 resistor which is in parallel with the

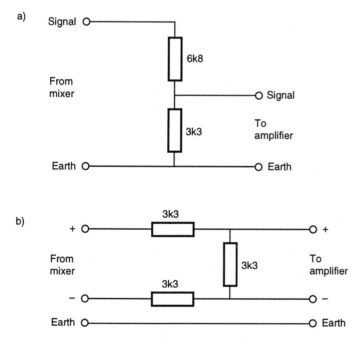

Figure 12.1 (a) An unbalanced resistive attenuator. (b) A balanced resistive attenuator

amplifier's input can be increased in value for less attenuation, and decreased in value for greater attenuation. With care, the resistors can be built into connecting plugs, the latter then needing to be clearly labelled.

12.3.2 Power output

A manufacturer will state the maximum power a particular model can provide into a given load, e.g.: '200 watts into 8 ohms', often with 'both channels driven' written after it. This last means that both channels of a stereo amplifier can deliver this simultaneously. When one channel only is being driven, the maximum output is often a bit higher, say 225 watts, because the power supply is less heavily taxed. Thus 200 watts into 8 ohms means that the amplifier is capable of delivering 40 volts into this load, with a current of 5 amps. If the load is now reduced to 4 ohms then the same amplifier should produce 400 watts. A theoretically perfect amplifier should then double its output when the impedance it drives is halved. In practice, this is beyond the great majority of power amplifiers and the 4 ohm specification of the above example may be more like 320 watts, but this only around 1 dB below the theoretically perfect value. A 2 ohm load is very punishing for an amplifier, and should be avoided even though a manufacturer sometimes claims a model is capable of, say, 800 watts of short-term peaks into 2 ohms. This at least tells us that the amp should be able to drive 4 ohm loads without any trouble.

Because 200 watts is only 3 dB higher than 100 watts, then, other things being equal, the exact wattage of an amplifier is less important than factors such as its ability to drive difficult reactive loads for long periods. Often, 'RMS' will be seen after the wattage rating. This stands for root-mean-square, and defines the raw 'heating' power of an amplifier, rather than its peak output. All amplifiers should be specified RMS so that they can easily be compared. The RMS value is 0.707 times the instantaneous peak capability, and it is unlikely that one would encounter a professional amplifier with just a peak power rating.

Power bandwidth is not the same as power rating, as discussed in Fact File 12.2.

FACT FILE 12.2   Power bandwidth

Power bandwidth is a definition of the frequency response limits within which an amplifier can sustain its specified output. Specifically, a 3 dB drop of output power is allowed in defining a particular amplifier's power bandwidth. For example, a 200 watt amplifier may have a power bandwidth of 10 Hz to 30 kHz, meaning that it can supply 200 watts − 3 dB (= 100 watts) at 10 Hz and 30 kHz, compared with the full 200 watts at mid frequencies. Such an amplifier would be expected to deliver the full 200 watts at all frequencies between about 30Hz and 20 kHz, and this should also be looked for in the specification. Often, though, the power rating of an amplifier is much more impressive when measured using single sine-wave tones than with broad-band signals, since the amplifier may be more efficient at a single frequency.

Power bandwidth can indicate whether a given amplifier is capable of driving a subwoofer at high levels in a PA rig, as it will be called upon to deliver much of its power at frequencies below 100 Hz or so. The driving of high-frequency horns also needs good high-frequency power bandwidth so that the amplifier never clips the high frequencies, which easily damages horns as has been said.

12.3.3 Frequency response

Frequency response, unlike power bandwidth, is simply a measure of the limits within which an amplifier responds equally to all frequencies when delivering a very low power. The frequency response is usually measured with the amplifier delivering 1 watt into 8 ohms. A specification such as '20 Hz – 20 kHz ± 0.5 dB' should be looked for, meaning that the response is virtually flat across the whole of the audible band. Additionally, the –3 dB points are usually also stated, e.g.: '–3 dB at 12 Hz and 40 kHz', indicating that the response falls away smoothly below and above the audio range. This is desirable as it gives a degree of protection for the amp and speakers against subsonic disturbances and RF interference.

12.3.4 Distortion

Distortion should be 0.1% THD (see section A1.3) or less across the audio band, even close to maximum-rated output. It often rises slightly at very high frequencies, but this is of no consequence. Transient distortion, or transient intermodulation distortion (TID), is also a useful specification. It is usually assessed by feeding both a 19 kHz and a 20 kHz sine wave into the amplifier and measuring the relative level of 1 kHz difference tone. The 1 kHz level should be at least 70 dB down, indicating a well-behaved amplifier in this respect. The test should be carried out with the amplifier delivering at least two-thirds of its rated power into 8 ohms. Slew rate distortion is also important (see Fact File 12.3).

12.3.5 Crosstalk

Crosstalk figures of around –70 dB at mid frequencies should be a reasonable minimum, degrading to around –50 dB at 20 kHz, and by perhaps the same amount at 25 Hz or so. 'Dynamic crosstalk' is sometimes specified, this manifesting itself mainly at low frequencies because the power supply works hardest when it is called upon to deliver high currents during high-level, low-frequency drive. Current demand by one channel can modulate the power supply voltage rails, which gets into the other channel. A number of amplifiers have completely separate power supplies for each channel, which eliminates such crosstalk, or at least separate secondary windings on the mains transformer plus two sets of rectifiers and reservoir capacitors which is almost as good.

12.3.6 Signal-to-noise ratio

Signal-to-noise ratio is a measure of the output residual noise voltage expressed as a decibel ratio between that and the maximum output voltage, when the input is short-circuited. Noise should never be a problem with a modern power amplifier and signal-to-noise ratios of at least 100 dB are common. High-powered models (200 watts upwards) should have signal-to-noise ratios correspondingly greater (e.g.: 110 dB or so) in order that the output residual noise remains below audibility.

12.3.7 Impedance

The input impedance of an amplifier ought to be at least 10 kΩ, so that if a mixer is required to drive, say, ten amplifiers in parallel, as is often the case with PA rigs,

FACT FILE Slew rate

12.3

Slew rate is a measure of the ability of an amplifier to respond accurately to high-level transients. For instance, the leading edge of a transient may demand that the output of an amplifier swings from 0 to 120 watts in a fraction of a millisecond. The slew rate is defined in V μs⁻¹ (volts per microsecond) and a power amplifier which is capable of 200 watts output will usually have a slew rate of at least 30 V μs⁻¹. Higher-powered models require a greater slew rate simply because their maximum output voltage swing is greater. A 400 watt model might be required to swing 57 volts into 8 ohms as compared with the 200 watt model's 40, so its slew rate needs to be at least:

$$30 \times (57 \div 40) = 43 \text{ V μs}^{-1}$$

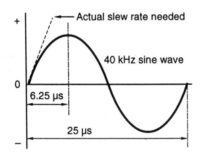

In practice, modern power amplifiers achieve slew rates comfortably above these figures.

An absolute minimum can be estimated by considering the highest frequency of interest, 20 kHz, then doubling it for safety, 40 kHz, and considering how fast a given amplifier must respond to reproduce this accurately at full output. A sine wave of 40 kHz reaches its positive-going peak in 6.25 μs, as shown in the diagram. A 200 watt model delivers a peak voltage swing of 56.56 volts peak to peak (1.414 times the RMS voltage). It may seem then that it could therefore be required to swing from 0 V to +28.28 V in 6.25 μs, thus requiring a slew rate of 28.28 ÷ 6.25, or 4.35 V μs⁻¹. But the actual slew rate requirement is rather higher because the initial portion of the sine wave rises steeply, tailing off towards its maximum level.

Musical waveforms come in all shapes and sizes of course, including near-square waves with their almost vertical leading edges, with a minimum slew rate of around eight times this (i.e.: 30 V μs⁻¹) might be considered as necessary. It should be remembered, though, that the harmonics of an HF square wave are well outside the audible spectrum, and thus slew rate distortion of such waves at HF is unlikely to be audible. Extremely high slew rates of several hundred volts per microsecond are sometimes encountered. These are achieved in part by a wide frequency response and 'fast' output transistors, which are not always as stable into difficult speaker loads as are their 'ordinary' counterparts. Excessive slew rates are therefore to be viewed with scepticism.

the total load will be 10 k ÷ 10, or 1 k , which is still a comfortable load for the mixer. Because speakers are of very low impedance, and because their impedance varies greatly with frequency, the amplifier's output impedance must not be greater than a fraction of an ohm, and a value of 0.1 ohms or less is needed. A power amplifier needs to be a virtually perfect 'voltage source', its output voltage remaining substantially constant with different load impedances.

The output impedance does, however, rise a little at frequency extremes. At LF, the output impedance of the power supply rises and therefore so does the amplifier's. It is common practice to place a low-valued inductor of a couple of microhenrys in series with a power amp's output which raises its output impedance a little at HF, this being to protect the amp against particularly reactive speakers or excessively capacitive cables, which can provoke HF oscillation.

12.3.8 Damping factor

Damping factor is a numerical indication of how well an amplifier can 'control' a speaker. There is a tendency for speaker cones and diaphragms to go on vibrating a little after the driving signal has stopped, and a very low output impedance virtually short-circuits the speaker terminals which 'damps' this. Damping factor is the ratio between the amplifier's output impedance and the speaker's rated imped- ance, so a damping factor of '100 into 8 ohms' means that the output impedance of the amplifier is 8 ÷ 100 ohms, or 0.08 ohms. One hundred is quite a good figure (the higher the better, but a number greater than 200 *could* imply that the amplifier is insufficiently well protected from reactive loads and the like), but it is better if a frequency is given. Damping factor is most useful at low frequencies because it is the bass cones which vibrate with greatest excursion, requiring the tightest control. A damping factor of '100 at 40 Hz' is therefore a more useful specification than '100 at 1 kHz'.

12.3.9 Phase response

Phase response is a measurement of how well the frequency extremes keep in step with mid frequencies. At very low and very high frequencies, 15° phase leads or phase lags are common, meaning that in the case of phase lag, there is a small delay of the signal compared with mid frequencies, and phase lead means the opposite. At 20 Hz and 20 kHz, the phase lag or phase lead should not be greater than 15°, otherwise this may imply a degree of instability when difficult loads are being driven, particularly if HF phase errors are present.

The absolute phase of a power amplifier is simply a statement of whether the output is in phase with the input. The amplifier should be non-phase-inverting overall. One or two models do phase invert, and this causes difficulties when such models are mixed with non-inverting ones in multi-speaker arrays when phase cancellations between adjacent speakers, and incorrect phase relationships between stereo pairs and the like, crop up. The cause of these problems is not usually apparent and can waste much time.

12.4 Coupling

The vast majority of power amplifier output stages are 'direct coupled', that is the output power transistors are connected to the speakers with nothing in between beyond perhaps a very low-valued resistor and a small inductor. The DC voltage operating points of the circuit must therefore be chosen such that no DC voltage appears across the output terminals of the amplifier. In practice this is achieved by using 'split' voltage rails of opposite polarity (e.g.: +46 volts DC) between which the symmetrical output stage 'hangs', the output being the mid-point of the voltage rails (i.e.: 0 V). Small errors are always present, and so 'DC offsets' are produced which means that several millivolts of DC voltage will always be present across the output terminals. This DC flows through the speaker, causing its cone to deflect either forwards or backwards a little from its rest position. As low a DC offset as possible must therefore be achieved, and a value of ±40 mV is an acceptable maximum. Values of 15 mV or less are quite common.

Chapter 13

Lines and interconnection

This chapter is concerned with the interconnection of analogue audio signals, and the solving of problems concerned with analogue interfacing. It is not intended to cover digital interfacing systems here, since this subject is adequately covered elsewhere (see the Further reading list at the end of Chapter 10). The proper interconnection of analogue audio signals, and an understanding of the principles of balanced and unbalanced lines, is vital to the maintenance of high quality in an audio system, and will remain important for many years notwithstanding the growing usage of digital systems.

13.1 Transformers

Mains transformers are widely used throughout the electrical and electronics industries, usually to convert the 240 V AC mains voltage to a rather lower voltage. Audio transformers are widely used in audio equipment for balancing and isolating purposes, and whereas mains transformers are required only to work at 50 Hz, audio transformers must give a satisfactory performance over the complete audio spectrum. Fortunately most audio transformers are only called upon to handle a few volts at negligible power, so they are generally much smaller than their mains counterparts. The principles of transformer operation are outlined in Fact File 13.1.

13.1.1 Transformers and impedances

Consider Figure 13.1(a). The turns ratio is 1:2, so the square of the turns ratio (used to calculate the impedance across the secondary) is 1:4, and therefore the impedance across the secondary will be found to be $10 \times 4 = 40$ k. Another example is shown in Figure 13.1(b). The turns ratio is 1:4. 0.7 volts is applied across the primary and gives 2.8 volts across the secondary. The square of the turns ratio is 1:16, so the impedance across the secondary is 2 k $\times 16 = 32$ k. The transformer also works backwards, as shown in Figure 13.1(c). A 20 k resistor is now placed across the secondary. The square of the turns ratio is 1:16, and therefore the impedance across the primary is 20 k $\div 16 = 1$k25.

Consider now a microphone transformer that is loaded with an impedance on both sides, as shown in Figure 13.2. The transformer presents the 2 k impedance of the mixer to the microphone, and the 200 ohm impedance of the microphone to the

FACT FILE 13.1 The transformer

From the diagrams it can be seen that the transformer consists of a laminated core (i.e.: a number of thin sheets of metal 'laminated' together to form a single thick core) around which is wound a 'primary' winding and a 'secondary' winding. If an alternating current is passed through the primary winding, magnetic flux flows in the core (in a similar fashion to the principle of the tape head; see Fact File 8.1), and thus through the secondary winding. Flux changes in the secondary winding cause a current to be induced in it. The voltage across the secondary winding compared with that across the primary is proportional to the ratio between the number of turns on each coil. For example, if the primary and secondary windings each have the same number of turns, then 1 volt across the primary will also appear as 1 volt across the secondary. If the secondary has twice the number of turns as the primary then twice the voltage will appear across it. The transformer also works in reverse – voltage applied to the secondary will be induced into the primary in proportion to the turns ratio.

The current flowing through the secondary is in *inverse* proportion to the turns ratio, such that equal *power* exists on the primary and secondary sides of the transformer (it is not magic – the increased voltage across the secondary of a step-up transformer is traded off against reduced current!).

It is important to remember that the principle of operation of the transformer depends on AC in the windings inducing an alternating field into the core (i.e.: it is the *change* in direction of magnetic flux which induces a current in the secondary, not simply the presence of constant flux). A DC signal, therefore, is not passed by a transformer.

Impedances are proportional to the *square* of the turns ratio, as discussed in the main text. A transformer will 'reflect' the impedances between which it works. In the case of a 1:1 transformer the impedance across the secondary is equal to the impedance across the primary, but in the case of a 1:2 transformer the impedance seen across the secondary would be four times that across the primary.

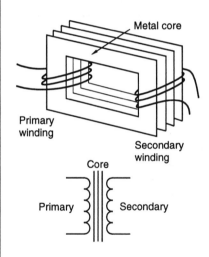

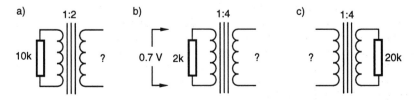

Figure 13.1 Examples of transformer circuits. (a) What is the impedance across the secondary? (b) What are the impedance and voltage across the secondary? (c) What is the impedance across the primary?

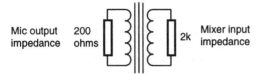

Mic output   200   2k   Mixer input
impedance   ohms   impedance

Figure 13.2 The input impedance of the mixer is seen by the microphone, modified by the turns ratio of the transformer, and vice versa

mixer. With a step-up ratio of 1:4 the square of the turns ratio would be 1:16. The microphone would be presented with an impedance of 2 k ÷ 16 = 125 ohms, whereas the mixer would be presented with an impedance of 200 × 16 = 3200 ohms. In this particular case a 1:4 step-up transformer is unsuitable because microphones like to work into an impedance five times or more their own impedance, so 125 ohms is far too low. Similarly, electronic inputs work best when driven by an impedance considerably lower than their own, so 3200 ohms is far too high.

13.1.2 Limitations of transformers

Earlier it was mentioned that an audio transformer must be able to handle the complete audio range. At very high and very low frequencies this is not easy to achieve, and it is usual to find that distortion rises at low frequencies, and also to a lesser extent at very high frequencies. The frequency response falls away at the frequency extremes, and an average transformer may well be 3 dB down at 20 Hz and 20 kHz compared with mid frequencies. Good (= expensive) transformers have a much better performance than this. All transformers are designed to work within certain limits of voltage and current, and if too high a voltage is applied a rapid increase in distortion results.

The frequency response and distortion performance is affected by the impedances between which the transformer works, and any particular model will be designed to give its optimum performance when used for its intended application. For example, a microphone transformer is designed to handle voltages in the millivolt range up to around 800 mV or so. The primary winding will be terminated with about 200 ohms, and the secondary will be terminated with around 1–2 k (or rather more if a step-up ratio is present). A line level transformer on the other hand must handle voltages up to 8 volts or so, and will probably be driven by a source impedance of below 100 ohms, and will feed an impedance of 10 k or more. Such differing parameters as these require specialised designs. There is no 'universal' transformer.

Transformers are sensitive to electromagnetic fields, and so their siting must be given consideration. Place an audio transformer next to a mains transformer and hum will be induced into it, and thus into the rest of the audio circuit. Most audio transformers are built into metal screening cans which considerably reduce their susceptibility to radio-frequency interference and the like.

13.2 Unbalanced lines

'Unbalanced' in this context does not mean unstable or faulty. The unbalanced audio line is to be found in virtually all domestic audio equipment, much semi-

---

FACT FILE 13.2   Earth loops

It is possible to wire cables such that the screening braid of a line is connected to earth at both ends. In many pieces of audio equipment the earth side of the audio circuit is connected to the mains earth. When two or more pieces of equipment are connected together this creates multiple paths to the mains earth, and low-level mains currents can circulate around the screening braids of the connecting leads if the earths are at even slightly different potentials. This induces 50 Hz mains hum into the inner conductor. A

common remedy for this problem is to disconnect the earth wires in the mains plugs on all the interconnected pieces of equipment except one, the remaining connection providing the earth for all the other pieces of equipment via the audio screening braids. This, though, is potentially dangerous, since if a piece of equipment develops a fault and the mains plug with the earth connection is unplugged, then the rest of the system is now unearthed and the fault could in serious cases place a mains voltage on the metal parts of the equipment. A lot of units are now 'double insulated', so that internal mains wiring cannot place mains voltage on the metal chassis. The mains lead is just two core, live and neutral.

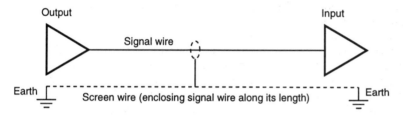

Figure 13.3 Simple unbalanced interconnection

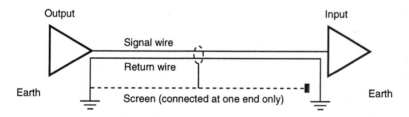

Figure 13.4 Alternative unbalanced interconnection

professional and some professional audio equipment as well. It consists of a 'send' and 'return' path for the audio signal, the return path being an outer screening braid which encloses the send wire and screens it from electromagnetic interference, shown in Figure 13.3. The screening effect considerably reduces interference such as hum, RF, and other induction, without eliminating it entirely. If the unbalanced line is used to carry an audio signal over tens of metres, the cumulative effect of interference may be unacceptable. Earth loops can also be formed (see Fact File 13.2). Unbalanced lines are normally terminated in connectors such as phono plugs, DIN plugs and quarter-inch 'A-gauge' jack plugs.

An improved means of unbalanced interconnection is shown in Figure 13.4. The connecting lead now has *two* wires inside the outer screen. One is used as the signal wire, and instead of the return being provided by the outer screen, it is provided by the second inner wire. The screening braid is connected to earth *at one end only*, and so it merely provides an interference screen without affecting the audio signal.

13.3 Cable effects with unbalanced lines

13.3.1 Cable resistance

'Loop' resistance is the total resistance of both the send and return paths for the signal, and generally, as long as the loop resistance of a cable is a couple of orders of magnitude (i.e.: a factor of 100) lower than the input impedance of the equipment it is feeding, it can be ignored. For example, the output impedance of a tape recorder might be 200 ohms. The input impedance of the amplifier it would be connected to would normally be 10 k or more. The DC resistance of a few metres of connecting cable would only be a fraction of an ohm and so would not need to be considered. But what about 100 metres of microphone cable? The input impedance of a microphone amplifier would normally be at least 1000 ohms. Two orders of magnitude lower than this is 10 ohms. Even 100 metres of mic lead will have a lower resistance than this unless very thin cheap wire is used, and so again the DC resistance of microphone cables can be ignored.

Speaker cables *do* need to be watched, because the input impedance of loud-speakers is of the order of 8 ohms. Wiring manufacturers quote the value of DC resistance per unit length (usually 1 metre) of cable, and a typical cable suitable for speaker use would be of 6 amp rating and about 12 milliohms (0.012 ohms) resistance per metre. Consider a 5 metre length of speaker cable. Its total loop resistance then would be 10 metres multiplied by 0.012 ohms = 0.12 ohms. This is a bit too high to meet the criterion stated above, an 8 ohm speaker requiring a cable of around 0.08 ohm loop resistance. In practice, though, this would probably be perfectly adequate, since there are many other factors which will affect sound quality. Nevertheless, it does illustrate that quite heavy cables are required to feed speakers, otherwise too much power will be wasted in the cable itself before the signal reaches the speaker.

If the same cable as above were used for a 40 m feed to a remote 8 ohm loudspeaker, the loop resistance would be nearly 1 ohm and nearly one-eighth of the amplifier power would be dissipated in heat in the cable. The moral here is to use the shortest length of cable as is practicable, or if long runs are required use the 100 volt line system (see section 13.8).

13.3.2 Cable and transformer inductance

The effect of cable inductance (see section 1.8) becomes more serious at high frequencies, but at audio frequencies it is insignificant even over long runs of cable. Conversely, inductance is extremely important in transformers. The coils on the transformer cores consist of a large number of turns of wire, and the electromagnetic field of each turn works against the fields of the other turns. The metallic core greatly enhances this effect. Therefore, the inductance of each transformer coil is very high and presents a high impedance to an audio signal. For a given frequency, the higher the inductance the higher the impedance in ohms.

13.3.3 Cable capacitance

The closer the conductors in a cable are together, the greater the capacitance (see section 1.8). The surface area of the conductors is also important. Capacitance is the opposite of inductance in that, for a given frequency, the greater the capacitance the *lower* is the impedance in ohms. In a screened cable the screening braid entirely encloses the inner conductor and so the surface area of the braid, as seen by this inner conductor, is quite large. Since large surface area implies high-capacitance, screened cable has a much higher capacitance than ordinary mains wiring, for example. So when an audio signal looks into a connecting cable it sees the capacitance between the conductors and therefore a rather less-than-infinite impedance between them, especially at high frequencies. A small amount of the signal can therefore be conducted to earth via the screen.

In the diagram in Figure 13.5 there are two resistors of equal value. A voltage V_1 is applied across the two. Because the value of the resistors is the same, V_1 is divided exactly in half, and V_2 will be found to be exactly half the value of V_1. If the lower resistor were to be increased in value to 400 ohms then twice the voltage would appear across it than across the upper resistor. The ratio of the resistors equals the ratio of the voltages across them.

Consider a 200 ohm microphone looking into a mic lead, as shown in Figure 13.6(a). C is the capacitance between the screening braid and the inner core of the cable. The equivalent of this circuit is shown in Figure 13.6(b). Manufacturers quote the capacitance of cables in picofarads (pF) per unit length. A typical value for screened cable is 200 pF (0.0002 μF) per metre. A simple formula exists for determining the frequency at which 3 dB of signal is lost for a given capacitance and source resistance:

$$f = 159\ 155/RC$$

where f = frequency in hertz (Hz), R = resistance in ohms, and C = capacitance in microfarads (μF).

To calculate the capacitance which will cause a 3 dB loss at 40 kHz, putting it safely out of the way of the audio band, the formula must be rearranged to give the maximum value of acceptable capacitance:

$$C = 159\ 155/R\,f$$

Thus, if $R = 200$ (mic impedance), $f = 40\ 000$:

$$C = 159\ 155 \div (200 \times 40\ 000) \approx 0.02\ \mu F$$

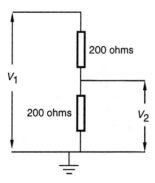

Figure 13.5 The voltage V_2 across the output is half the input voltage (V_1)

a)

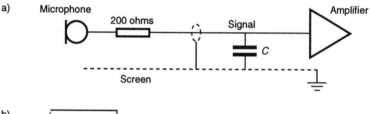

b)

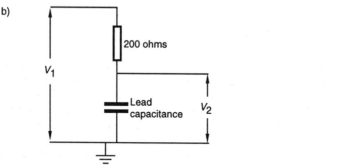

Figure 13.6 (a) A microphone with a 200 ohm output impedance is connected to an amplifier. Lead capacitance conducts high frequencies to ground more than low frequencies, and thus the cable introduces HF roll-off. V_2 is lower at HF than at LF

So a maximum value of 0.02 µF of lead capacitance is acceptable for a mic lead. Typical lead capacitance was quoted as 0.0002 µF per metre, so 100 metres will give 0.02 µF, which is the calculated acceptable value. Therefore one could safely use up to 100 metres of typical mic cable with a standard 200 ohm microphone without incurring significant signal loss at high frequencies.

The principle applies equally to other audio circuits, and one more example will be worked out. A certain tape recorder has an output impedance of 1 k . How long a cable can it safely drive? From the above formula:

$$C = 159\ 155 \div (1000 \times 40\ 000) \approx 0.004\ \mu F$$

In this case, assuming the same cable capacitance, the maximum safe cable length is 0.004 ÷ 0.0002 = 20 metres. In practice, modern audio equipment generally has a low enough source impedance to drive long leads, but it is always wise to check up on this in the manufacturer's specification. Probably of greater concern will be the need to avoid long runs of unbalanced cable due to interference problems.

13.4 Balanced lines

The balanced line is better at rejecting interference than the unbalanced line, and improvements upon the performance of the unbalanced line in this respect can be 80 dB or more for high-quality microphone lines.

As shown in Figure 13.7, the connecting cable consists of a pair of inner conductors enclosed by a screening braid. At each end of the line is a 'balancing' transformer. The output amplifier feeds the primary of the output transformer and its voltage appears across the secondary. The send and return paths for the audio

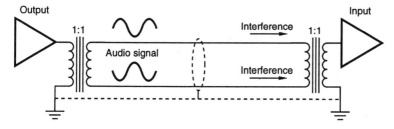

Figure 13.7 A balanced interconnection using transformers

signal are provided by the two inner conductors, and the screen does not form part of the audio circuit. If an interference signal breaks through the screen it is induced equally into both signal lines. At the secondary transformer's primary the induced interference current, flowing in the same direction in both legs of the balanced line, cancels out, thus rejecting the interference signal. Two identical signals, flowing in opposite directions, cancel out where they collide.

Such an interfering signal is called a 'common mode' signal because it is equal and common to both audio lines. The rejection of this in the transformer is termed 'common mode rejection' (CMR). A common mode rejection ratio (CMRR) of at least 80 dB may be feasible. Meanwhile, the legitimate audio signal flows through the primary of the transformer as before, because the signal appears at each end of the coil with equal strength but *opposite* phase. Such a signal is called a 'differential signal', and the balanced input is also termed a 'differential input' because it accepts differential mode signals but rejects common mode signals.

So balanced lines are used for professional audio connections because of their greatly superior rejection of interference, and this is particularly useful when sending just a few precious millivolts from a microphone down many metres of cable to an amplifier.

13.5 Working with balanced lines

In order to avoid earth loops (see Fact File 13.2) with the balanced line, the earth screen is often connected at one end only, as shown in Figure 13.8(a), and still acts as a screen for the balanced audio lines. There is now no earth link between the two pieces of equipment, and so both can be safely earthed at the mains without causing an earth loop. The transformers have 'isolated' the two pieces of equipment from each other. The one potential danger with this is that the connecting lead with its earth disconnected in the plug at one end may later be used as a microphone cable. The lack of earth continuity between microphone and amplifier will cause inadequate screening of the microphone, and will also prevent a phantom power circuit being made (see section 4.9), so such cables and tie-lines should be marked 'earth off' at the plug without the earth connection.

Unfortunately, not all pieces of audio equipment have balanced inputs and outputs, and one may be faced with the problem of interfacing a balanced output with an unbalanced input, and an unbalanced output with a balanced input. A solution is shown in Figure 13.8(b), where the output transformer is connected to the signal and earth of the unbalanced input to give signal continuity. Because the

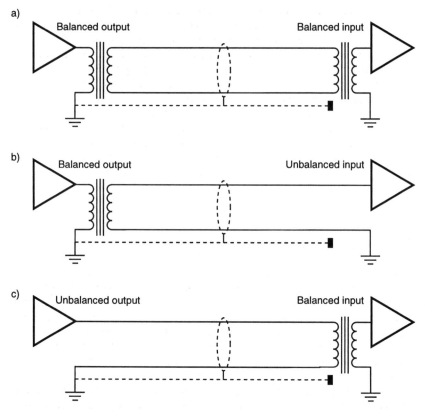

Figure 13.8 (a) Balanced output to balanced input with screen connected to earth only at output. (b) Balanced output to unbalanced input. (c) Unbalanced output to balanced input

FACT FILE

13.3

XLR-3 connectors

The most common balanced connector in professional audio is the XLR-3. This

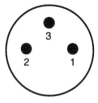

Viewed from end of
male pins

connector has three pins (as shown in the diagram), carrying respectively:

Pin 1   Screen
Pin 2   Signal (Live or 'hot')
Pin 3   Signal (Return or 'cold')

It is easy to remember this configuration, since X–L–R stands for Xternal, Live, Return. Unfortunately, an American convention still hangs on in some equipment which reverses the roles of pins 2 and 3, making pin 2 return and pin 3 live (or 'hot'). The result of this is an apparent absolute phase reversal in signals from devices using this convention when compared with an identical signal leaving a standard device. Modern American equipment mostly uses the European convention, and American manufacturers have now agreed to standardise on this approach.

input is unbalanced, there is no common mode rejection and the line is as susceptible to interference as an ordinary unbalanced line is. But notice that the screen is connected at one end only, so at least one can avoid an earth loop.

Figure 13.8(c) illustrates an unbalanced output feeding a balanced input. The signal and earth from the output feed the primary of the input transformer. Again the screen is not connected at one end so earth loops are avoided. Common mode rejection of interference at the input is again lost, because one side of the transformer primary is connected to earth. A better solution is to use a balancing transformer as close to the unbalanced output as possible, preferably before sending the signal over any length of cable. In the longer term it would be a good idea to fit balancing transformers inside unbalanced equipment with associated three-pin XLR-type sockets (see Fact File 13.3) if space inside the casing will allow. (Wait until the guarantee has elapsed first!)

13.6 Star-quad cable

Two audio lines can never occupy exactly the same physical space, and any interference induced into a balanced line may be slightly stronger in one line than in the other. This imbalance is seen by the transformer as a small differential signal which it will pass on, so a small amount of the unwanted signal will still get through. To help combat this, the two audio lines are twisted together during manufacture so as to present, on average, an equal face to the interference along both lines. A further step has been taken in the form of a cable called 'star-quad'. Here, four audio lines are incorporated inside the screen, as shown in Figure 13.9.

It is connected as follows. The screen is connected as usual. The four inner cores are connected in pairs such that two of the opposite wires (top and bottom in the figure) are connected together and used as one line, and the other two opposite wires are used as the other. All four are twisted together along the length of the cable during manufacture. This configuration ensures that for a given length of cable, both audio lines are exposed to an interference signal as equally as possible so that any interference is induced as equally as possible. So the input transformer sees the interference as a virtually perfect common mode signal, and efficiently rejects it. This may seem like taking things to extremes, but star-quad is in fact quite widely used for microphone cables. When multicore cables are used which contain many separate audio lines in a single thick cable, the balanced system gives good immunity from crosstalk, due to the fact that a signal in a particular wire will be induced equally into the audio pair of the adjacent line, and will therefore be a common mode signal. Star-quad multicores give even lower values of crosstalk.

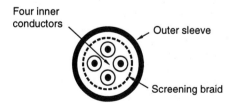

Figure 13.9 Four conductors are used in star-quad cable

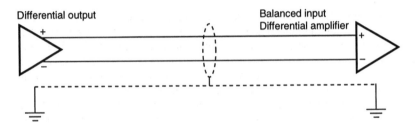

Figure 13.10 An electronically balanced interconnection using differential amplifiers

13.7 Electronic balancing

Much audio equipment uses an electronically balanced arrangement instead of a transformer, and it is schematically represented as in Figure 13.10. The transformers have been replaced by a differential amplifier. The differential amplifier is designed to respond only to differential signals, as is the case with the transformer, and has one positive and one negative input. Electronically balanced and transformer-balanced equipment can of course be freely intermixed. Reasons for dispensing with transformers include lower cost (a good transformer is rather more expensive than the electronic components which replace it), smaller size (transformers take up at least a few cubic centimetres, the alternative electronics rather less), less suscepti-bility to electromagnetic interference, and rather less sensitivity to the impedances between which they work with respect to distortion, frequency response and the like.

Good electronic balancing circuitry is, however, tricky to design, and the use of high-quality transformers in expensive audio equipment may well be a safer bet than electronic balancing of unknown performance. The best electronic balancing is usually capable of equal CMR performance to the transformer. Critics of trans-former balancing cite factors such as the low-frequency distortion performance of a transformer, and its inability to pass extremely low frequencies, whereas critics of electronic balancing cite the better CMR available from a transformer when compared with a differential amplifier, and the fact that only the transformer provides true isolation between devices. Broadcasters often prefer to use trans-former balancing because signals are transferred over very long distances and isolation is required, whereas recording studios often prefer electronic balancing, claiming that the sound quality is better.

13.8 100 volt lines

13.8.1 Principles

In section 13.3.1 it was suggested that the resistance of even quite thick cables was still sufficient to cause signal loss in loudspeaker interconnection unless short runs were employed. But long speaker lines are frequently unavoidable, examples being: back-stage paging and show relay speakers in theatres; wall-mounted speakers in lecture theatres and halls; paging speakers in supermarkets and factories; and open-air 'tannoy' horns at fairgrounds and fêtes. All these require long speaker runs, or alternatively a separate power amplifier sited close to each speaker, each amplifier being driven from the line output of a mixer or microphone amplifier. The latter solution will in most cases be considered an unnecessarily expensive and complicated solution. So the '100 volt line' was developed so that long speaker cable runs could be employed without too much signal loss along them.

The problem in normal speaker connection is that the speaker cable has a resistance comparable with, or even greater than, the speaker's impedance over longer runs. It was shown in section 13.1.2 that a transformer reflects impedance according to the square of the turns ratio. Suppose a transformer with a turns ratio of 5:1 is connected to the input of an 8 ohm speaker, as shown in Figure 13.11. The square of the turns ratio is 25:1 so the impedance across the primary of the transformer is $25 \times 8 = 200$ ohms. Now, the effective impedance of the speaker is much greater than the cable resistance, so most of the voltage will now reach the primary of the transformer and thence to the secondary and the speaker itself. But the transformer also transforms voltage, and the voltage across the secondary will only be a fifth of that across the primary. To produce 20 volts across the speaker then, one must apply 100 volts to the primary.

In a 100 V line system, as shown in Figure 13.12, a 50 W power amplifier drives a transformer with a step-up ratio of 1:5. Because the output impedance of a power amplifier is designed to be extremely low, the impedance across the secondary is also low enough to be ignored. The 20 volts, 2.5 amp output of the 50 watt amplifier is stepped up to 100 volts. The current is correspondingly stepped down to 0.5 amps (see Fact File 13.1), so that the total power remains the same. Along the speaker line there is a much higher voltage than before, and a much lower current. The voltage drop across the cable resistance is proportional to the current flowing through it, so this reduction in current means that there is a much smaller voltage drop due to the line. At the speaker end, a transformer restores the voltage to 20 volts and the current to 2.5 amps, and so the original 50 watts is delivered to the speaker.

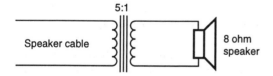

Figure 13.11 Transformer coupling to a loudspeaker as used in 100 volt line systems

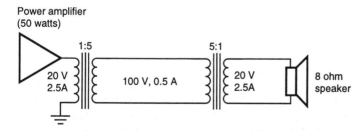

Figure 13.12 Voltage/current relationships in an example of 100 volt line operation

A 50 watt amplifier has been used in the discussion. Any wattage of amplifier can be used, the transformer being chosen so that the step-up ratio gives the standard 100 volts output when the amplifier is delivering its maximum power output. For example, an amplifier rated at 100 watts into 8 ohms produces about 28 V. The step-up ratio of the line transformer would then have to be 28:100, or 1:3.6, to give the standard 100 volt output when the amplifier is being fully driven.

Returning to the loudspeaker end of the circuit. What if the speaker is only rated at 10 watts? The full 100 watts of the above amplifier would burn it out very quickly, and so a step-down ratio of the speaker transformer is chosen so that it receives only 10 watts. As 10 watts across 8 ohms is equivalent to around 9 volts, the required speaker transformer would have a step-down ratio of 100:9, or approximately 11:1.

13.8.2 Working with 100 V lines

Speaker line transformers usually have a range of terminals labelled such that the primary side has a choice of wattage settings (e.g.: 30 W, 20 W, 10 W, 2 W) and the secondary gives a choice of speaker impedance, usually 15 ohms, 8 ohms and 4 ohms. This choice means that a number of speaker systems can be connected along the line (a transformer being required for each speaker enclosure), the wattage setting being appropriate to the speaker's coverage. For example, a paging system in the back-stage area of a theatre could be required to feed paging to six dressing rooms, a large toilet, and a fairly noisy green room. The dressing rooms are small and quiet, and so small speakers rated at 10 watts are employed with line transformers wired for 2 watts output. The large toilet requires greater power from the speaker, so one could use a 10 W speaker with a line transformer wired for 10 W output. The noisy green room could have a rather larger 20 W speaker with a line transformer wired for 20 W output. In this way, each speaker is supplied only with the wattage required to make it loud enough to be clearly heard in that particular room. A 20 watt speaker in a small dressing room would be far too loud, and a 2 watt speaker in a larger, noisier room would be inadequate.

As a string of loudspeakers is added to the system, one must be careful that the *total* wattage of the speakers does not exceed the output wattage of the power amplifier, or the latter will be overloaded. In the example, the six dressing rooms were allocated 2 W each, total 12 W. The toilet was allocated 10 W, the green room 20 W. The total is therefore 42 W, and a 50 W amplifier and line transformer would be adequate. In practice, a 100 W amplifier would be chosen to allow for both a good safety margin and plenty of power in hand if extra speakers need to be connected at a later date.

From the foregoing, it might well be asked why the 100 volt line system is not automatically used in all speaker systems. One reason is that 100 volts is high enough to give an electric shock and so is potentially dangerous in the domestic environment and other places where members of the public could interfere with an inadequately installed system. Secondly, the ultimate sound quality is compromised by the presence of transformers in the speaker lines – they are harder to design than the microphone and line level transformers already discussed, because they have to handle high voltages as well as several amps – and whereas they still give a perfectly adequate and extremely useful performance in paging and background music applications, they are not therefore used in high-quality PA systems or hi-fi and studio monitor speakers.

13.9 600 ohms

One frequently sees 600 ohms mentioned in the specifications of mixers, micro-phone amplifiers, and other equipment with line level outputs. Why is 600 ohms so special? The short answer is: it is not.

13.9.1 Principles

As has been shown, the output impedances of audio devices are low, typically 200 ohms for microphones and the same value or rather less for line level outputs. The input impedances of devices are much higher, at least 1–2 k for microphone inputs and 10 k or more for line inputs. This is to ensure that virtually the whole of the output voltage of a piece of equipment appears across the input it is feeding. Also, lower input impedances draw a higher current for a given voltage, the obvious example of this being an 8 ohm loudspeaker which draws several amps from a power amplifier. So a high input impedance means that only very small currents need to be supplied by the outputs, and one can look upon microphone and line level signals as being purely voltage without considering current.

This works fine, unless you are a telephone company that needs to send its signals along miles of cable. Over these distances, a hitherto unmentioned parameter comes into play which would cause signal loss if not dealt with, namely the wavelength of the signal in the cable. Now the audio signal is transmitted along a line at close to the speed of light (186 000 miles per second or 3×10^8 metres per second). The shortest signal wavelength will occur at the upper limit of the audio spectrum, and will be around 9.3 miles (14.9 km) at 20 kHz.

When a cable is long enough to accommodate a whole wavelength or more, the signal can be reflected back along the line and cause some cancellation of the primary signal. Even when the cable run is somewhat less than a wavelength some reflection and cancellation still occurs. To stop this from happening the cable must be terminated correctly, to form a so-called 'transmission line', and input and output impedances are chosen to be equal. The value of 600 ohms was chosen many decades ago as the standard value for telecommunications, and therefore the '600 ohm balanced line' is used to send audio signals along lines which need to be longer than a mile or so. It was chosen because comparatively little current needs to be supplied to drive this impedance, but on the other hand it is not high enough to allow much interference, as it is much easier for interference to affect a high-impedance circuit than a low one. Thus, professional equipment began to appear which boasted '600 ohms' to make it compatible with these lines. Unfortunately, many people did not bother to find out, or never understood, why 600 ohms was sometimes needed, and assumed that this was a professional audio standard *per se*, rather than a telecommunications standard. It was used widely in broadcasting, which has parallels with telecommunications, and may still be found in many cases involving older equipment.

The 600 ohm standard also gave rise to the standard reference level unit of 0 dBm, which corresponds to 1 mW of power dissipated in a resistance of 600 ohms. The corresponding voltage across the 600 ohm resistance at 0 dBm is 0.775 volts, and this leads some people still to confuse dBm with dBu, but 0 dBu refers simply to 0.775 volts with no reference to power or impedance: dBu is much more appropriate in modern equipment, since, as indicated above, the current flowing in most interfaces is negligible and impedances vary; dBm should only correctly be used in 600 ohm systems, unless an alternative impedance is quoted (e.g. dBm (75 ohms) is sometimes used in video equipment where 75 ohm termination is common).

13.9.2 Problems with 600 ohm equipment

A 600 ohm output impedance is too high for normal applications. With 600 ohms, 200 pF per metre cable, and an acceptable 3 dB loss at 40 kHz, the maximum cable length would be only around 33 m, which is inadequate for many installations. (This HF loss does not occur with the properly terminated 600 ohm system because the cable assumes the properties of a transmission line.) Furthermore, consider a situation where a mixer with a 600 ohm output impedance is required to drive five power amplifiers, each with an input impedance of 10 k. Five lots of 10 k in parallel produce an effective impedance of 2 k, as shown in Figure 13.13(a). Effectively then 600 ohms is driving 2 k, as shown in Figure 13.13(b). If V_1 is 1 volt then V_2 (from Ohm's law) is only 0.77 volts. Almost a quarter of the audio signal has been lost, and only a maximum of 33 metres of cable can be driven anyway.

Despite this, there are still one or two manufacturers who use 600 ohm impedances in order to appear 'professional'. It actually renders their equipment *less* suitable for professional use, as has been shown. One specification that is frequently encountered for a line output is something like: 'capable of delivering +20 dBu into 600 ohms'. Here +20 dBu is 7.75 volts, and 600 ohms is quite a low impedance, thus

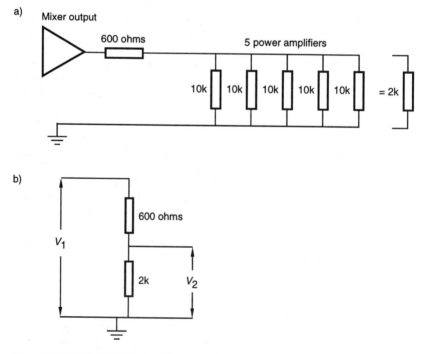

Figure 13.13 (a) Considerable signal loss can result if a 600 ohm output is connected to a number of 10 k inputs in parallel. (b) Electrical equivalent

drawing more current from the source for a given voltage than, say, 10 kΩ does. The above specification is therefore useful, because it tells the user that the equipment can deliver 7.75 volts even into 600 ohms, and can therefore safely drive, say, a stack of power amplifiers, and/or a number of tape recorders, etc., without being overloaded. A domestic cassette recorder, for instance, despite having a low output impedance, could not be expected to do this. Its specification may well state '2 volts into 10 kΩ or greater'. This is fine for domestic applications, but one should do a quick calculation or two before asking it to drive all the power amplifiers in the building at once.

13.10 DI boxes

13.10.1 Overview

A frequent requirement is the need to interface equipment that has basically non-standard unbalanced outputs with the standard balanced inputs of mixers, either at line level or microphone level. An electric guitar, for example, has an unbalanced output of fairly high impedance – around 10 kΩ or so. The standard output socket is the 'mono' quarter-inch jack, and output voltage levels of around a volt or so (with the guitar's volume controls set to maximum) can be expected. Plugging the guitar directly into the mic or line level input of a mixer is unsatisfactory for several reasons: the input impedance of the mixer will be too low for the guitar, which likes to drive impedances of 500 kΩ or more; the guitar output is unbalanced so the interference-rejecting properties of the mixer's balanced input will be lost; the high output impedance of the guitar renders it incapable of driving long studio tie-lines; and the guitarist will frequently wish to plug the instrument into an amplifier as well as the mixer, and simply using the same guitar output to feed both via a splitter lead electrically connects the amplifier to the studio equipment which causes severe interference and low-frequency hum problems. Similar problems are encountered with other instruments such as synthesisers, electric pianos, and pickup systems for acoustic instruments.

To connect such an instrument with the mixer, a special interfacing unit known as a DI box (DI = direct injection) is therefore employed. This unit will convert the instrument's output to a low-impedance balanced signal, and also reduce its output level to the millivolt range suitable for feeding a microphone input. In addition to the input jack socket, it will also have an output jack socket so that the instrument's unprocessed signal can be passed to an amplifier as well. The low-impedance balanced output appears on a standard three-pin XLR panel-mounted plug which can now be looked upon as the output of a microphone. An earth-lift switch is also provided which isolates the earth of the input and output jack sockets from the XLR output, to trap earth loop problems.

13.10.2 Passive DI boxes

The simplest DI boxes contain just a transformer, and are termed 'passive' because they require no power supply. Figure 13.14 shows the circuit. The transformer in this case has a 20:1 step-down ratio, converting the fairly high output of the instrument to a lower output suitable for feeding microphone lines. Impedance is converted according to the square of the turns ratio (400:1), so a typical guitar output

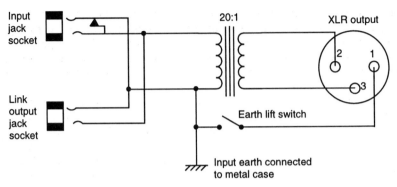

Figure 13.14 A simple passive direct-injection box

impedance of 15 kΩ will be stepped down to about 40 ohms which is comfortably low enough to drive long microphone lines. But the guitar itself likes to look into a high impedance. If the mixer's microphone input impedance is 2 kΩ, the transformer will step this up to 800 kΩ) which is adequately high for the guitar. The 'link output jack socket' is used to connect the guitar to an amplifier if required. Note the configuration of the input jack socket: the make-and-break contact normally short-circuits the input which gives the box immunity from interference, and also very low noise when an instrument is not plugged in. Insertion of the jack plug opens this contact, removing the short-circuit. The transformer isolates the instrument from phantom power on the microphone line.

This type of DI box design has the advantages of being cheap, simple, and requiring no power source – there are no internal batteries to forget to change. On the other hand, its input and output impedances are entirely dependent on the reflected impedances each side of the transformer. Unusually low microphone input impedances will give insufficiently high impedances for many guitars. Also, instruments with passive volume controls can exhibit output impedances as high as 200 kΩ with the control turned down a few numbers from maximum, and this will cause too high an impedance at the output of the DI box for driving long lines. The fixed turns ratio of the transformer is not equally suited to the wide variety of instruments the DI box will encounter, although several units have additional switches which alter the transformer tapping giving different degrees of attenuation.

13.10.3 Active DI boxes

The active DI box replaces the transformer with an electronic circuit which presents a constant very high impedance to the instrument and provides a constant low-impedance output. Additionally, the presence of electronics provides scope for including other features such as several switched attenuation values (say –20 dB, –40 dB, –60 dB), high and low filters and the like. The box is powered either by internal batteries, or preferably by the phantom power on the microphone line. If batteries are used, the box should include an indication of battery status; a 'test' switch is often included which lights an LED when the battery is good. Alternatively, an LED comes on as a warning when the voltage of the battery drops below a certain level. The make-and-break contacts of the input jack socket are often configured so that insertion of the jack plug automatically switches the unit on. One should be mindful of this because if the jack plug is left plugged into the unit overnight, for instance, this will waste battery power. Usually the current consumption of the DI box is just a few milliamps, so the battery will last for perhaps a hundred hours. Some guitar and keyboard amplifiers offer a separate balanced output on an XLR socket labelled 'DI' or 'studio' which is intended to replace the DI box, and it is often convenient to use this instead.

DI boxes are generally small and light, and they spend much of their time on the floor being kicked around and trodden on by musicians and sound engineers. Therefore, rugged metal boxes should be used (not plastic) and any switches, LEDs, etc. should be mounted such that they are recessed or shrouded for protection. Switches should not be easily moved by trailing guitar leads and feet. The DI box can also be used for interfacing domestic hi-fi equipment such as cassette recorders and radio tuners with balanced microphone inputs.

13.11 Splitter boxes

The recording or broadcasting of live events calls for the outputs of microphones and instruments to be fed to at least two destinations, namely the PA mixer and the mixer in the mobile recording or outside broadcast van. The PA engineer can then balance the sound for the live audience, and the recording/broadcast balancer can independently control the mix for these differing requirements. It is possible of course to use two completely separate sets of microphones for this, but when one considers that there may be as many as ten microphones on a drum kit alone, and a vocalist would find the handling of two mics strapped together with two trailing leads rather unacceptable, a single set of microphones plugged into splitter boxes is the obvious way to do it. Ten or fifteen years ago recording studios and broadcasting companies would have frowned on this because the quality of some of the microphones then being used for PA was insufficient for their needs; but today PA mics tend to be every bit as good as those found in studios, and indeed many of the same models are common to both environments.

A microphone cannot be plugged into two microphone inputs of two separate mixers directly because on the one hand this will electrically connect one mixer with

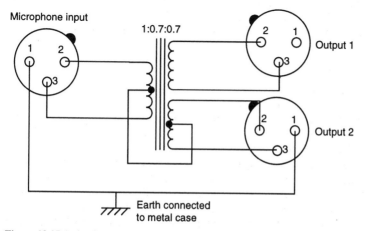

Figure 13.15 A simple passive splitter box

the other causing ground loop and interference problems, not to mention the fact that one phantom power circuit will be driving directly into the other phantom power circuit; and on the other hand the impedance seen by the microphone will now be the parallel result of the two mixers resulting in impedances of as low as 500 ohms which is too low for many microphones. A splitter box is therefore used which isolates the two mixers from each other and maintains a suitable impedance for the microphone. A splitter box will often contain a transformer with one primary winding for the microphone and two separate secondary windings giving the two outputs, as shown in Figure 13.15.

The diagram requires a bit of explanation. Firstly, phantom power must be conveyed to the microphone. In this case, output 2 provides it via the centre tap of its winding which conveys the power to the centre tap of the primary. The earth screen, on pin 1 of the input and output XLR sockets, is connected between the input and output 2 only, to provide screening of the microphone and its lead and also the phantom power return path. Note that pin 1 of output 1 is left unconnected so that earth loops cannot be created between the two outputs.

The turns ratio of the transformer must be considered next. The 1:0.7:0.7 indicates that each secondary coil has only 0.7 times the windings of the primary, and therefore the output voltage of the secondaries will each be 0.7 times the microphone output, which is about 3 dB down. There is therefore a 3 dB insertion loss in the splitter transformer. The reason for this is that the impedance as seen by the microphone must not be allowed to go too low. If there were the same number of turns on each coil, the microphone would be driving the impedance across each output directly in parallel. Two mixers with input impedances of 1 kΩ would therefore together load the microphone with 500 ohms, which is too low. But the illustrated turns ratio means that each 1 kΩ impedance is stepped up by a factor of $1:(0.7)^2 \approx 1:0.5$, so each mixer input appears as 2 kΩ to the microphone, giving a resultant parallel impedance of 1 kΩ, equal to the value of each mixer on its own which is fine. The 3 dB loss of signal is accompanied by an effective halving of the microphone impedance as seen by each mixer, again due to the transformer's impedance conversion according to the square of the turns ratio, so there need not be a signal-to-noise ratio penalty.

Because of the simple nature of the splitter box, a high-quality transformer and a metal case with the necessary input and output sockets are all that is really needed. Active electronic units are also available which eliminate the insertion loss and can even provide extra gain if required. The advantages of an active splitter box over its transformer counterpart are, however, of far less importance than, say, the advantages that an active DI box has over its passive counterpart.

13.12 Jackfields (patchbays)

13.12.1 Overview

A jackfield (or patchbay) provides a versatile and comprehensive means of interconnecting equipment and tie-lines in a non-permanent manner such that various source and destination configurations can quickly and easily be set up to cater for any requirements that may arise.

For example, a large mixing console may have microphone inputs, line inputs, main outputs, group outputs, auxiliary outputs, and insert send and returns for all input channels and all outputs. A jackfield, which usually consists of banks of 19 inch (48 cm) wide rack-mounting modules filled with rows of quarter-inch 'GPO'- (Telecom) type balanced jack sockets, is used as the termination point for all of the above facilities so that any individual input or output can be separately accessed. There are usually 24 jack sockets to a row, but sometimes 20 or 28 are encountered. Multicore cables connect the mixer to the jackfield, multipin connectors normally being employed at the mixer end. At the jackfield end multipin connectors can again be used, but as often as not the multicores will be hard wired to the jack socket terminals themselves. The layout of a mixer jackfield was discussed in section 6.4.9.

In addition to the mixer's jackfield there will be other jackfields either elsewhere in the rack or in adjacent racks which provide connection points for the other equipment and tie-lines. In a recording studio control room there will be such things as multitrack inputs and outputs, mic and line tie-lines linking control room to studio, outboard processor inputs and outputs, and tie-lines to the other control rooms and studios within a studio complex. A broadcasting studio will have similar arrangements, and there may also be tie-lines linking the studio with nearby concert halls or transmitter distribution facilities. A theatre jackfield will in addition carry tie-lines leading to various destinations around the auditorium, back stage, in the wings, in the orchestra pit, and almost anywhere else. There is no such thing as too many tie-lines in a theatre.

13.12.2 Patch cords

Patch cords (screened leads of around 1 metre in length terminated at each end with a 'B'-gauge jack plug (small tip) giving tip/ring/sleeve connections) are used to link two appropriate sockets. The tip is live (corresponding to pin 2 on an XLR), the ring is return (pin 3 on an XLR), and the sleeve is earth (pin 1 on an XLR), providing balanced interconnection. The wire or 'cord' of a normal patch cord is coloured red. Yellow cords should indicate that the patch cord reverses the phase of the signal (i.e.: tip at one end is connected to ring of the other end) but this convention is not followed rigidly, leading to potential confusion. Green cords indicate that the earth is left unconnected at one end, and such cords are employed if an earth loop occurs when two separately powered pieces of equipment are connected together via the jackfield.

A large-tipped 'A'-gauge jack plug, found on the end of most stereo headphone leads and often called a stereo jack (although it was originally used as a single-line balanced jack before stereo came into vogue), may sometimes be inserted into a GPO-type socket. This works, but the spring-loaded tip connector is positioned so as to make secure contact with the small 'B'-type tip and the large 'A' tip can sometimes bend the contact. When the correct type of jack plug is later inserted there will be intermittent or no contact between the tip and the socket's tip connector. The insertion of large tipped jack plugs should therefore be discouraged.

13.12.3 Normalling

Normally, jackfield insertion points will be unused. Therefore the insertion send socket must be connected to the insertion return socket so that signal continuity is achieved. When an outboard unit is to be patched in, the insertion send socket is used to feed the unit's input. The unit's output is fed to the insertion return socket on the jackfield. This means that the send signal must be disconnected from the return

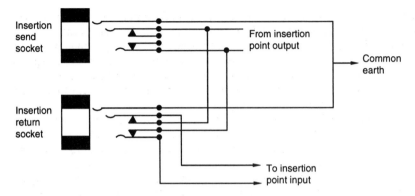

Figure 13.16 Normalling at jackfield insertion points

socket and replaced by the return from the processor. To effect this requirement, extra make-and-break 'normalling' contacts on the jack socket are employed. Figure 13.16 shows how this is done. The signal is taken from the top jack socket to the bottom jack socket via the black triangle make-and-break contactors on the bottom socket. There is signal continuity. If a jack plug is now inserted into the bottom socket the contacts will be moved away from the make-and-break triangles, disconnecting the upper socket's signal from it. Signal from that jack plug now feeds the return socket. The make-and-break contacts on the upper jack socket are left unused, and so insertion of a jack plug into this socket alone has no effect on signal continuity. The send socket therefore simply provides an output signal to feed the processor.

Sometimes these make-and-break contacts are wired so that the signal is also interrupted if a jack plug is inserted into the send socket alone. This can be useful if, for instance, an alternative destination is required for, say, a group output. Insertion of a jack plug into the send socket will automatically mute the group's output and allow its signal to be patched in elsewhere, without disturbing the original patching arrangement. Such a wiring scheme is, however, rather less often encountered.

In addition to providing insert points, normalling can also be used to connect the group outputs of a mixer to the inputs of a multitrack recorder or a set of power amplifiers. The recorder or amplifier inputs will have associated sockets on the jackfield, and the mixer's group output sockets will be normalled to these as described. If need be, the input sockets can be overplugged in order to drive the inputs with alternative signals, automatically disconnecting the mixer's outputs in the process.

13.12.4 Other jackfield facilities

Other useful facilities in a jackfield include multiple rows of interconnected jacks, or 'mults'. These consist of, say, six or eight or however many adjacent sockets which are wired together in parallel so that a mixer output can be patched into one of the sockets, the signal now appearing on all of the sockets in the chain which can then be used to feed a number of power amplifier inputs in parallel, or several tape machines. The disadvantage of this poor man's distribution amplifier is that if there is a short-circuit or an interference signal on any of the inputs it is feeding this will affect all of the sockets on the mult. There is no isolation between them.

Outboard equipment will generally be equipped with XLR-type input and output connectors (or sometimes 'mono' jack sockets). It is useful therefore to have a panel of both male and female XLR sockets near to where such outboard gear is usually placed, these being wired to a row of sockets on the jackfield to facilitate connection between the mixer or tie-lines and these units.

Special additional make-and-break contacts can be included in the jack socket which are operated in the usual manner by jack plug insertion but have no contact with any part of the audio signal. These can be used to activate warning indicators to tell operators that a certain piece of equipment is in use, for example.

Since most if not all of the interconnections in a rig pass through the jackfield it is essential that the contacts are of good quality, giving reliable service. Palladium metal plating is employed which is tough, offering good resistance to wear and oxidation. This should always be looked for when jackfield is being ordered. Gold or silver plating is not used because it would quickly wear away in the face of professional use. The latter also tarnishes rather easily.

There is a miniature version known as the Bantam jack. This type is frequently employed in the control surface areas of mixers to give convenient access to the patching. Very high-density jackfields can be assembled, which has implications for the wiring arrangements on the back. Several earlier examples of Bantam-type jackfields were unreliable and unsuited to professional use. Later examples are rather better, and of course palladium contacts should always be specified.

Electronically controlled 'jackfields' dispense with patch cords altogether. Such systems consist of a digitally controlled 'stage box' which carries a number of input and output sockets into which the mixer inputs and outputs and any other tie-lines, processor and tape machine inputs and outputs are plugged. The unit is controlled by a keypad, and a VDU displays the state of the patch. Information can also be entered identifying each input and output by name according to the particular plug-up arrangement of the stage box. Any output can be routed to any input, and an output can be switched to drive any number of inputs as required. Various patches can be stored in the system's memory and recalled; rapid repatching is therefore possible, and this facility can be used in conjunction with timecode to effect automatic repatches at certain chosen points on a tape during mixdown for instance. MIDI control is also a possibility.

13.13 Distribution amplifiers

A distribution amplifier is an amplifier used for distributing one input to a number of outputs, with independent level control and isolation for each output. It is used widely in broadcast centres and other locations where signals must be split off and routed to a number of independent locations. This approach is preferable to simple parallel connections, since each output is unaffected by connections made to the others, preventing one from loading down or interfering with the others.

Chapter 14

Outboard equipment

Outboard equipment includes such things as parametric equalisers, graphic equal-isers, delay lines, echo devices, dynamics units, multi-effects processors, gates, and room simulators. They provide extra facilities which will not normally be present in the mixing console itself, although several consoles incorporate such things as parametric equalisation and dynamics control.

14.1 The graphic equaliser

The graphic equaliser, pictured in Figure 14.1, consists of a row of faders (or sometimes rotary controls), each of which can cut and boost a relatively narrow band of frequencies. Simple four- or five-band devices exist which are aimed at the electronic music market, these being multiband tone controls. They perform the useful function of expanding existing simple tone controls on guitars and amplifiers, and several amplifiers actually incorporate them.

The professional rack-mounting graphic equaliser will have at least ten fre-quency bands, spaced at octave or one-third-octave intervals. The ISO (Interna-tional Standards Organisation) centre frequencies for octave bands are 31 Hz, 63 Hz, 125 Hz, 250 Hz, 500 Hz, 1 kHz, 2 kHz, 4 kHz, and 16 kHz. Each fader can cut or boost its band by typically 12 dB or more. Figure 14.2 shows two possible types of filter action. The 1 kHz fader is chosen, and three levels of cut and boost are illustrated. Maximum cut and boost of both types produces very similar Q (see Fact File 6.6). A high Q result is obtained by both types when maximum cut or boost is applied. The action of the first type is rather gentler when less cut or boost

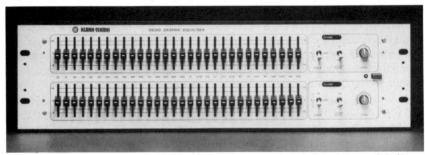

Figure 14.1 A typical two-channel graphic equaliser. (Courtesy of Klark-Teknik Research Ltd.)

a)

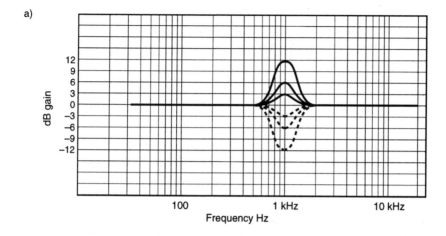

b)

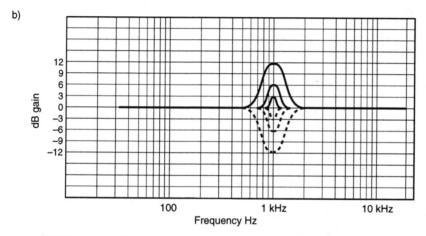

Figure 14.2 Two types of filter action shown with various degrees of boost and cut. (a) Typical graphic equaliser with Q dependent upon degree of boost/cut. (b) Constant Q filter action

is applied, and the Q varies according to the degree of deviation from the fader's central position.

Many graphic equalisers conform to this type of action, and it has the disadvantage that a relatively broad band of frequencies is affected when moderate degrees of boost or cut are applied. The second type maintains a tight control of frequency bandwidth throughout the cut and boost range, and such filters are termed constant Q, the Q remaining virtually the same throughout the fader's travel. This is particularly important in the closely spaced one-third-octave graphic equaliser which has 30 separate bands, so that adjacent bands do not interact with each other too much. The ISO centre frequencies for 30 bands are 25 Hz, 31 Hz, 40 Hz, 50 Hz, 63 Hz, 80 Hz, 100 Hz, 125 Hz, 180 Hz, 200 Hz, 250 Hz, 315 Hz, 400 Hz, 500 Hz, 630 Hz, 800 Hz, 1 kHz, 1k25 Hz, 1k8 Hz, 2 kHz, 2k5 Hz, 3k15 Hz, 4 kHz, 5 kHz, 6k3 Hz, 8 kHz, 10 kHz, 12k5 Hz, 16 kHz, and 20 kHz. The value of using standard centre frequencies is that complementary equipment such as spectrum analysers which will often be used in conjunction with graphic equalisers have their scales centered on the same frequencies.

Even with tight constant Q filters, the conventional analogue graphic equaliser still suffers from adjacent filter interaction. If, say, 12 dB of boost is applied to one frequency and 12 dB of cut applied to the next, the result will be more like a 6 dB boost and cut, the response merging in between to produce an ill-defined Q value. Such extreme settings are, however, unlikely. The digital graphic equaliser applies cut and boost in the digital domain, and extreme settings of adjacent bands can be successfully accomplished without interaction if required.

Some graphic equalisers are single channel, some are stereo. All will have an overall level control, a bypass switch, and many also sport separate steep-cut LF filters. A useful facility is an overload indicator – usually an LED which flashes just before the signal is clipped – which indicates signal clipping anywhere along the circuit path within the unit. Large degrees of boost can sometimes provoke this. Some feature frequency cut only, these being useful as notch filters for getting rid of feedback frequencies in PA/microphone combinations. Some can be switched between cut/boost, or cut only. It is quite possible that the graphic equaliser will be asked to drive very long lines, if it is placed between mixer outputs and power amplifiers for example, and so it must be capable of doing this. The '+20 dBu into 600 ohms' specification should be looked for as is the case with mixers. It will be more usual though to patch the graphic equaliser into mixer output inserts, so that the mixer's output level meters display the effect on level the graphic equaliser is having. Signal-to-noise ratio should be at least 100 dB.

The graphic equaliser can be used purely as a creative tool, providing tone control to taste. It will frequently be used to provide overall frequency balance correction for PA rigs. It has formerly been used to equalise control room speakers, but poor results are frequently obtained due to the fact that a spectrum analyser's microphone samples the complete room frequency response whereas the perceived frequency balance is a complex combination of direct and reflected sound arriving at different times. The graphic equaliser can also change the phase response of signals, and there has been a trend away from their use in the control room for monitor EQ, adjustments being made to the control room acoustics instead.

The parametric equaliser was fully described in section 6.4.4.

14.2 The compressor/limiter

The compressor/limiter (see Fact File 14.1) is used in applications such as dynamics control and as a guard against signal clipping. Such a device is pictured in Figure 14.3. The three main variable parameters are attack, release and threshold. The

Figure 14.3 A typical compressor/limiter. (Courtesy of Drawmer Distribution Ltd.)

FACT FILE

14.1

Compression and limiting

A compressor is a device whose output level can be made to change at a different rate to input level. For example, a compressor with a ratio of 2:1 will give an output level that changes by only half as much as the input level above a certain threshold (see diagram). For example, if the input level were to change by 6 dB the output level would change by only 3 dB. Other compression ratios are available such as 3:1, 5:1, etc. At the higher ratios, the output level changes only a very small amount with changes in input level, which makes the device useful for reducing the dynamic range of a signal. The threshold of a compressor determines the

signal level above which action occurs.

A limiter is a compressor with a very high compression ratio. A limiter is used to ensure that signal level does not rise above a given threshold. A 'soft' limiter has an action which comes in only gently above the threshold, rather than acting as a brick wall, whereas a 'hard' limiter has the effect almost of clipping anything which exceeds the threshold.

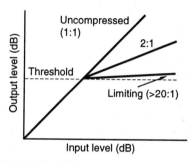

attack time, in microseconds (μs) and milliseconds (ms), is the time taken for a limiter to react to a signal. A very fast attack time of 10 μs can be used to avoid signal clipping, any high-level transients being rapidly brought under control. A fast release time will rapidly restore the gain so that only very short-duration peaks will be truncated. A ducking effect can be produced by using rapid attack plus a release of around 2–300 ms. A threshold level is chosen which causes the limiting to come in at a moderate signal level so that peaks are pushed down before the gain is quickly reinstated. Such a ducking effect is ugly on speech, but is useful for overhead cymbal mics for example.

A slow release time of several seconds, coupled with a moderate threshold, will compress the signal dynamics into a narrower window, allowing a higher mean signal level to be produced. Such a technique is often used in vocals to obtain consistent vocal level from a singer. AM radio is compressed in this way so as to squeeze wide dynamic range material into this narrow dynamic range medium. It is also used on FM radio to a lesser extent, although very bad examples of its application are frequently heard on pop stations. An oppressive, raspy sound is the result, and in the pauses in between items or speech one hears the system gain creeping back up, causing pumping noises. Background noise rapidly ducks back down when the presenter again speaks. The effect is like listening to a tape with noise reduction encoding, the reciprocal decoding being absent.

Many units offer separate limiting and compressing sections, their attack, release and threshold controls in each section having values appropriate to the two applications. Some also include gates (see section 9.5.2) with variable threshold and ratio, rather like an upside-down limiter, such that *below* a certain level the level drops faster than would be expected. 'Gain make-up' is often available to compensate for the overall level-reducing effect of compression. Meters may indicate the amount of level reduction occurring.

14.3 Echo and reverb devices

14.3.1 Echo chamber

Echo, or more properly reverb, used to be produced literally by setting aside a suitable reflective room in which was placed a loudspeaker and two microphones, the mics being placed a fair distance apart from each other, and from the speaker (see

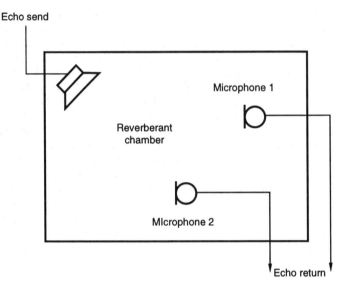

Figure 14.4 In a reverberation chamber the send is fed to a loudspeaker which in turn is used to excite the room. The returns are fed from two microphones placed in the reverberant field

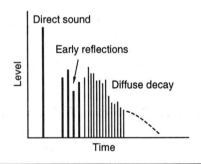

FACT FILE
14.2
Simulating reflections

Pre-delay in a reverb device is a means of delaying the first reflection to simulate the effect of a large room with distant surfaces. Early reflections may then be programmed to simulate the first few reflections from the surfaces as the reverberant field builds up, followed by the general decay of reverberant energy in the room as random reflections lose their energy (see diagram).

Pre-delay and early reflections have an important effect on one's perception of the size of a room, and it is these first few milliseconds which provide the brain with one of its main clues as to room size. Rever-

beration time (RT) alone is not a good guide to room size, since the RT is affected both by room volume and absorption (see Fact Files 1.5 and 1.6); thus the same RT could be obtained from a certain large room and a smaller, more reflective room. Early reflections, though, are dictated only by the distance of the surfaces.

Figure 14.4). Signal was sent to the speaker at a fairly high level, which excited reflections in the room. The two microphones provided the 'stereo' echo return signals. The technique was undoubtedly successful, quality varying according to room size, shape and acoustic treatment. Pre-delay (see Fact File 14.2) could be used to simulate reflections from more distant surfaces.

14.3.2 Echo plate

The echo plate consists of a large, thin metal plate, several square metres in area. It is enclosed in an acoustically isolated housing. A driving transducer is attached close to one edge, and two or more receiving transducers are positioned at various other locations on the plate (see Figure 14.5). Signal is applied to the driving transducer which acoustically excites the plate. Its resonant properties ensure that vibrations persist for a usefully long period of time. The receiving transducers pick up the signal to provide the echo signal. The several receiving transducers can give various types of echo according to their positions, and more than one driving transducer can also be used to excite different resonant modes. A pre-delay signal (tape machine or solid-state delay) may be used to simulate a larger room. The plate is capable at its best of quite pleasing effects, and the twangy 'plate-echo' character of some settings is deliberately simulated in digital reverb devices in some of the programs because it suits some instruments well. The characteristics of plate reverb can be altered using a graphic equaliser if necessary. The plate must be well isolated acoustically because it is sensitive to both air-borne and structure-borne vibrations. If it is mounted in a noisy area the echo return may contain echoey versions of the sound in the plate's surroundings! A separate floor from the studio is often chosen.

14.3.3 Spring reverb

The spring reverb, popularised by the American Hammond organ company many years ago, consists basically of a coil of springy wire several millimetres in diameter and up to a metre or so long. It is acoustically excited by a transducer at one end, and other transducers, both at the far end and sometimes along the length, pick up the

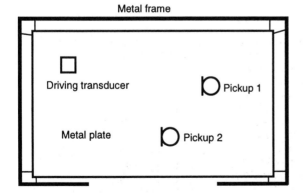

Figure 14.5 A reverberation plate is a metal sheet suspended from a frame, with transducers mounted at appropriate locations. Often the damping of the plate may be varied to alter the decay time of vibrations

vibrations. Some are fairly sophisticated, utilising several springs and a number of transducers. The modest quality of their performance does not, however, prevent them from having an application in guitar amplifiers, and quite a pleasing reverb quality can be obtained. The springs are acoustically sensitive, and one frequently hears loud twangs and crashes through the speakers when guitar amplifiers are being moved about.

14.3.4 Digital reverb

The present-day digital reverb, such as that pictured in Figure 14.6, can be quite a sophisticated device. Research into path lengths, boundary and atmospheric absorption, and the physical volume and dimensions of real halls, have been taken into account when algorithms have been designed. Typical front panel controls will include selection of internal pre-programmed effects, labelled as 'large hall', 'medium hall', 'cathedral', 'church', etc., and parameters such as degree of pre-delay, decay time, frequency balance of delay, dry-to-wet ratio (how much direct untreated sound appears with the effect signal on the output), stereo width, and relative phase between the stereo outputs can often be additionally altered by the user. A small display gives information about the various parameters.

Memory stores generally contain a volatile and a non-volatile section. The non-volatile section contains factory pre-set effects, and although the parameters can be varied to taste the alterations can not be stored in that memory. Settings can be stored in the volatile section, and it is usual to adjust an internal pre-set to taste and then transfer and store this in the volatile section. For example, a unit may contain 100 pre-sets. The first 50 are non-volatile, and can not be permanently altered. The last 50 can store settings arrived at by the user by transferring existing settings to this section. The method of doing this varies between models, but is usually a simple two- or three-button procedure, for example by pressing 5, 7 and 'store'. This means that a setting in the non-volatile memory which has been adjusted to taste will be stored in memory 57. Additional adjustments can then be made later if required.

Several units provide a lock facility so that stored effects can be made safe against accidental overwriting. An internal battery backup protects the memory contents when the unit is switched off. Various unique settings can be stored in the memories, although it is surprising how a particular model will frequently stamp its

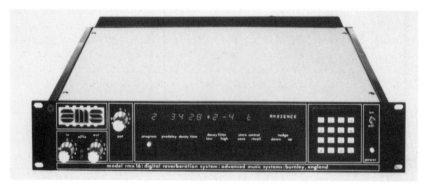

Figure 14.6 A typical digital reverberation unit: the AMS RMX-16. (Courtesy of AMS Industries PLC)

Figure 14.7 The tc 2290 multi-effects processor. (Courtesy of tc Electronics)

own character on the sound however it is altered. This can be both good or bad of course, and operators may have a preference for a particular system. Certain processors sound a bit like clangy spring reverbs whatever the settings. Some sound dull due to limited bandwidth, the frequency response extending only up to around 12 kHz or so. This must be carefully looked for in the specification. Sometimes the bandwidth reduces with increasing reverberation or echo decay times. Such a shortcoming is sometimes hard to find in a unit's manual.

In all the above reverb devices it should be noted that the input is normally mono and the output stereo. In this way 'stereo space' can be added to a mono signal, there being a degree of decorrelation between the outputs. Occasionally, a reverb device may have stereo inputs, so that the source can be assumed to be other than a point.

14.4 Multi-effects processors

Digital multi-effects processors such as that shown in Figure 14.7 can offer a great variety of features. Parametric equalisation is available, offering variations in Q, frequency and degree of cut and boost. Short memory capacity can store a sample, the unit being able to process this and reproduce it according to the incoming signal's command. MIDI interfacing (see Chapter 15) has become popular for the selection of effects under remote control, as has the RS 232 computer interface, and a floppy-disk drive is sometimes encountered for loading information. Repeat echo, autopan, phase, modulation, flange, high and low filters, straight signal delay, pitch change, gating, and added harmony may all be available in the pre-sets, various multifunction nudge buttons being provided for overall control. Many units are only capable of offering one type of effect at a time. Several have software update options so that a basic unit can be purchased and updates later incorporated internally to provide, say, longer delay times, higher sample storage capacity, and new types of effect as funds allow and as the manufacturer develops them. This helps to keep obsolescence at bay in an area which is always rapidly changing.

14.5 Frequency shifter

The frequency shifter shifts an incoming signal by a few hertz. It is used for acoustic feedback control in PA work, and operates as follows. Feedback is caused by sound from a speaker re-entering a microphone to be reamplified and reproduced again by the speaker, forming a positive feedback loop which builds up to a continuous loud howling noise at a particular frequency. The frequency shifter is placed in the signal path such that the frequencies reproduced by the speakers are displaced by several hertz compared with the sound entering the microphone, preventing additive effects when the sound is recycled, so the positive feedback loop is broken. The very small frequency shift has minimal effect on the perceived pitch of the primary sound.

14.6 Miscellaneous devices

The Aphex company of America introduced a unit called the *Aural Exciter* in the 1970s, and for a time the mechanisms by which it achieved its effect were shrouded in a certain amount of mystery. The unit made a signal 'sparkle', enhancing its overall presence and life, and it was usually applied to individual sounds in a mix such as solo instruments and voices, but sometimes also to the complete stereo signal. Such devices succeed entirely by their subjective effect, and several companies later produced similar units. They achieve their psycho-acoustic effect by techniques such as comb filtering, selective boosting of certain frequencies, and by introducing relatively narrow-band phase shifts between stereo channels.

The de-esser cleans up closely miced vocals. Sibilant sounds can produce a rasping quality, and the de-esser dynamically filters the high-level, high-frequency component of the sound to produce a more natural vocal quality.

14.7 Connection of outboard devices

A distinction needs to be made between processors which need to interrupt a signal for treatment, and those which basically add something to an existing signal. Graphic and parametric equalisers, compressors, de-essers and gates need to be placed in the signal path. One would not normally wish to mix, say, an uncompressed signal with its compressed version, or an ungated with the gated sound. Such processors will generally be patched in via the mixer's channel insertion send and returns (see Figure 14.8), or patched in ahead of the incoming signal or immediately after an output. Devices such as echo, reverb, chorus, flange are generally used to add something to an existing signal and usually a channel aux send will be used to drive them. Their outputs will be brought back to additional input channels and these signals mixed to taste with the existing dry signal (see Figure 14.9).

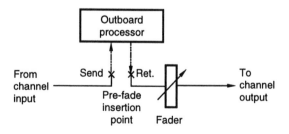

Figure 14.8 Outboard processors such as compressors are normally patched in at an insertion point of the required mixer channel

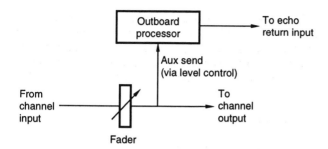

Figure 14.9 Reverberation and echo devices are usually fed from a post-fader auxiliary send, and brought back to a dedicated echo return or another channel input

Sometimes just the effects signal will be required, in which case either the aux send will be switched to pre-fade and that channel's fader closed, or the channel will simply be derouted from the outputs. The channel is then used merely to send the signal to the effects unit via the aux. The returns will often contain a degree of dry signal anyway, the ratio of dry to effect being adjusted on the processor.

Chapter 15

MIDI

MIDI, the Musical Instrument Digital Interface, now forms an integral part of many studio and multimedia operations. It can be used for controlling musical instruments such as samplers and synthesisers, internal sound cards within PCs (on which may be provided FM and wavetable synthesis functions, as well as audio sampling), and a range of other studio equipment such as mixers, effects devices and recording devices. MIDI software is often integrated with digital audio software, so as to produce a package capable of recording, editing and replaying both musical performance data and sampled digital audio data, combining the benefits of both approaches.

MIDI has become popular as a means of controlling and storing sound information in an economical way. As previously explained, sampled digital audio can occupy very large amounts of memory space (around 5 Mbytes per minute for high quality sound), whereas the MIDI data for a minute of music might only consume 5 kbytes. Whilst the two are not directly comparable, the widespread adoption of General MIDI and the availability of cheap sound cards for PCs has meant that many high quality computer sound applications can be handled using MIDI control instead of digital sound files.

This chapter will provide an introduction to MIDI and its applications. For more detailed information the reader is referred to the book *MIDI Systems and Control* by Francis Rumsey, as detailed at the end of this chapter.

15.1 What is MIDI?

MIDI is principally a means of remote control for electronic musical instruments. It integrates timing and system control functions with note triggering and other performance information. It is possible to control a number of musical instruments polyphonically from one controller using a single MIDI data stream. The interface is serial (see Fact File 15.1) which offers a number of advantages.

15.2 MIDI and digital audio contrasted

For many the distinction between MIDI and digital audio may be a clear one, but those new to the subject often confuse the two. Any confusion is often due to both MIDI and digital audio equipment appearing to perform the same task – that is the

FACT FILE	A serial
15.1	**interface**

A serial interface is one in which the bits of data making up each byte are transferred one after another. It is thus different from a parallel interface in which a number of bits may be transferred together, using a number of lines in parallel. A serial link between two devices requires a single communications channel, whereas a parallel link requires many; so the serial link is useful in cases where economy in wiring is paramount, or where only a single channel exists (over a telephone link, for example). Most computers, and this includes those inside MIDI devices, use parallel data

internally, requiring some conversion to and from serial form for MIDI communication. Parallel data is converted to serial form using a device known as a shift register, illustrated below. The rate of the clock signal determines the rate at which the bits of the parallel input word are shifted out of the serial output. The reverse process can be used for converting serial inputs to parallel form.

The adoption of a simple serial standard for MIDI was dictated largely by economic and practical considerations, as it was intended that the interface should be installed on relatively cheap items of equipment, and that it should be available to as wide a range of users as possible. The simplicity and ease of installation of a MIDI system has been largely responsible for its rapid proliferation as an international standard.

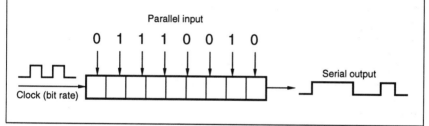

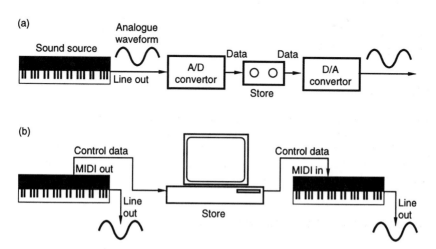

Figure 15.1 (a) Digital audio recording and (b) MIDI contrasted. In (a) the sound waveform is converted into digital data and stored, whereas in (b) only control information is stored, and a MIDI-controlled sound generator is required during replay

recording of multiple channels of music using digital equipment – and is not helped by the way in which some manufacturers refer to MIDI sequencing as digital recording.

Digital audio involves a process whereby an audio waveform (such as the line output of a musical instrument) is sampled regularly and then converted into a series of binary words which actually represent the sound itself (see Chapter 10). A digital audio recorder stores this sequence of data and can replay it by passing the original data through a digital-to-analogue convertor which turns the data back into a sound waveform, as shown in Figure 15.1. MIDI, on the other hand, handles digital information which controls the generation of sound. The data represents events such as notes pressed, wheels moved and switches operated. When a music recording is made using a MIDI sequencer this event data is stored, and can be replayed by transmitting it to a collection of MIDI-controlled musical instruments. It is the instruments which actually reproduce the recording.

A digital audio recorder, then, allows any sound to be stored and replayed without the need for additional hardware. It is useful for recording acoustic sounds such as voices, where MIDI is not much help. A MIDI sequencer is almost useless without a collection of sound generators, and the replayed sound quality and timbre are not fixed – they depend on the voices of the sound generators. An advantage of the MIDI recording is that the stored data represents the events in a piece of music, making it possible to change the music by changing the event data.

15.3 Basic MIDI principles

15.3.1 System specifications

The MIDI standard specifies a uni-directional asynchronous serial interface running at a data rate of 31.25 kbits ± 1% (see Fact File 15.2). The rate had to be slow enough to be carried over simple cables and interface hardware, but fast enough to allow musical information to be transferred from one instrument to another without noticeable delays, giving the impression of real-time control.

There is a defined hardware interface which should be incorporated in all MIDI equipment, which ensures electrical compatibility within a system. This is shown in Fact File 15.3. Concerning any potential for delay between IN and THRU connectors, it should be stated at the outset that the delay in data passed through this circuit is only a matter of microseconds, so this contributes little to any audible delays perceived in the musical outputs of some instruments in a large system. The bulk of any perceived delay will be due to other factors, which are covered in later sections.

15.3.2 Simple interconnection

Before going further it is necessary to look at a simple practical application of MIDI, to see how various aspects of the standard message protocol have arisen. For the present it will be assumed that we are dealing with a music system, although, as will be seen, MIDI systems can easily incorporate non-musical devices in various ways.

In the simplest MIDI system, one instrument could be connected to another using a standard MIDI cable (see Fact File 15.4) as shown in Figure 15.2. Here, instrument 1 sends information relating to actions performed on its own controls (notes pressed,

FACT FILE 15.2 — Asynchronous data transmission

MIDI uses what is known as asynchronous serial communication. In asynchronous communication data is transmitted without an accompanying clock signal, and the sending device expects the receiver to lock on to the incoming data whenever it arrives. This makes it very easy to interlink two devices, as there is only really a need for a single wire to carry the data (although in practice there are two to complete the circuit, and a screen to prevent interference).

In an asynchronous interface the clocks of the transmitter and the receiver must be running at exactly the same rate, otherwise data may be lost due to timing errors. This close tolerance in clock rates is fundamental to the satisfactory operation of an asynchronous interface, as it is vital that the receiver can rely on a particular bit of a received word occurring at a particular time. The tolerance in clock frequency allowed for MIDI is ± 1%. Also, data may be transmitted at any time, and there may be long gaps in between messages. Thus there is a need for a means of signalling the start and end of a data word. This is achieved using start and stop bits.

A start bit precedes every byte transmitted, and a stop bit follows every byte. On recognition of the leading edge of the start bit, the receiver adjusts the phase of its clock such that it thereafter accepts the bits of the following word in their correct time slots, and the stop bit terminates the word (see diagram). A device known as a UART (Universal Asynchronous Receiver/Transmitter), which is really just a more advanced form of shift register, automatically handles the timing of the serial interface and converts data to and from the parallel form.

The MIDI standard specifies a unidirectional serial interface running at 31.25 kbaud. The baud rate is the maximum number of bits per second that can be transferred over the interface. Data is transmitted unidirec-tionally, that is it has only one direction: from the transmitter to the receiver. Standard MIDI messages typically consist of 1, 2, or 3 bytes, although there are longer messages for special purposes.

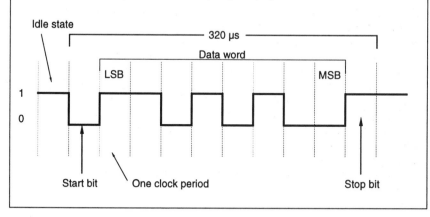

The MIDI standard specifies a unidirectional serial interface. Diagram showing idle state, 320 µs, Data word, LSB, MSB, 1, 0, Start bit, One clock period, Stop bit.

pedals pressed, etc.) to instrument 2, which imitates these actions as far as it is able. This type of arrangement can be used for 'doubling-up' sounds, 'layering' or 'stacking', such that a composite sound can be made up from two synthesisers' outputs. (It should be pointed out that the audio outputs of the two instruments would have to be mixed together for this effect to be heard.) Larger MIDI systems could be built up by further 'daisy-chaining' of instruments, such that instruments further down the chain all received information generated by the first (see Figure 15.3),

FACT FILE

15.3

MIDI hardware interface

Most equipment using MIDI has three interface connectors: IN, OUT and THRU (sic), as shown below. The OUT connector carries data which the device itself has generated, relating to keys pressed, controls altered, etc. It is transmitted from the device's UART. The IN connector receives data from other devices and relays it to the device's UART. Data messages received will be acted upon by the instrument concerned if it is capable of handling the particular message, and if it is destined for that instrument (see text).

The THRU connector is a direct relay of the data that is present at the IN. It is simply a buffered feed of the input data, and it has not been processed in any way. A few cheaper devices do not have THRU connectors, but it is possible to buy 'MIDI THRU boxes' which provide a number of THRUs from one input. Occasionally, devices without a THRU socket allow the OUT socket to be switched between OUT and THRU. The THRU socket can be used to 'daisy-chain' MIDI devices together, so that transmitted information from one controller can be sent to a number of receivers without the need for multiple outputs from the controller.

The opto-isolator (an encapsulated device containing a light-emitting diode and a photo-transistor) between the MIDI IN and the UART couples the data optically, and means that there is no direct electrical link between devices. It helps to reduce the effects of any problems which might occur if one instrument in a system were to develop an electrical fault.

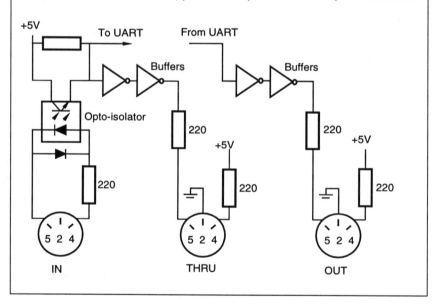

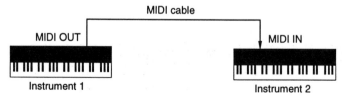

Figure 15.2 In the simplest form of MIDI interconnection, the OUT of one instrument is connected to the IN of another

FACT FILE 15.4 Connectors and cables

The connectors used for MIDI interfaces are like the five-pin DIN plugs used in some hi-fi systems. The specification also allows for the use of professional XLR connectors, although these are rarely encountered in practice. Only three of the pins of a five-pin DIN plug are actually used in most equipment (the three innermost pins). The cable should be a shielded twisted pair with the shield connected to pin 2 at both ends, although within the receiver itself the MIDI IN does not have pin 2 connected to earth. This is to avoid earth loops (see Fact File 13.2), but makes it possible to use a cable either way round. In the cable, pin 5 at one end should be connected to pin 5 at the other, and likewise pin 4 to pin 4, and pin 2 to pin 2. Unless DIN cables follow this convention they will not work.

A current loop is created between a MIDI OUT or THRU and a MIDI IN, when connected with the appropriate cable, as shown below, and data is represented by the sending device turning this current on and off – a binary zero is represented by a current flowing through the loop whereas a one is represented by a lack of current.

It is recommended that no more than 15 m of cable is used for a single cable run in a simple MIDI system, although deterioration of the signal is gradual and depends on the conditions, cable, and equipment in use. Longer distances may be accommodated with the use of buffer or 'booster' boxes. Professional microphone cable terminated in DIN connectors may be used as a higher quality solution, because domestic cables will not always be a shielded twisted pair and thus will be more susceptible to external interference, as well as radiating more themselves which could interfere with adjacent audio signals.

It is recommended that the correct cable is used in professional installations where MIDI cables are installed in the same trunking as audio cables, to avoid any problems with crosstalk.

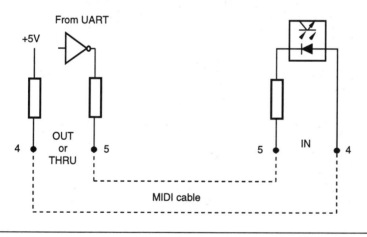

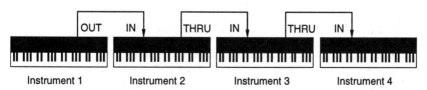

Figure 15.3 Devices can be 'daisy-chained' using the THRU connector, in which case all subsequent devices receive commands from the first device in the chain

although this is not a very satisfactory way of building a large MIDI system as described later in this chapter.

15.3.3 MIDI channels

MIDI messages are made up of a number of bytes. Each part of the message has a specific purpose, and one of these is to define the receiving channel to which the message refers. In this way, a controlling device can make a message device-specific—in other words it can define which receiving instrument will act on the data. This is most important in large systems which use a computer sequencer as a master controller, when a large amount of information will be present on the MIDI data bus, not all of which is intended for every instrument. If a device is set to receive on a specific channel or on a number of channels it will act only on information which is 'tagged' with its own channel numbers. It will usually ignore everything else. There are sixteen basic MIDI channels and instruments can usually be set to receive on any specific channel or channels (omni off mode), or to receive on all channels (omni on mode).

Later it will be seen that the limit of sixteen MIDI channels can be exceeded easily by using multiport MIDI interfaces connected to a computer. In such cases it is important not to confuse the MIDI data channel with the physical port to which a device may be connected, since each physical port will be capable of transmitting on all sixteen data channels.

15.4 MIDI communications

15.4.1 Message format

There are two basic types of MIDI message byte: the status byte and the data byte. Status bytes always begin with a binary one to distinguish them from data bytes, which always begin with a zero. As shown in Figure 15.4, the first half of the status byte denotes the message type and the second half denotes the channel number. Because the most significant bit (MSB) of each byte is reserved to denote the type (status or data) there are only seven active bits per byte which allows 2^7 (that is 128) possible values. (Some basic information about binary data was given in Chapter 10 and Fact File 15.5 gives further details of the method of binary shorthand known as hexadecimal.)

The first byte in a MIDI message is normally a status byte, which contains information about the channel number to which the message applies. It can be seen that four bits of the status byte are set aside to indicate the channel number, which allows for 2^4 (or 16) possible channels. The status byte is the label that denotes which

Figure 15.4 General format of a MIDI message. The 'sss' bits are used to define the message type, the 'nnnn' bits define the channel number, whilst the 'xxxxxxx' and 'yyyyyy' bits carry the message data

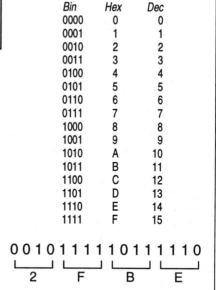

FACT FILE

Hexadecimal

15.5

Long binary numbers can be fairly unwieldy to write down, so various forms of shorthand are used to make them more manageable. The most common of these is hexadecimal. The hexadecimal system (base 16) represents decimal values from 0 to 15 using the sixteen symbols 0–9 and A–F, as shown in the table. Each hex character corresponds to a group of four binary digits (bits), or one 'nibble' of a binary word. An example showing how a long binary word may be written in hex is shown in the diagram – it is simply a matter of breaking the word up into 4 bit nibbles and representing each nibble as the appropriate hex character. Hex can be converted back to binary using the reverse process.

Bin	Hex	Dec
0000	0	0
0001	1	1
0010	2	2
0011	3	3
0100	4	4
0101	5	5
0110	6	6
0111	7	7
1000	8	8
1001	9	9
1010	A	10
1011	B	11
1100	C	12
1101	D	13
1110	E	14
1111	F	15

0010 1111 1011 1110
  2   F   B   E

Table 15.1 MIDI messages summarised

Message	Status	Data 1	Data 2
Note off	&8n	Note number	Velocity
Note on	&9n	Note number	Velocity
Polyphonic aftertouch	&An	Note number	Pressure
Control change	&Bn	Controller number	Data
Program change	&Cn	Program number	–
Channel aftertouch	&Dn	Pressure	–
Pitch wheel	&En	LSbyte	MSbyte
System exclusive			
System exclusive start	&F0	Manufacturer ID	Data, (Data), (Data)
End of SysEx	&F7	–	
System common			
Quarter frame	&F1	Data	–
Song pointer	&F2	LSbyte	MSbyte
Song select	&F3	Song number	–
Tune request	&F6	–	
System realtime			
Timing clock	&F8	–	–
Start	&FA	–	–
Continue	&FB	–	–
Stop	&FC	–	–
Active sensing	&FE	–	–
Reset	&FF	–	–

receiver or part of a receiver the message is intended for, and it also denotes which type of message is to follow (e.g. a note on message). There are only three bits to denote the message type because the first bit must always be a one. This theoretically allows for eight message types, but there are some special cases in the form of system messages, all of which begin with binary 1111 (&F).

Standard MIDI messages can be one, two or three bytes long, but System Exclusive messages are special exceptions which often exceed this length for a variety of reasons. Table 15.1 shows the format and content of the main MIDI message types. The letter 'n' is used to denote any channel number from 1-16, although these are actually encoded as 0-15 in the message. A short description of some of the most commonly encountered MIDI messages is given in the following sections, although there is not space here to go into detail on every type.

15.4.2 Channel and system messages contrasted

Two primary classes of message exist: those which relate to specific MIDI channels and those which relate to the system as a whole. Channel messages start with status bytes in the range &8n to &En (they start at hexadecimal eight because the MSB must be a one for a status byte). System messages all begin with &F, and do not contain a channel number. Instead the second part of the status byte (that which normally would contain the channel number) is used for further identification of the system message, allowing for sixteen possible system messages running from &F0 to &FF.

System messages are themselves split into three groups: system common, system exclusive and system realtime. The common messages may apply to any device on the MIDI bus, depending only on the device's ability to handle the message. The exclusive messages apply to whichever manufacturer's devices are specified later in the message (see below), and the realtime messages are intended for devices which are to be synchronised to the controller's musical tempo.

15.4.3 Note on and note off messages

Much of the musical information sent over a typical MIDI interface will consist of these two message types. As indicated by the titles, the note on message turns on a musical note, and the note off message turns it off. Note on takes the general format:

[&8n] [Note number] [Velocity]

and note off takes the form:

[&9n] [Note number] [Velocity]

A MIDI instrument will generate note on messages at its MIDI OUT corresponding to whatever notes are pressed on the keyboard, on whatever channel the instrument is set to transmit. Also, any note which has been turned on must subsequently be turned off in order for it to stop sounding, thus if one instrument receives a note on message from another and then loses the MIDI connection for any reason, the note will continue sounding ad infinitum. This situation can occur if a MIDI cable is pulled out during transmission.

MIDI note numbers relate directly to the western musical chromatic scale, and the format of the message allows for 128 note numbers which cover a range of a little over ten octaves – adequate for the full range of most musical material. This

Table 15.2 MIDI note numbers related to the musical scale

Musical note	MIDI note number	
C–2	0	
C–1	12	
C0	24	
C1	36	
C2	48	
C3 (middle C)	60	(Yamaha convention)
C4	72	
C5	84	
C6	96	
C7	108	
C8	120	
G8	127	

quantisation of the pitch scale is geared very much towards keyboard instruments, and is perhaps less suitable for other instruments and cultures where the definition of pitches is not so black-and-white. None the less, means have been developed of adapting control to situations where unconventional tunings are required. Note numbers normally relate to the musical scale as shown in Table 15.2, although there is a lack of consistency here. Yamaha established the use of C3 for middle C, whereas others have used C4. Some software allows the user to decide which convention will be used for display purposes.

15.4.4 Velocity information

Note messages are associated with a velocity byte, and this is used to represent the speed with which a key was pressed or released. The former will correspond to the force exerted on the key as it is depressed: in other words, 'how hard you hit it' (called 'note on velocity'). It is applied to control parameters such as the volume or timbre of a note, and can be used to scale the effect of one or more of the envelope generators in a synthesiser. This velocity value has 128 possible states, but not all MIDI instruments are able to generate or interpret the velocity byte, in which case they will set it to a value half way between the limits, i.e.: 64_{10}. Some instruments may act on velocity information even if they are unable to generate it themselves. Note off velocity (or 'release velocity') is not widely used. None the less it is available for special effects if a manufacturer decides to implement it.

The note on, velocity zero value is reserved for the special purpose of turning a note off, for reasons which will become clear under 'Running status' below. If an instrument sees a note number with a velocity of zero, its software should interpret this as a note off message.

15.4.5 Running status

When a large amount of information is transmitted over a single MIDI interface, delays naturally arise due to the serial nature of transmission wherein data such as the concurrent notes of a chord must be sent one after the other. It will be advantageous, therefore, to reduce the amount of data transmitted as much as possible, in order to keep the delay as short as possible and to avoid overloading the devices on the bus with unnecessary data.

Running status is an accepted method of reducing the amount of data transmitted, and one which all MIDI software should understand. It involves the assumption that once a status byte has been asserted by a controller there is no need to reiterate this status for each subsequent message of that status, so long as the status has not changed in between. Thus a string of note on messages could be sent with the note on status only sent at the start of the series of note data, for example:

[&9n] [Data] [Velocity] [Data] [Velocity] [Data] [Velocity]

It will be appreciated that for a long string of note data this could reduce the amount of data sent by nearly one third. But as in most music each note on is almost always followed quickly by a note off for the same note number, this method would clearly break down, as the status would be changing from note on to note off very regularly, thus eliminating most of the advantage gained by running status. This is the reason for the adoption of note on, velocity zero as equivalent to a note off message, because it avoids a change of status during running status, allowing a string of what appear to be note on messages, but which represent both note on and note off.

Running status is not used at all times for a string of same-status messages, and will often only be called upon by an instrument's software when the rate of data exceeds a certain point. Indeed, an examination of the data from a typical synthesiser indicates that running status is not used during a large amount of ordinary playing.

15.4.6 Control change messages

As well as note information, a MIDI device may be capable of transmitting control information which corresponds to the various switches, control wheels and pedals associated with it. The controller messages have proliferated enormously since the early days of MIDI, and not all devices will implement all of them. It should be noted that there are two distinct kinds of controller: that is, the switch type, and the continuous type. The continuous controller is any variable wheel, lever, slider or pedal that might have one of a number of positions, whereas the switch normally only has two states.

Most of the continuous controllers in real instruments are only represented to 7 bit resolution, giving 2^7 or 128 possible steps over their range, but there is provision for a second message to be sent carrying a further seven bits to add finer control. This is why there are two controller numbers for each 14 bit controller, as exemplified in the table, one for coarse control and one for fine control.

On/off switches can be represented easily in binary form (0 for OFF, 1 for ON), and it would be possible to use just a single bit for this purpose, but, in order to conform to the standard format of the message, switch states are normally represented by data values between &00 and &3F for OFF, and &40-&7F for ON. In other words switches are now considered as 7 bit continuous controllers, and it may be possible on some instruments to define positions in between off and on in order to provide further degrees of control, such as used in some 'sustain' pedals (although this is not common in the majority of equipment). In older systems it may be found that only &00 = OFF and &7F = ON.

15.4.7 Channel modes

Although grouped with the controllers under the same status, the channel mode

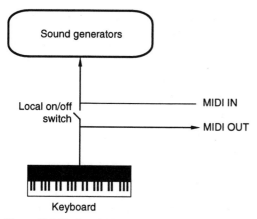

Figure 15.5 The Local Off switch disconnects a keyboard from its associated sound generators in order that the two parts may be treated independently in a MIDI system

messages set the mode of operation of the instrument receiving on that particular channel.

'Local on/off' is used to make or break the link between an instrument's keyboard and its own sound generators. Effectively there is a switch between the output of the keyboard and the control input to the sound generators which allows the instrument to play its own sound generators in normal operation when the switch is closed (see Figure 15.5). If the switch is opened, the link is broken and the output from the keyboard feeds the MIDI OUT while the sound generators are controlled from the MIDI IN. In this mode the instrument acts as two separate devices: a keyboard without any sound, and a sound generator without a keyboard. This configuration can be useful when the instrument in use is the master keyboard for a large sequencer system, where it may not always be desired that everything played on the master keyboard results in sound from the instrument itself.

'Omni off' ensures that the instrument will only act on data tagged with its own channel number(s), as set by the instrument's controls. 'Omni on' sets the instrument to receive on all of the MIDI channels. In other words, the instrument will ignore the channel number in the status byte and will attempt to act on any data that arrives, whatever its channel. Devices should power-up in this mode according to the original specification, but more recent devices will tend to power- up in the mode that they were left. Mono mode sets the instrument such that it will only reproduce one note at a time, as opposed to 'Poly' (phonic) in which a number of notes may be sounded together.

In poly mode the instrument will sound as many notes as it is able at the same time. Instruments differ as to the action taken when the number of simultaneous notes is exceeded: some will release the first note played in favour of the new note, whereas others will refuse to play the new note. The more intelligent of them may look to see if the same note already exists in the notes currently sounding, and only accept a new note if is not already sounding. Even more intelligently, some devices may release the quietest note (that with the lowest velocity value), or the note furthest through its velocity envelope, to make way for a later arrival. It is also common to run a device in poly mode on more than one receive channel, provided that the software can handle the reception of multiple polyphonic channels. A multitimbral sound generator may well have this facility, commonly referred to as 'multi' mode, making it act as if it were a number of separate instruments each receiving on a separate channel. In multi mode a device may be able to dynamically assign its polyphony between the channels and voices in order that the user does not need to assign a fixed polyphony to each voice.

15.4.8 Program change

The program change message is used most commonly to change the 'patch' of an instrument or other device. A patch is a stored configuration of the device, describing the set-up of the tone generators in a synthesiser, and the way in which they are interconnected. It often corresponds to a preprogrammed voice on a synthesiser. Program change is channel-specific, and there is only a single data byte associated with it, specifying to which of 128 possible stored programs the receiving device should switch. On non-musical devices such as effects units, the program change message is often used to switch between different effects, and the different effects programs may be mapped to specific program change numbers.

15.4.9 System exclusive

A system exclusive message is one which is unique to a particular manufacturer, and often a particular instrument. The only thing that is defined about such messages is how they are to start and finish, with the exception of the use of system exclusive messages for standardised universal information, which makes use of the three highest numbered manufacturer IDs.

System exclusive messages generated by a device will naturally be produced at the MIDI OUT, not at the THRU, so a deliberate connection must be made between the transmitting device and the receiving device before data transfer may take place. Occasionally it is necessary to make a return link from the OUT of the receiver to the IN of the transmitter so that two-way communication is possible, and so that the receiver can control the flow of data to some extent by telling the transmitter when it is ready to receive and when it has received correctly (a form of handshaking).

The message takes the general form:

&[F0] [ident.] [data] [data] ... [F7]

where [ident.] identifies the relevant manufacturer ID, which is a number defining which manufacturer's message is to follow. Originally, manufacturer IDs were a single byte but the number of IDs has been extended by setting aside the [00] value of the ID to indicate that two further bytes of ID follow. Manufacturer IDs are therefore either one or three bytes long.

Data of virtually any sort can follow the ID. It can be used for a variety of miscellaneous purposes which have not been defined in the MIDI standard, and the message can have virtually any length that the manufacturer requires, although it is often split into packets of a manageable size in order not to cause receiver memory buffers to overflow. Exceptions are data bytes which look like other MIDI status bytes (except realtime messages), as they will naturally be interpreted as such by any receiver, which might terminate reception of the system exclusive message. The message should be terminated with &F7.

Examples of applications for such messages can be seen in the form of sample data dumps (from a sampler to a computer and back again for editing purposes), although this is painfully slow, and voice data dumps (from a synthesiser to a computer for storage and editing of user-programmed voices). There are now an enormous number of uses of system exclusive messages, both in the universal categories and in the manufacturer categories.

15.5 Synchronisation

MIDI information is also used to synchronise devices within a system. It depends on the type of device as to whether or not MIDI synchronisation data is recognised. Sequencers normally need a timing reference, either internal or external, to determine the rate at which music and other data are replayed. Drum machines also need timing information because a drum machine usually contains a sequencer which stores the patterns of rhythm which a player may lay down. If two or more sequencers or drum machines are connected together using MIDI cables it is possible to run them in synchronism with each other, stopping and starting in the same places, and autolocating to the same positions in the song. In such cases one device acts as the master and all the others as slaves. Wherever the master goes the slave follows.

A normal synthesiser, effects unit or sampler is not concerned with timing information, because it has no functions which would be affected by a timing clock. Such devices do not normally store rhythm patterns, although there are some keyboards with onboard sequencers which ought to handle timing data.

As MIDI has become more involved with the professional recording studio it has become possible for MIDI timing information to be locked to a pre-recorded timecode which may be derived from a video or audio recorder.

15.5.1 System Realtime

A group of system messages called the System Realtime messages (see Table 15.1)) control the execution of timed sequences in a MIDI system, and these are often used in conjunction with the Song Pointer (which is really a System Common message) for autolocation. The MIDI clock byte, a single status byte (&F8), is transmitted by the master device six times per MIDI beat. A MIDI beat is equivalent to a musical semiquaver or sixteenth note. At any one musical tempo a MIDI beat could be said to represent a fixed amount of time, but this would change if the tempo changed. Any slaves lock to the tempo rate indicated by these clock bytes.

START (&FA), transmitted by the master, should result in the playback of the sequence from the very beginning. STOP (&FC) is used to halt the execution of a song which is running on a synchronised instrument. CONTINUE (&FB) is used to restart the execution from the point at which the sequence was stopped, not from the beginning.

15.5.2 Song position pointers (SPPs)

SPPs are used when the master device needs to tell slaves where it is in a song. For example, one might 'fast-forward' through a song and start again twenty bars later, in which case the other timed devices in the system would have to know where to restart. An SPP would be sent followed by 'continue' and then regular clocks.

The SPP represents the position in a stored song in terms of number of MIDI beats (not clocks) from the start of the song. Thus again it relates directly to musical time rather than real time, and does not necessarily represent the number of seconds into the song. It uses two data bytes, and the MSB of data bytes must be zero, so the resulting fourteen bits can specify up to 16 384 MIDI beats. SPP is often used in conjunction with &F3 (Song Select), which is used to define one of a collection of stored songs to be recalled.

SPPs are fine for directing the movements of an entirely musical system, in which every action is related to a particular beat or subdivision of a beat, but not so fine when actions must occur at a particular point in real time. If, for example, one was using a MIDI system to dub music and effects to a picture in which an effect was intended to be triggered at a particular visual occurrence, that effect would have to maintain its position in time no matter what happened to the music. If the effect was to be triggered by a sequencer at a particular number of beats from the beginning of the song, this point could change in real time if the tempo of the music was altered slightly to fit a particular visual scene. Clearly some means of real-time synchronisation is required either instead of, or as well as, the clock and song pointer arrangement, such that certain events in a MIDI-controlled system may be triggered at specific times in hours, minutes and seconds.

15.5.3 MIDI TimeCode (MTC)

MIDI Timecode (MTC) is used widely as a means of synchronising MIDI-controlled equipment to a real time reference. SMPTE/EBU longitudinal timecode (LTC) is described in Chapter 16, and MTC is a means of transferring LTC around a MIDI system.

In an LTC timecode frame two binary data groups are allocated to each of hours, minutes, seconds and frames, these groups representing the tens and units of each, so there are eight binary groups in total representing the time value of a frame. In order to transmit this information over MIDI, it has to be turned into a format which is compatible with other MIDI data (i.e. a status byte followed by relevant data bytes). There are two types of MTC synchronising message: one which updates a receiver regularly with running timecode, and another which transmits one-time updates of the timecode position for situations such as exist during the high-speed spooling of tape machines, where regular updating of each single frame would involve too great a rate of transmitted data. The former is known as a quarter-frame message (see Fact File 15.6), denoted by the status byte (&F1), whilst the latter is known as a full-frame message and is transmitted as a universal realtime system exclusive (SysEx) message, the details of which will not be given here.

15.6 Interfacing a computer to a MIDI system

In order to use a computer as a central controller for a MIDI system it must have at least one MIDI interface, consisting of at least an IN and an OUT port. (THRU is not strictly necessary in most cases.) Unless the computer has a built-in interface, as found on old Atari machines, some form of third-party hardware interface must be added, and there are many ranging from simple single ports to complex multiple port products.

FACT FILE

15.6

Quarter-frame MTC messages

One timecode frame is represented by too much information to be sent in a standard three byte MIDI message, so it is broken down into eight separate messages called quarter-frame messages. Four of these quarter-frame messages are transmitted in the period of one timecode frame (in order to limit the message rate), so it takes two frames to transmit a complete frame value and the receiver is updated every two frames. Receivers therefore need to maintain a two frame offset between their displayed timecode and the last decoded value, because by the time a frame value has been completely transmitted two frames will have elapsed. Internally, the timing resolution of software can be made higher than that of the timecode messages, with the messages being used as a reference point every so often.

Each message of the group of eight represents a part of the timecode frame value, as shown in the diagram, and takes the general form:

&[F1] [DATA]

The data byte begins with zero (as always), and the next seven bits of the data word are made up of a 3 bit code defining whether the message represents hours, minutes, seconds

or frames, MSnibble or LSnibble, followed by the four bits representing the binary value of that nibble. In order to reassemble the correct timecode value from the eight quarter-frame messages, the LS and MS nibbles of hours, minutes, seconds and frames are each paired within the receiver to form 8 bit words as follows:

Frames: rrr qqqqq

where rrr is reserved for future use and qqqqq represents the frames value from 0 to 29;

Seconds: rr qqqqqq

where rr is reserved for future use and qqqqqq represents the seconds value from 0 to 59;
Minutes: rr qqqqqq
as for seconds;

Hours: r qq ppppp

where r is undefined, qq represents the timecode type (see below), and ppppp is the hours value from 0 to 23.
The timecode frame rate is denoted as follows in the hours count:

00 = 24 fps

01 = 25 fps

10 = 30 fps drop-frame

11 = 30 fps non-drop-frame

Unassigned bits should be set to zero..

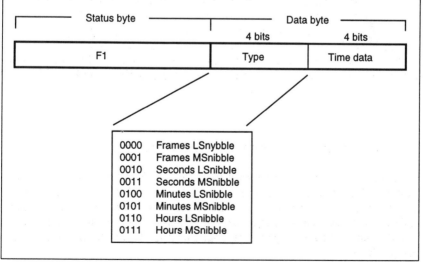

Status byte		Data byte	
		4 bits	4 bits
F1		Type	Time data

0000	Frames LSnybble
0001	Frames MSnibble
0010	Seconds LSnibble
0011	Seconds MSnibble
0100	Minutes LSnibble
0101	Minutes MSnibble
0110	Hours LSnibble
0111	Hours MSnibble

15.6.1 Single port MIDI interfaces

A typical single port MIDI interface will be connected either to one of the spare I/O ports of the computer, or plugged into an expansion slot. On the Macintosh, for example, a MIDI interface is normally connected directly to one of the two serial ports as shown in Figure 15.6. Provided that electrical conditions were properly controlled it would be possible to connect a number of receiving devices to a single port, either in a chain or using a splitter box or router. Some examples of simple setups are shown in Figure 15.7. As shown in one of these figures, some sound generators also double as a serial MIDI interface for the host computer, avoiding the need for a separate box.

The limitation of single port MIDI interfaces is that they can only address the basic sixteen channels of the MIDI protocol, and so one can run short of channels when controlling large systems with many voices and devices.

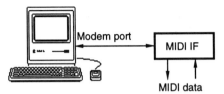

Figure 15.6 A spare general purpose serial interface, such as the Macintosh modem port, can be used to connect an external MIDI interface

(a)

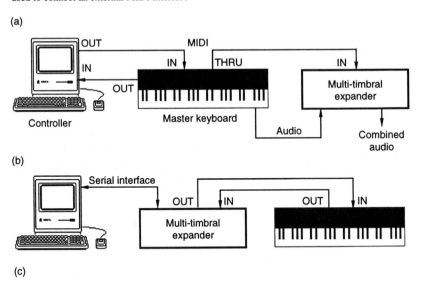

(b)

(c)

Insert MIDI Sys and Contr. Fig 5.3 (without box)

Figure 15.7 (a) Simple single port system with MIDI IF integrated with computer. (b) Using sound module as a MIDI IF. (c) Rear panel of Yamaha TG100, which will act as a MIDI IF for a computer, showing 'TO HOST' port for serial connection to computer

15.6.2 Multiport interfaces

Multiport interfaces have a number of independent MIDI OUT ports. Such an interface is connected to the host computer using either a parallel or serial I/O port, or using an expansion card, as shown in Figure 15.8. Most multiport systems allow up to 16 channels to be addressed by each of the MIDI OUTs, thus expanding the total number of channels addressable to 16 times the number of ports. In such systems it is common for each instrument to be connected to its own port, both IN and OUT, which is especially useful with multi-timbral sound modules capable of operating on all 16 channels simultaneously. A number of possibilities are illustrated in Figure 15.9. It also allows any instrument to be used as a controller, and

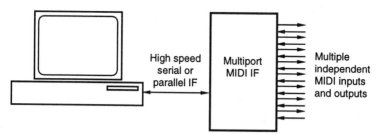

Figure 15.8 A multiport MIDI interface allows interconnection via a large number of independent MIDI ports

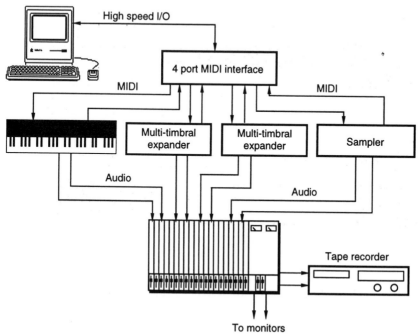

Figure 15.9(a) Example of an intermediate-level computer-controlled MIDI system for music production, using a 4 port MIDI interface. All devices can both receive and send MIDI information from and to the computer

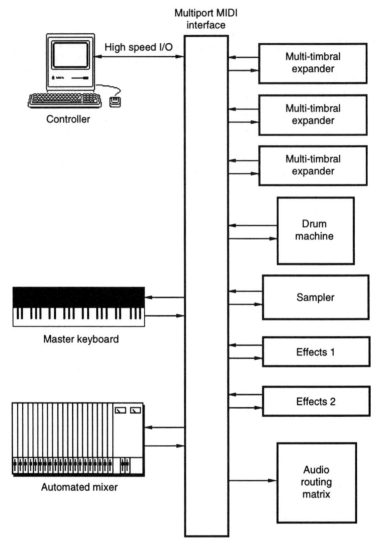

Figure 15.9(b) Example of a large computer-controlled MIDI system, incorporating MIDI-controlled audio effects and an automated mixer. The audio routing is not shown here

makes it possible for instruments to send system exclusive information back to the computer without replugging. Multiport interfaces will often allow data received from more than one port to be merged, or allow recording from more than one source at a time.

The more advanced multiport interfaces may also incorporate a timecode reader and generator, allowing for the synchronisation of video and audio tape recorders with the MIDI system. The conversion of SMPTE/EBU timecode to MIDI timecode or an equivalent then takes place within the interface before being transferred to the host computer. MIDI Machine Control (MMC), as described below, can be used to control the tape machine from a MIDI port. Some examples of system configurations are shown in Figure 15.10.

(a)

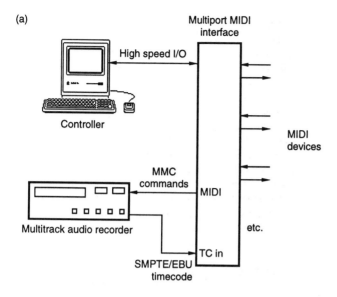

(b)

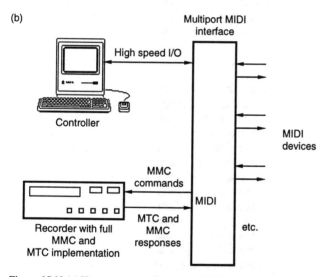

Figure 15.10 (a) Here a tape recorder is controlled using MMC commands from the sequencer, with positional information being fed back to the sequencer via the SMPTE/EBU timecode reader in the MIDI interface. (b) A tape recorder with a built-in timecode reader and a full MMC/MTC implementation does not require a timecode reader in the MIDI interface, because positional information is returned to the sequencer in the form of MTC messages

15.6.3 Interface driver software

Quite commonly, a software driver is provided for an external MIDI interface or expansion card. This driver is used by the computer's operating system to address the MIDI interface concerned, and it takes care of handling the various routines necessary to carry data to and from the physical I/O ports. The MIDI application therefore simply talks to the driver, and it follows that you must have the correct driver installed for the interface which you intend to use. High-end MIDI software usually comes complete with a number of drivers and configuration documents for the most popular MIDI interfaces.

15.7 An overview of software for MIDI

Sequencers are probably the most ubiquitous of the available MIDI software packages. A sequencer will be capable of storing a number of 'tracks' of MIDI information, editing it and otherwise manipulating it for musical composition purposes. It is also capable of storing MIDI events for non-musical purposes such as studio automation, and may be equipped with digital audio recording capabilities in some cases. Some of the more advanced packages are available in modular form (allowing the user to buy only the functional blocks required) and in cut-down or 'entry-level' versions for the new user.

The dividing line between sequencer and music notation software is a grey one, since there are features common to each. Music notation software is designed to allow the user control over the detailed appearance of the printed musical page, rather as desktop publishing packages work for typesetters, and such software often provides facilities for MIDI input and output. MIDI input is used for entering note pitches during setting, whilst output is used for playing the finished score in an audible form. Most major packages will read and write standard MIDI files (see Fact File 15.7), and can therefore exchange data with sequencers, allowing sequenced music to be exported to a notation package for fine tuning of printed appearance. It is also common for sequencer packages to offer varying degrees of music notation capability, although the scores which result are rarely as professional in appearance as those produced by dedicated notation software.

Librarian and editor software is used for managing large amounts of voice data for MIDI-controlled instruments. Such packages communicate with MIDI instruments using system exclusive messages in order to exchange parameters relating to voice programs. The software may then allow these voice programs or 'patches' to be modified using an editor, offering a rather better graphical interface than those usually found on the front panels of most sound modules. Banks of patches may be stored on disk by the librarian, in order that libraries of sounds can be managed easily, and this is often cheaper than storing patches in the various 'memory cards' offered by synth manufacturers. Banks of patch information may be accessed by sequencer software in order that the operator may choose voices for particular tracks by name, rather than by program change numbers. Sample editors are also available, offering similar facilities, although sample dumps using system exclusive are not really recommended, unless they are short, since the time taken can be excessive. Sample data can be transferred to a computer using a faster interface than MIDI (such as SCSI) and the sample waveforms can be edited graphically.

Amongst other miscellaneous software packages available for the computer are MIDI mixer automation systems, guitar sequencers, MIDI file players, multimedia authoring systems and alternative user interfaces. Development software is also available for MIDI programmers, providing a programming environment for the writing of new MIDI software applications. There are also a number of software applications designed principally for research purposes or for experimental music composition.

Sequencers and notation packages typically store data on disk in their own unique file formats. Occasionally it is possible for a file from one package to be read by others, especially if the packages are from the same manufacturer, but this is rare. The standard MIDI file was developed in an attempt to make interchange of information between packages more straightforward, and is now used widely in the industry in addition to manufacturers' own file formats.

Three types of standard MIDI file exist to encourage the interchange of sequencer data between software packages. The MIDI file contains data representing events on individual sequencer tracks, as well as containing labels such as track names, instrument names and time signatures. It is the intention that not only should software packages running on the same computer type be able to read these universal files, but that such files may be transferred to other computers (either over a network, or on a disk that can be read by the other computer), so that a package running under a completely different operating system may read the data.

File type 0 is the simplest, and is used for single-track data, whilst file type 1 supports multiple tracks which are 'vertically' synchronous with each other (such as the parts of a song), and file type 2 contains multiple tracks which have no direct timing relationship and may thus be asynchronous. Type 2 could be used for transferring song files which are made up of a number of discrete sequences, each with a multiple track structure.

15.8 General MIDI

If MIDI files are to be exchanged between systems and replayed on different sound generating hardware, one needs a means of ensuring that the music will sound at least similar on the two systems. One of the problems with MIDI-controlled sound generators used to be that although voice programs could be selected using MIDI program change commands, there was no standard mapping of voices to program numbers. In other words, program change 3 might correspond to 'alto sax' on one instrument and 'grand piano' on another. Consequently a music sequence could sound completely different when replayed on two different multitimbral sound generators. General MIDI was introduced to standardise some basic elements of sound generator control, so that MIDI files could be exchanged more easily between systems, ensuring that an approximation to the sounds and instruments intended by the composer would be heard no matter what General MIDI sound generator was used for replay. Currently, General MIDI is specified at Level 1, although there are proposals to extend the concept to further levels.

General MIDI specifies a number of other things as well as standard sounds. For example, it specifies a minimum degree of polyphony, and requires that a sound generator should be able to receive MIDI data on all 16 channels simultaneously and polyphonically, with a different voice on each channel. There is also a requirement that the sound generator should support percussion sounds in the form of drum kits, so that a General MIDI sound module is capable of acting as a complete 'band in a box'. Some multimedia computers now incorporate General MIDI synthesisers as part of the sound hardware, and have operating systems capable of addressing MIDI instruments.

Dynamic voice allocation is the norm in GM sound modules, with a requirement either for at least 24 dynamically allocated voices in total, or 16 for melody and 8

Table 15.3 General MIDI program number ranges (except channel 10)

Program change (decimal)	Sound type
0–7	Piano
8–15	Chromatic percussion
16–23	Organ
24–31	Guitar
32–39	Bass
40–47	Strings
48–55	Ensemble
56–63	Brass
64–71	Reed
72–79	Pipe
80–87	Synth lead
88–95	Synth pad
96–103	Synth effects
104–111	Ethnic
112–119	Percussive
121–128	Sound effects

for percussion. In order to ensure compatibility between sequences that are replayed on GM modules, percussion sounds are always allocated to MIDI channel 10. Program change numbers are mapped to specific voice names, with ranges of numbers allocated to certain types of sounds, as shown in Table 15.3. Precise voice names may be found in the GM documentation (see the end of the chapter). Channel 10, the percussion channel, has a defined set of note numbers on which particular sounds are to occur, so that the composer may know for example that key 39 will always be a 'hand clap'.

15.9 MIDI Machine Control (MMC)

MIDI may be used for remotely controlling tape machines and other studio equipment. MMC uses universal realtime SysEx messages, and has a lot in common with a remote control protocol known as 'ESbus' which was devised by the EBU and SMPTE as a universal standard for the remote control of tape machines, VTRs and other studio equipment.

There are a number of levels of complexity at which MMC can be made to operate, and communication is possible in both closed and open loop modes (either with or without a return connection to the controlling computer). By allowing this flexibility, MMC makes it possible for designers to implement it at anything from a very simple level (i.e. cheaply) to a very complicated level involving all the finer points. MMC is gaining increasing popularity in semi-professional equipment, because it is somewhat cheaper to implement than ESbus and allows equipment to be integrated easily with a MIDI-based studio system. There are a number of tape machines and synchronisers now on the market with MIDI interfaces, and some sequencer packages handle the remote control of studio machines using the MMC protocol. Studio machines may be connected to the main studio computer by connecting them to one port on a multiport MIDI interface, as was shown in Figure 15.10.

MMC could be used simply to control the basic transport functions of an audio tape recorder. In such a case only a very limited set of commands would need to be implemented in the tape recorder, and very little would be needed in the way of responses. In fact it would be quite feasible to operate the transport of a tape recorder using an open-loop approach – simply sending 'play', 'stop', 'rewind', etc. as required by the controlling application.

Recommended further reading

MMA (1983) *MIDI 1.0 Detailed Specification.* MIDI Manufacturers Association
MMA (1991) *General MIDI System Level 1.* MIDI Manufacturers Association
MMA (1993) *4.2 Addendum to MIDI 1.0 Specification.* MIDI Manufacturers Association
Rumsey, F. (1994) *MIDI Systems and Control*, 2nd edn., Focal Press
Yavelow, C. (1992) *Macworld Music and Sound Bible.* IDG Books Worldwide, Inc., San Mateo, CA, USA

Chapter 16

Timecode and synchronisation

The boundaries between audio and video operations are less clear these days, and subjects such as timecode which used to be almost universally the domain of the video engineer are now as pertinent to the audio engineer. Timecode is used widely in the audio post-production industry for synchronising machines and providing a real-time positional reference on tapes. It is used in video editing and in the editing of digital audio recordings, and it is used in hard-disk recording systems for compiling edit lists and for synchronisation. Many modern analogue tape recorders have timecode facilities as do professional digital recorders, some even being equipped with 'chase' synchronisers.

In the following chapter the basics of timecode and machine synchronisation are discussed, but omitting discussion of the many systems which have been used in the past (and are still used in certain cases) for the synchronisation of film systems. MIDI Timecode (MTC) is discussed in Chapter 15.

16.1 SMPTE/EBU timecode

The American Society of Motion Picture and Television Engineers proposed a system to facilitate the accurate editing of video tape in 1967. This became known as SMPTE ('simpty') code, and it is basically a continuously running eight-digit clock registering time from an arbitrary start point (which may be the time of day) in hours, minutes, seconds and frames, against which the programme runs. The clock information is encoded into a signal which can be recorded on the audio track of a tape. Every single frame on a particular video tape has its own unique number called the timecode address and this can be used to pinpoint a precise editing position.

A number of frame rates are used, depending on the television standard to which they relate, the frame rate being the number of still frames per second used to give the impression of continuous motion: 30 frames per second (fps), or true SMPTE, was used for monochrome American television, and is now only used for CD mastering in the Sony 1630 format; 29.97 fps is used for colour NTSC television (mainly USA, Japan and parts of the Middle East), and is called 'SMPTE drop-frame' (see Fact File 16.1); 25 fps is used for PAL and SECAM TV and is called 'EBU' (Europe, Australia, etc.); and 24 fps is used for some film work.

Each timecode frame is represented by an 80 bit binary 'word', split principally into groups of 4 bits, with each 4 bits representing a particular parameter such as tens

---

FACT FILE 16.1   Drop-frame timecode

When colour TV (NTSC standard) was introduced in the USA it proved necessary to change the frame rate of TV broadcasts slightly in order to accommodate the colour information within the same spectrum. The 30 fps of monochrome TV, originally chosen so as to lock to the American mains frequency of 60 Hz, was thus changed to 29.97 fps, since there was no longer a need to maintain synchronism with the mains owing to improvements in oscillator stability. In order that 30 fps timecode could be made synchronous with the new frame rate it became necessary to drop two frames every minute, except for every tenth minute, which resulted in minimal long-term drift between timecode and picture (75 ms over 24 hours). The drift in the short term gradually increased towards the minute boundaries and was then reset.

A flag is set in the timecode word to denote NTSC drop-frame timecode. This type of code should be used for all applications where the recording might be expected to lock to an NTSC video programme.

---

of hours, units of hours, and so forth, in BCD (binary-coded decimal) form (see Figure 16.1). Sometimes, not all four bits per group are required — the hours only go up to '23', for example — and in these cases the remaining bits are either used for special control purposes or set to zero (unassigned): 26 bits in total are used for time address information to give each frame its unique hours, minutes, seconds, frame value; 32 are 'user bits' and can be used for encoding information such as reel number, scene number, day of the month and the like; bit 10 can denote drop-frame mode if a binary 1 is encoded there, and bit 11 can denote colour frame mode if a binary 1 is encoded. The end of each word consists of 16 bits in a unique sequence, called the 'sync word', and this is used to mark the boundary between one frame and the next. It also allows the reader to tell in which direction the code is being read, since the sync word begins with 11 in one direction and 10 in the other.

This binary information cannot be recorded to tape directly, since its bandwidth would be too wide, so it is modulated in a simple scheme known as 'bi-phase mark', or FM, such that a transition from one state to the other (low to high or high to low) occurs at the edge of each bit period, but an additional transition is forced within the period to denote a binary 1 (see Figure 16.2). The result looks like a square wave with two frequencies, depending on the presence of ones and zeros in the code. Depending on the frame rate, the maximum frequency of square wave contained within the timecode signal is either 2400 Hz (80 bits × 30 fps) or 2000 Hz (80 bits × 25 fps), and the lowest frequency is either 1200 Hz or 1000 Hz, and thus it may easily be recorded on an audio machine. The code can be read forwards or backwards, and phase inverted. Readers are available which will read timecode over a very wide range of speeds, from around 0.1 to 200 times play speed. The rise-time of the signal, that is the time it takes to swing between its two extremes, is specified as 25 µs ± 5 µs, and this requires an audio bandwidth of about 10 kHz.

There is another form of timecode known as VITC (Vertical Interval Timecode), used widely in VTRs. VITC is recorded not on an audio track, but in the vertical sync period of a video picture, such that it can always be read when video is capable of being read, such as in slow-motion and pause modes. This code will not be covered further here.

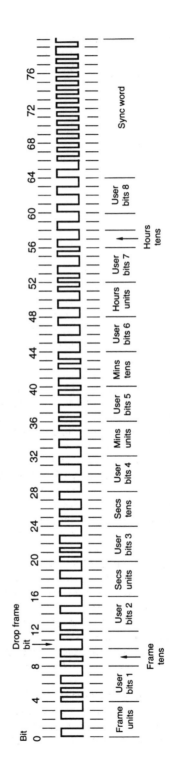

Figure 16.1 The data format of a SMPTE/EBU longitudinal timecode frame

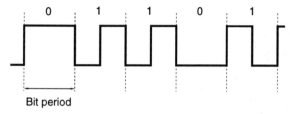

Bit period

Figure 16.2 Linear timecode data is modulated before recording using a scheme known as 'bi-phase mark' or FM (frequency modulation). A transition from high to low or low to high occurs at every bit-cell boundary, and a binary '1' is represented by an additional transition within a bit cell

16.2 Recording timecode

Timecode may be recorded or 'striped' on to tape before, during or after the programme material is recorded, depending on the application. In many cases the timecode must be locked to the same speed reference as that used to lock the speed of the tape machine, otherwise a long-term drift can build up between the passage of time on the tape and the measured passage in terms of timecode. Such a reference is usually provided in the form of a video composite sync signal, and video sync inputs are increasingly provided on digital tape recorders for this purpose.

Timecode generators are available in a number of forms, either as stand-alone devices (such as that pictured in Figure 16.3), as part of a synchroniser or editor, or integrally within a tape recorder. In large centres timecode is sometimes centrally distributed and available on a jackfield point. When generated externally, timecode normally appears as an audio signal on an XLR connector or jack, and this should be routed to the track required for timecode on the tape recorder. Most generators allow the user to preset the start time and the frame-rate standard.

Timecode is often recorded on to an outside track of a multitrack tape machine (usually track 24), or a separate timecode or cue track will be provided on digital machines. The signal is recorded at around 10 dB below reference level, and crosstalk between tracks or cables is often a problem due to the very audible mid-frequency nature of timecode. Some quarter-inch analogue machines have a facility for recording timecode in a track which runs down the centre of the guard band in the NAB track format (see section 8.4.1). This is called 'centre-track timecode', and a head arrangement similar to that shown in Figure 16.4 may be used for recording and replay. Normally separate heads are used for recording timecode to those for audio, to avoid crosstalk, although some manufacturers seem to have circumvented this problem and use the same heads. In the former case a delay line is used to synchronise timecode and audio on the tape.

Professional R-DAT machines are often capable of recording timecode, this being converted internally into a DAT running-time code which is recorded in the subcode area of the digital recording. On replay, any frame rate of timecode can be derived, no matter what was used during recording, which is useful in mixed-standard environments.

In mobile film and video work which often employs separate machines for recording sound and picture it is necessary to stripe timecode on both the camera's tape or film and on the audio tape. This can be done by using the same timecode generator to feed both machines, but more usually each machine will carry its own

Figure 16.3 A stand-alone timecode generator. (Courtesy of Avitel Electronics Ltd.)

a)

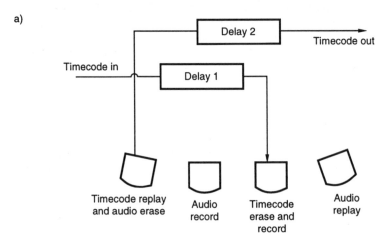

b)

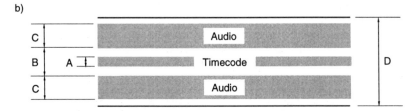

A = 0.8mm, B = 2.1mm, C = 2.0mm, D = 6.3mm

Figure 16.4 The centre-track timecode format on quarter-inch tape. (a) Delays are used to record and replay a timecode track in the guard band using separate heads. (Alternatively, specially-engineered combination heads may be used.) (b) Physical dimensions of the centre-track timecode format

generator and the clocks will be synchronised at the beginning of each day's shooting, both reading absolute time of day. Highly stable crystal control ensures that sync between the clocks will be maintained throughout the day, and it does not then matter whether the two (or more) machines are run at different times or for different lengths of time because each frame has a unique time of day address code which enables successful post-production syncing.

The code should run for around 20 seconds or more before the programme begins in order to give other machines and computers time to lock in. If programme is spread over several reels, the timecode generator should be set and run such that no number repeats itself anywhere throughout the reels, thus avoiding confusion during post-production. Alternatively the reels can be separately numbered.

16.3 Synchronisers

16.3.1 Overview

A synchroniser is a device which reads timecode from two or more machines and controls the speeds of 'slave' machines so that their timecodes run at the same rate as the 'master' machine. It does so by modifying the capstan speed of the slave machines, using an externally applied speed reference signal, usually in the form of a 19.2 kHz square wave whose frequency is used as a reference in the capstan servo circuit (see Figure 16.5). The synchroniser is microprocessor controlled, and can incorporate offsets between the master and slave machines, programmed by the user. It may also be able to store pre-programmed points for such functions as record drop-in, drop-out, looping and autolocation, for use in post-production.

16.3.2 Chase synchroniser

A simple chase synchroniser could simply be a box with a timecode input for master and slave machines and a remote control interface for each machine (see Figure 16.6). Such a synchroniser is designed to cause the slave to follow the master wherever it goes, like a faithful hound. If the master goes into fast forward so does

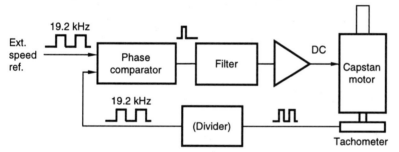

Figure 16.5 Capstan speed control is often effected using a servo circuit similar to this one. The frequency of a square wave pulse generated by the capstan tachometer is compared with an externally generated pulse of nominally the same frequency. A signal based on the difference between the two is used to drive the capstan motor faster or slower

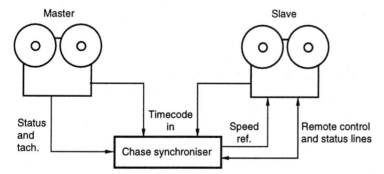

Figure 16.6 A simple chase synchroniser will read timecode, direction and tachometer information from the master, compare it with the slave's position and control the slave accordingly until the two timecodes are identical (plus or minus any entered offset)

the slave, the synchroniser keeping the position of the slave as close as possible to the master, and when the master goes back into play the synchroniser parks the slave as close as possible to the master position and then drops it into play, adjusting the capstan speed to lock the two together. A full-featured chase synchroniser is pictured in Figure 16.7.

In fast wind modes, a chase synchroniser will tend not to read timecode, since the tape is not normally in contact with the heads and the timecode reader may not be able to read code at wind speeds, so it reads tachometer pulses from the tape machine's roller guide, transferred over the remote interface. The synchroniser will be programmed so as to count the correct number of tach pulses per second for each machine (they tend to differ considerably) or it may be able to work this out automatically during the first few seconds of operation. When the machine goes back into play it reads timecode again and adjusts its estimation of its position, which should be fairly close to that worked out from the tach pulses. The synchroniser then uses the difference between the master and slave timecode values, plus or minus any offset, to speed up or slow down the slave in order to lock it closely to the master.

Figure 16.7 A modular chase synchroniser with serial bus control facilities. (Courtesy of Audio Kinetics UK Ltd.)

FACT FILE 16.2 Types of lock

Frame lock or absolute lock
This term or a similar term is used to describe the mode in which a synchroniser works on the absolute time values of master and slave codes. If the master jumps in time, due to a discontinuous edit for example, then so does the slave, often causing the slave to spool off the end of the reel if it does not have such a value on the tape.

Phase lock or sync lock
These terms are often used to describe a mode in which the synchroniser initially locks to the absolute value of the timecode on

master and slaves, switching thereafter to a mode in which it simply locks to the frame edges of all machines, looking at the sync word in the timecode and ignoring the absolute value. This is useful if discontinuities in the timecode track are known or anticipated, and ensures that a machine will not suddenly drop into a fast spool mode during a programme.

Slow and fast relock
After initial lock is established, a synchroniser may lose lock due to a timecode drop-out or discontinuity in timecode phase. In fast relock mode the synchroniser will attempt to relock the machines as quickly as possible, with no concern for the audible effects of pitch slewing. In slow relock mode, the machines will relock more slowly at a rate intended to be inaudible.

Such a synchroniser could be used to lock two multitrack recorders together, for example in order to increase the number of available tracks, or it could be used to slave a quarter-inch machine to a VTR for laying off or laying back stereo sound tracks in video editing. It should act as an almost invisible link between the machines and should require little attention. The initiation for chasing should not need to come from the user; the slave should start to move as soon as it sees timecode move from the master. Some chase synchronisers of this sort will even work if no remote connection is made to the master, simply chasing the timecode presented to its input (which could have come from anywhere). Systems vary as to what they will do if the master timecode drops out or jumps in time. In the former case most synchronisers wait a couple of seconds or so before stopping the slave, and in the latter case they may try to locate the slave to the new position (this depends on the type of lock employed, as discussed in Fact File 16.2).

Occasionally a machine may be fitted with an internal chase synchroniser which locks to a timecode input on the rear of the machine. It may also have a built-in timecode generator.

16.3.3 Full-featured synchroniser

In post-production operations a controller is often required which offers more facilities than the simple chase synchroniser, such as the example pictured in Figure 16.8. Such a device may allow for multiple machines to be controlled from a single controller, perhaps using a computer network link to communicate commands from the controller to the individual tape machines. In some 'distributed intelligence' systems, each tape machine has a local chase synchroniser which communicates with the controller, the controller not being a synchroniser but a 'command centre' (see Figure 16.9). The ES-bus is a remote control bus used increasingly in such applications, designed to act as a remote control bus for audio and video equipment.

The sync controller in such a system will offer facilities for storing full edit decision lists (EDLs) containing the necessary offsets for each slave machine and

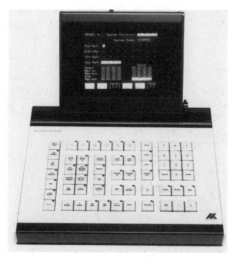

Figure 16.8 A full-featured sync controller: the Audio Kinetics *Eclipse*. (Courtesy of Audio Kinetics UK Ltd.)

FACT FILE

16.3

Synchroniser terminology

Pre-roll

The period prior to the required lock point, during which machines play and are synchronised. Typically machines park about 5 seconds before the required lock point and then pre-roll for 5 seconds, after which it is likely that the synchroniser will have done its job. It is rare not to be able to lock machines in 5 seconds, and often it can be faster.

Post-roll

The period after a programmed record drop-out point during which machines continue to play in synchronised fashion.

Loop

A programmed section of tape which is played repeatedly under automatic control, including a pre-roll to lock the machines before each pass over the loop.

Drop-in and drop-out

Points at which the controller or synchroniser executes a pre-programmed record drop-in or drop-out on a selected slave machine. This may be at the start and end of a loop.

Offset

A programmed timecode value which offsets the position of a slave with relation to the master, in order that they lock at an offset. Often each slave may have a separate offset.

Nudge

Occasionally it is possible to nudge a slave's position frame by frame with relation to the master once it has gained lock. This allows for small adjustments to be made in the relative positions of the two machines.

Bit offset

Some synchronisers allow for offsets of less than one frame, with resolution down to one-eightieth of a frame (one timecode bit).

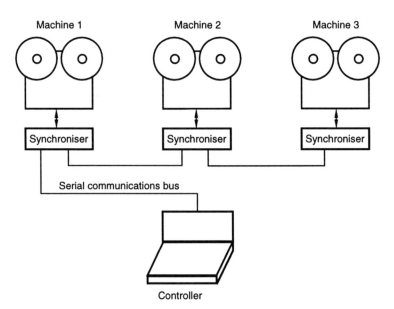

Figure 16.9 In an advanced synchronised system each machine is equipped with its own modular synchroniser, receiving commands and offsets from a serially-linked controller

the record drop-in and drop-out points for each machine. This can be used for jobs such as automatic dialogue replacement (ADR), in which sections of a programme can be set to loop with a pre-roll (see Fact File 16.3) and drop-in at the point where dialogue on a film or video production is to be replaced. A multitrack recorder may be used as a slave, being dropped in on particular tracks to build up a sound master tape. Music and effects can then be overdubbed.

In locked systems involving video equipment the master machine is normally the video machine, and the slaves are audio machines. This is because it is easier to synchronise audio machines, and because video machines may need to be locked to a separate video reference which dictates their running speed. In cases involving multiple video or digital audio machines, none of the machines is designated the master, and all machines slave to the synchroniser which acts as the master. Its timecode generator is locked to the house video or audio reference, and all machines lock to its timecode generator. This technique is also used in video editing systems.

Recommended further reading

Amyes, T. (1990) *The Technique of Audio Post-Production in Video and Film.* Focal Press

Ratcliff, J. (1995) *Timecode: A User's Guide.* Focal Press

Chapter 17

Stereo recording and reproduction

This chapter investigates the principles and practice of stereo recording and reproduction. The discussion of 'stereo' is not limited to conventional two loud-speaker reproduction, but considers it in the true sense of the Greek stereo, meaning 'solid', or 'three-dimensional'. Stereo techniques cannot be considered from a purely theoretical point of view, neither can the theory be ignored, the key being in a proper synthesis of theory and subjective assessment. Some techniques which have been judged subjectively to be good do not always stand up to rigorous theoretical analysis, and those which are held up as theoretically 'correct' are sometimes judged subjectively to be poorer than others. Part of the problem is that the mechanisms of directional perception are not yet entirely understood. Further-more, most commercial stereo reproduction uses only two loudspeakers, and thus the listening situation is already a distortion of reality (since real sonic experience involves sound arriving from all around the head), perhaps leading listeners to prefer distorted images because of the pleasing artefacts such as 'spaciousness' which are often side-effects, just as many listeners 'prefer' sound which is distorted in other ways. Most of the stereo techniques used today combine aspects of imaging accuracy with an attempt to give the impression of spaciousness in the sound field, and to many theorists these two are almost mutually exclusive.

In the following chapters stereo pickup and reproduction is considered from both a theoretical and a practical point of view, recognising that theoretical rules may have to be bent or broken for operational and subjective reasons. Since the subject is far too large even to be summarised in the short space available, a list of recommended further reading is given at the end of the chapter to allow the reader greater scope for personal study.

17.1 Relationships between directional perception and stereo sound techniques

The subject of stereo recording and reproduction is closely linked to the subject of directional perception in human hearing, since it is the aim of any stereo pickup and reproduction technique to create the illusion of directionality and space in repro-duced sound. A study of directional hearing also helps the reader considerably in appreciating the differences between alternative techniques such as binaural stereo, coincident-pair stereo, spaced-pair stereo, pan-potted stereo and various surround sound approaches. Chapter 2 of this book provides an overview of the main mechanisms involved in directional perception, and the further reading recommendations at the end of that chapter go into greater depth if that is required.

To summarise, the ability of the human hearing mechanism to detect the location of sounds depends on a combination of frequency-dependent amplitude difference between the ears (due principally to the shadowing effect of the head) at mid-to-high frequencies, on frequency-dependent phase differences for continuous low frequency sounds, on time-of-arrival differences for complex transient sounds, and on the subtle spectral shaping effects of the outer ear. Reflections from other parts of the body and from the floor may also serve to alter the received spectrum. For every angle of incidence of sound a unique 'template' of notches and peaks in the frequency spectrum will modify the received sound signal, and this is one of the principal factors influencing the brain's ability to distinguish between front and rear, and between up and down in hearing. When considering stereo techniques there is a need to distinguish between the 'binaural effect' in which the two ears are independently stimulated by the same original wavefront (resulting in interaural delays of between 0 and about 600 µs) and the 'precedence' effect in which the two ears are both stimulated by two or more sources arriving at different times (effective with delays between the sources of up to about 50 ms, depending on the nature of the source). There is some evidence to suggest that the binaural hearing mechanism adapts itself to the direction suggested by the onset of a sound, ignoring subsequent information to some extent unless that information succeeds in retriggering the localisation process. It is possibly true, therefore, that *changes* in the sound have a more persuasive influence on localisation than continuous features.

It might reasonably be supposed that the best stereo sound system would be that which reproduced the sound signal to the ears as faithfully as possible, with all the original directional cues intact. Possibly that should be the aim, and indeed it is the aim of the so-called 'binaural' techniques discussed later in the chapter, but there are many stereo techniques that rely on loudspeakers for reproduction which only manage to provide *some* of the directional cues to the ears. Such techniques are compromises which have varying degrees of success, and they are necessary for the simple reason that they are reasonably straightforward from a recording point of view, the results can be reproduced in anyone's living room, and are demonstrably better than mono (single channel reproduction). Theoretical correctness is one thing, pragmatism is another, and the history of stereo could be characterised all along as being something of a compromise between the two.

17.2 Basic directional sound reproduction

Now that the mechanisms by which direction is perceived in sound have been introduced, the principles of basic directional sound reproduction in audio systems will be discussed as a precursor to the discussion of stereo signals and microphone configurations.

17.2.1 Historical development

We have become used to stereo sound as a two channel format, although a review of developments during the last century shows that two channels really only became the norm through economic and domestic necessity, and through the practical considerations of encoding directional sound easily for gramophone records and radio. A two loudspeaker arrangement is practical in the domestic environment, is reasonably cheap to implement, and provides good phantom images for a central listening position.

Early work on directional reproduction undertaken at Bell Labs in the 1930s involved attempts to recreate the 'sound wavefront' which would result from an infinite number of microphone/loudspeaker channels by using a smaller number of channels, as shown in Figure 17.1(a) and (b). In all cases, spaced pressure response (omnidirectional) microphones were used, each connected via a single amplifier to the appropriate loudspeaker in the listening room. Steinberg and Snow found that when reducing the number of channels from three to two, central sources appeared to recede towards the rear of the sound stage and that the width of the reproduced sound stage appeared to be increased. They attempted to make some calculated rather than measured deductions about the way that loudness differences between the channels affected directional perception, choosing to ignore the effects of time or phase difference between channels.

Some twenty years later Snow made comment on those early results, reconsidering the effects of time difference in a system with a small number of channels, since, as he pointed out, there was in fact a marked difference between the multiple-point-source configuration and the small-number-of-channels configuration. It was suggested that in fact the 'ideal' multi-source system recreated the original wavefront very accurately, allowing the ears to use exactly the same binaural perception mechanisms as used in the real-life sound field. It may be appreciated that the 'wall' of multiple loudspeakers acted as a source of wavelets, re-creating a new plane wave with its virtual source in the same relative place as the original source, thus resulting in a TOA difference between the listener's ears in the range 0–600 µs, depending on position. In the two or three channel system, far from this simply being a sparse approximation to the 'wavefront' system, the ears are subjected to two or three discrete arrivals of sound, the delays between which are likely to be considerably in excess of those normally experienced in binaural listening, being a number of milliseconds depending on the original spacing of the microphones. In this case, the effect of directionality relies much more on the precedence effect described in Chapter 2, and also on the relative levels of the channels. Snow therefore begs us to remember the fundamental difference between 'binaural' situations (see section 17.2.3), and what he calls 'stereophonic' situations (see section 17.2.2).

This difference was also recognised by Alan Blumlein, whose now famous patent specification of 1931 (accepted 1933) allows for the conversion of signals from a binaural format suitable for spaced pressure microphones to a format suitable for reproduction on loudspeakers, and for other formats of pickup which result in an approximation of the original time and phase differences at the ears when reproduced on loudspeakers. This will be discussed in more detail later on, but it is interesting historically to note how much writing on stereo reproduction even in the early 1950s appears unaware of Blumlein's most valuable work, which appears to have been ignored for some time.

A British paper presented by Clark, Dutton and Vanderlyn (of EMI) in 1957 revives the Blumlein theories, and shows in more rigorous mathematical detail than Blumlein's patent specification how a two-loudspeaker system may be used to create an accurate correlation between the original angle of offset of a sound source and the perceived angle of offset on reproduction by controlling only the relative signal amplitudes of the two loudspeakers (derived in this case from a pair of coincident figure-eight microphones). The authors discuss the three channel spaced

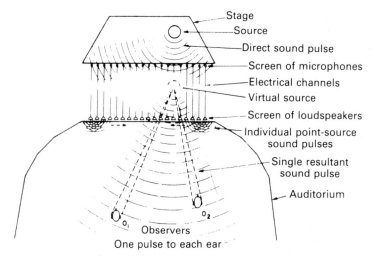

Figure 17.1(a) A good virtual sound stage could be constructed by an infinite number of microphone–loudspeaker channels, giving rise to correct binaural differences for listeners in any position (after Steinberg and Snow)

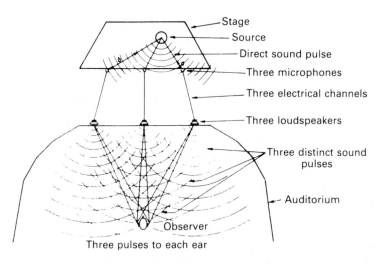

Figure 17.1(b) The reduction in number of channels to three gives rise to three distinct arrivals of sound from the three loudspeakers, dependent on the microphone spacing. The differences between the ears are not the same as in natural listening, and the precedence effect takes over (after Steinberg and Snow)

microphone system of Bell Labs, and suggest that although it produces convincing results in many listening situations it is uneconomical for domestic use, and that the two-channel simplification (again using spaced microphones, at about ten feet apart) has a tendency to result in the subjective 'hole-in-the-middle' effect with which many modern users of spaced microphones may be familiar (the sound appearing to come either from the left or the right but not anywhere in between). They concede, in a discussion, that the Blumlein method adapted by them does not take advantage of all the mechanisms of binaural hearing, especially the precedence effect, but that they have endeavoured to take advantage of, and recreate, a few of the directional cues which exist in the real-life situation.

There is therefore a historical basis for both the spaced microphone arrangement which makes use of the precedence effect (with only moderate level differences between channels), as well as the coincident microphone technique (or any other technique which results in only level differences between channels), with some evidence to show that the spaced technique is more effective with three channels than with only two. Later, we shall see that spaced techniques have a fundamental theoretical flaw from a point-of-view of 'correct' imaging of continuous sounds, which has not always been appreciated, although such techniques may result in subjectively acceptable sounds. Interestingly, three front channels are the norm in cinema sound reproduction, since the central channel has the effect of stabilising the important central image for off-centre listeners, having been used ever since the Disney film *Fantasia* in 1939. People have often misunderstood the intentions of Bell Labs in the 1930s, since it is not generally realised that they were working on a system suitable for auditorium reproduction with wide-screen pictures, as opposed to a domestic system.

Alternatives to three-channel stereo exist for the coverage of a wider listening area, these involving loudspeaker arrangements which are directional in such a way as to compensate off-centre listeners with a higher audio level from the more distant speaker to make up for the increased precedence effect of the closer speaker. One interesting example of this is the recent Canon 'Wide-Imaging Stereo' speaker system which makes use of shaped conical reflectors mounted above mid- and high-frequency drivers pointing upwards, in order to give the loudspeaker a polar response that favours listeners on the opposite side of the listening area, compensating for the increased precedence effect of the closer speaker.

17.2.2 Two channel stereo from loudspeakers

The principles of stereo reproduction using two channels on loudspeakers (as opposed to headphones) will now be discussed in more detail, since this is the most common listening arrangement in use today.

From the above discussions it will now be clear that in most cases stereo reproduction from two loudspeakers can only hope to achieve a modest illusion of the original soundfield, since reproduction is from the front quadrant only (although see section n below). It is possible to create an illusion of directionality and space in stereo sound using either time differences between the speaker outputs or using level differences between them. It is also possible to use a combination of the two, although there is a problem with 'time-difference' stereo in that contradictions may arise between transient and continuous sounds (see Section 6.8.1). The main point to be considered with loudspeaker reproduction is that both ears will receive the signals from both speakers, whereas in headphone listening each ear only receives one signal channel. The result of this is that the loudspeaker listener seated in a centre seat (see Figure 17.2) receives at his left ear the signal from the left speaker first, followed by that from the right speaker, and at his right ear the signal from the right speaker first, followed by that from the left speaker, the time t being the time taken for the sound to travel the extra distance from the more distant speaker. The basis on which 'level-difference' or 'Blumlein' stereo works is to convert level differences between two loudspeakers into low frequency phase differences be-

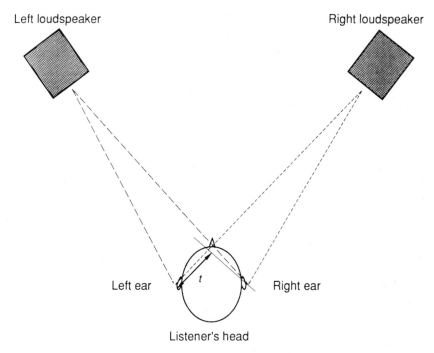

Figure 17.2 When listening to sound on two loudspeakers, both ears receive sound from both loudspeakers, the signal from the more distant loudspeaker arriving a time *t* later than that from the nearer speaker for each ear

tween the ears, based on the summation of the loudspeaker signals at the two ears, as described in Fact File 17.1.

Experiments by the author with wideband speech signals replayed on a standard loudspeaker arrangement have shown that a level difference of approximately 18 dB between channels is necessary to give the impression that a sound comes from either 'fully left' or 'fully right' in the image (see Figure 17.3), and that there is a measure of disagreement between listeners as to the positions of 'half left' and 'half right' (which might be expected). If a time difference also exists between the channels, then transient sounds will be 'pulled' towards the advanced speaker because of the precedence effect, the perceived position depending to some extent on the time delay. If the left speaker is advanced in time relative to the right speaker (or more correctly, the right speaker is delayed!) then the sound appears to come more from the left speaker, although this can be corrected by increasing the level to the right speaker. There is a trade-off between time and level difference, and it can be shown that if the left channel is, say, 2 ms earlier than the right, then the right must be made approximately 5 dB louder in order to compensate and bring the signal back to the centre. This was illustrated in Figure 2.5.

A time difference of between 2 and 4 ms (depending on the nature of the signal) appears to be required for a sound to appear either fully left or fully right, although this only holds true for transient signals and may be contradicted by the resulting LF phase differences for continuous sounds. Apparently the ear is able to resolve the conflicts which arise in such situations, often locking strongly onto the time

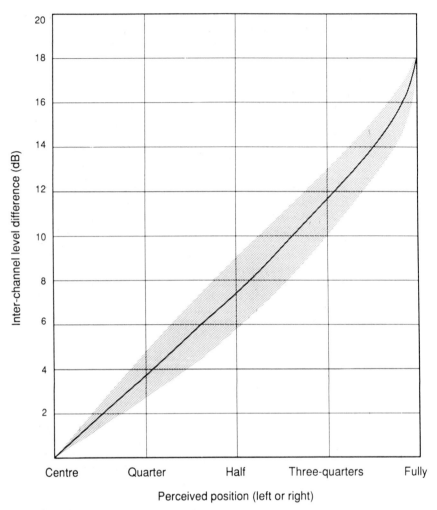

Figure 17.3 Speech sources appear at different positions on reproduction depending on the level difference between the channels. The shaded areas shows measured disagreement between listeners.

difference information in the signal in preference to other cues. Practical stereo microphone techniques operate using combinations of level and time difference between the channels.

The above theories of stereo have been termed 'summation localisation theories' by Günther Theile of the IRT (Institut für Rundfunktechnik) in Germany, who dismisses such theories as not allowing for the correct natural interaural attributes of sound signals required for spatial reproduction. Theile states that the correct way to derive signals suitable for natural stereo reproduction, even when using loudspeakers, is to generate 'head-referred' signals – that is signals similar to those used in binaural systems – since these contain the necessary information to reproduce a sound image in a 'simulation plane' between two loudspeakers. Subjective test results have shown that binaural signals equalised for a flat frontal frequency response are capable of producing convincing stereo from loudspeakers. There is often strong disagreement between those who adhere to this theory and those who adhere more closely to traditional summation theories of stereo.

FACT FILE 17.1 Stereo vector summation

If the outputs of the two speakers differ only in level and not in phase (time) then it can be shown (at least for low frequencies up to around 700 Hz) that the vector summation of the signals from the two speakers at each ear results in two signals which, for a given frequency, differ in phase angle proportional to the relative amplitudes of the two signals (the level difference between the ears being negligible at LF). For a given level difference between the speakers, the phase angle changes approximately linearly with frequency, which is the case when listening to a real point source. At higher frequencies the phase difference cue ceases to be useful but the shadowing effect of the head results in level differences between the ears.

If the amplitudes of the two channels are correctly controlled it is possible to produce resultant phase and amplitude differences for continuous sounds which are very close to those experienced with natural sources, thus giving the impression of virtual images anywhere between the left and right loudspeakers. This is the basis of Blumlein's 1931 stereophonic system 'invention', further developed by Clark, Dutton and Vanderlyn in 1957. The result of the phasor analysis is a simple formula which can be used to determine, for any angle subtended by the loudspeakers at the listener, what the apparent angle of the virtual image will be for a given difference between left and right levels. Firstly, referring to the diagram, it can be shown that:

$$\sin \alpha = ((L - R)/(L + R)) \sin \vartheta_0$$

where a is the apparent angle of offset from the centre of the virtual image, and J0 is the angle subtended by the speaker at the listener. Secondly, it can be shown that:

$$(L - R)/(L + R) = \tan \vartheta_t$$

where ϑ_t is the true angle of offset of a real source from the centre-front of a coincident pair of figure-eight velocity microphones. (L – R) and (L + R) are the difference (S) and sum (M) signals of a stereo pair, defined in Fact File 4.5.

It is a useful result since it shows that it is possible to use positioning techniques such as 'pan-potting' which rely on the splitting of a mono signal source into two components, with adjustment of the relative proportion fed to the left and right channels without affecting their relative timing. It also makes possible the combining of the two channels into mono without cancellations due to phase difference.

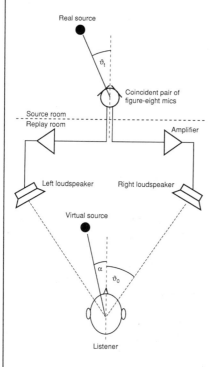

17.2.3 Two channel stereo from headphones

Headphone reproduction is different to loudspeaker reproduction since, as already stated, each ear is fed only with one channel's signal. This is therefore an example of the binaural situation and allows for the ears to be fed with signals which differ in time by up to the binaural delay (600 μs), and also differ in amplitude by amounts similar to those differences which result from the shadowing effects of the head. This suggests the need for a microphone technique which uses microphones spaced apart by the binaural distance, and baffled by an object similar to the human head, in order to produce signals with the correct differences.

Bauer has pointed out that if stereo signals designed for reproduction on loudspeakers were fed to headphones there would be a too-great level difference between the ears compared with the real life situation, and that the correct interaural delays would not exist. This results in an unnatural stereo image which does not have the expected sense of space. He therefore proposed a network which introduced a measure of delayed crosstalk between the channels to simulate the correct interaural level differences at different frequencies, as well as simulating the interaural time delays which would result from loudspeaker signals incident at 45° to the listener. He based the characteristics on research done by Weiner which produced graphs for the effects of diffraction around the human head for different angles of incidence. The characteristics of Bauer's circuit are shown in Figure 17.4 (with Weiner's results shown dotted). It may be seen that Bauer chooses to reduce the delay at HF, partially because the circuit design would have been too complicated, and partially because localisation relies more on amplitude difference at HF anyway.

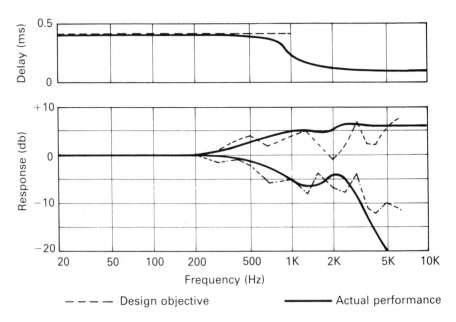

Figure 17.4 Bauer designed a circuit which introduced binaural delay and crosstalk into signals designed for loudspeaker reproduction so that they could be reproduced satisfactorily on headphones. The characteristics of his circuit are shown here. The upper graph shows the amount of delay introduced in the crossfeed between channels, whilst in the lower graph L_g and R_g are left and right channel gains to imitate the shadowing effects of the head

Bauer also suggests the reverse process (turning binaural signals into stereo signals for loudspeakers), pointing out that crosstalk must be removed between binaural channels for correct loudspeaker reproduction, since the cross-feed between the channels will otherwise occur twice (once between the pair of binaurally-spaced microphones, and again at the ears of the listener), resulting in poor separation and a narrow image. He suggests that this may be achieved using the subtraction of an anti-phase component of each channel from the other channel signal, although he does not discuss how the time difference between the binaural channels may be removed. Further work on a circuit for improving the stereo headphone sound image was done by Thomas in 1977, quoting listening tests which showed that all of his listening panel preferred stereo signals on headphones which had been subjected to the 'crossfeed with delay' processing.

Work by Theile on the frequency-response equalisation of headphones for optimum spatial impression has suggested that it is most important for the equalisation of dummy heads and headphones to be standardised, so as not to destroy the spectral cues which are vital for binaural localisation. It is also pointed out, nonetheless, that the ear/brain is able to adapt to the particular spectral characteristics of a reproduction system over a period of time, after which spatial information is more readily decoded.

17.3 Binaural signal processing

Recent work involving digital signal processing has resulted in systems which can be used for processing audio signals so as to introduce the correct 'head-related

---

FACT FILE 17.2 Transaural stereo

The basis of transaural stereo is in the preconditioning of loudspeaker signals with a component which represents the inverse of the crosstalk which occurs between the two ears when listening to loudspeakers. It assumes that the source material is binaurally recorded and allows the ears to be supplied with the correct binaural signal relationships.

Experiments by Atal and Schroeder in 1962 showed that it was possible to introduce a crosstalk-cancelling component into the signals fed to two loudspeakers which would compensate for the interaural crosstalk, although the system was so position critical on the part of the listener (one had to be within 75 mm of the 'correct' position), as well as requiring that the listening environment be anechoic, that the concept did not proceed into general acceptance. Nonetheless, the

results, for listeners in the correct environment and position, were said to be stunning, with directional information present in all dimensions.

The problem with transaural stereo, as with other forms of binaural processing, is that it is highly dependent on the head and pinna characteristics implemented in the filters. Recent work has shown an improvement in the success of transaurally processed binaural material, which owes much to the simplification of the filters. Filters which generalise many of the HRTF characteristics, and which are less specific to one person's hearing, tend to work for a greater number of listeners over a wider listening position. Nonetheless, it seems unlikely that any system which is highly dependent for its success on the listener's seating position will ultimately become popular. A commercial product embodying transaural principles has appeared in the form of *Roland Sound Space* – a binaural mixer and transaural crosstalk canceller designed principally as an audio effects unit.

transfer function' (HRTF) for every possible angle of incidence of a monophonic sound source. In practice, the system employs a series of digital filters, delays and mixers which simulate the effect of the pinnae on sound signals as well as simulating the delay and shadowing effects of the head. This has resulted in a system capable of acting as a 'binaural mixer', with the facility for positioning mono sounds at any elevation or offset angle around the listener. It is also possible to use the system as a 'room simulator' in order to test the effects of placing sources in virtual rooms of different shapes, sizes and surface characteristics, allowing a listener to place himself anywhere in this virtual room and receive the correct binaural signals. This could be of considerable benefit to acoustic designers.

A recent development in the field of stereo recording and reproduction exists in the form of 'transaural stereo', designed to allow the reproduction of binaural signals through loudspeakers such that all the original directional cues remain intact at the listener's ears. This is discussed in more detail in Fact File 17.2.

17.4 Multiple channel stereo on loudspeakers

Although it is not proposed to cover surround sound systems in great detail here, some of the most important approaches are outlined. Sound for pictures has long made use of more than two loudspeaker channels to cover a wide listening area, to 'fix' the central image where dialogue tends to reside, and to offer some additional spatial information such as rear images or 'surround sound'. Multi-channel repro-duction for domestic usage has had a chequered and not altogether successful history. Quadraphonic reproduction systems appeared in the 1970s, using a number of different standards, but they have not found continued favour and thus will not be discussed further here. The so-called 'Ambisonic' system, developed under the auspices of the NRDC in the UK during the 1970s and based mainly on work by Gerzon, Fellgett and Barton, has had a certain amount of success in sound-only applications. This system is based on well-researched psychoacoustic principles and aims to reproduce the original sound field as accurately as possible in a wide variety of possible configurations. Its principal child in commercial terms is the 'Soundfield' microphone which was mentioned in Section 4.7.

17.4.1 Cinema-style surround sound

The most popular multi-channel sound format for the cinema at the present time is based on four channels: left (L), centre (C), right (R) and 'surround' (S), with sound signals panned into positions around the listener during post-production by varying the relative levels between the channels. All dialogue is normally sent to the centre channel, with music and effects using the space of the other three channels. The S channel is delayed and band limited in order to reduce any detrimental effects of crosstalk, and to avoid the possibility of a precedence 'pull' towards the rear of the auditorium (epecially for listeners nearer the back than the front). The LCRS format for sound signals is inherent in the 'Dolby Stereo' format for cinema sound tracks, and 'Dolby Surround', its consumer equivalent. Dolby manufactures an encoder and decoder matrix which allows for the four channels to be combined into two channels for recording onto film or video with reasonable compatibility in the case of two channel listening. A block diagram of the consumer 'ProLogic' decoder matrix is shown in Figure 17.5.

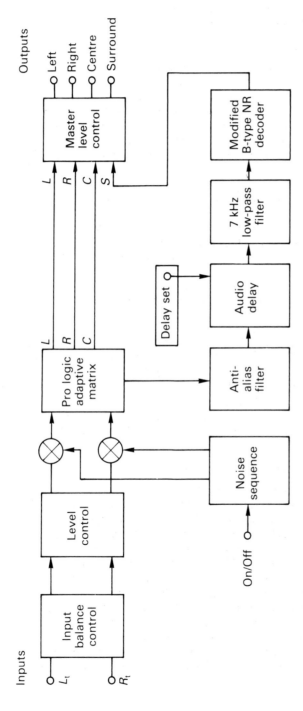

Figure 17.5 Dolby Surround ProLogic decoder block diagram. (Courtesy of Dolby Labs)

17.4.2 Ambisonics

Ambisonics aims to offer a complete hierarchical approach to directional sound pickup, storage or transmission and reproduction, which is equally applicable to mono, stereo, horizontal surround sound, or full 'periphonic' reproduction including height information. Depending on the number of channels employed it is possible to represent a smaller or greater number of dimensions in the reproduced sound. A number of formats exist for signals in the ambisonic system, and these are as follows: the A-format for microphone pickup (Figures 17.6 and 17.7), the B-format for studio equipment and processing (Figure 17.8), the C-format for transmission (Figure 17.9), and the D-format for decoding and reproduction (Figure 17.10). A format known as UHJ ('Universal HJ', 'HJ' simply being the letters denoting two earlier surround sound systems), described originally by Gaskell of the BBC Research Department, may be used for encoding surround-sound ambisonic information into two or three channels, whilst retaining good mono and stereo compatibility for 'non-surround' listeners.

Ambisonic sound should be distinguished from quadraphonic sound, since quadraphonics explicitly requires the use of four loudspeaker channels, and cannot be adapted to the wide variety of pickup and listening situations which may be encountered. Quadraphonics generally works by creating conventional stereo phantom images between each pair of speakers, and conventional stereo does not perform well when the listener is off-centre or when the loudspeakers subtend an angle larger than 60°. Since in quadraphonic reproduction the loudspeakers are angled at roughly 90° there is a tendency towards a hole-in-the-middle, as well as there being the problem that conventional stereo theories do not apply correctly for speaker pairs to the side of the listener. Ambisonics however, correctly encodes sounds from all directions in terms of pressure and velocity components, and correctly decodes these signals to a number of loudspeakers, with psychoacoustically optimised shelf filtering above 700 Hz to correct for the shadowing effects of the head, and an amplitude matrix which determines the correct levels to be fed to each speaker for the layout chosen.

The source of an ambisonic signal may be an ambisonic microphone such as the Calrec *Soundfield* (pictured earlier in Figure 4.12), or it may be an artificially panned mono signal, split into the correct B-format components and placed in a position around the listener by adjusting the ratios between the signals.

It is often the case that theoretical 'correctness' in a system does not automatically lead to widespread commercial adoption, and despite considerable coverage of ambisonic techniques, both in academic circles and in the trade journals, the

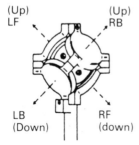

Figure 17.6 A-format capsule directions in an ambisonic microphone. The capsules have a sub-cardioid response $(2 + \cos\vartheta)$ and can be combined in pairs by subtraction to produce figure-eight responses

Figure 17.7 Capsules of the AMS Calrec *Soundfield* microphone arranged in the A format. (Courtesy of AMS)

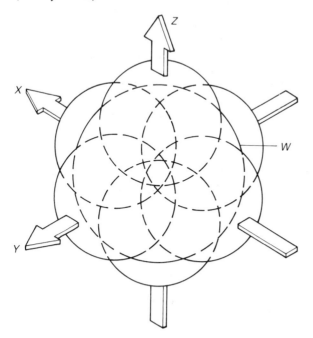

Figure 17.8 B-format components X, Y and Z are figure-eight patterns representing lateral position and height in the stereo image. They are derived from the A-format capsule pair signals using sum and difference processing. W is an omnidirectional pressure component created by adding the outputs of the A-format capsules in phase

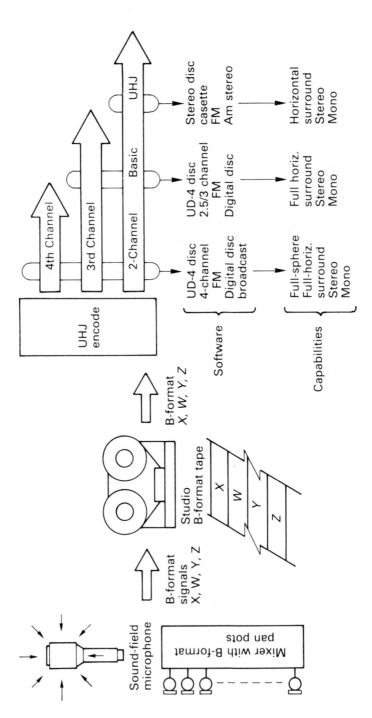

Figure 17.9 B-format signals can be converted to C-format (UHJ) for storage or transmission, the number of C-format channels dictating the amount of surround sound information recoverable upon decoding

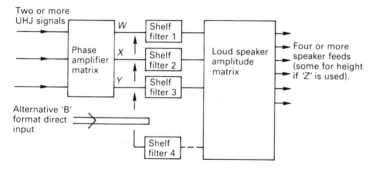

Figure 17.10 C-format signals are decoded to provide D-format signals for loudspeaker reproduction

system is still only used rarely in commercial recording and broadcasting. It is true that the Soundfield microphone is used widely, but this is principally because of its unusual capacity for being steered electrically so as to allow the microphone to be 'pointed' in virtually any direction without physically moving it, and set to any polar pattern between omni and figure-eight, simply by turning knobs on a control box. It is used in this respect as an extremely versatile stereo microphone for two channel recording and reproduction.

17.4.3 The 3–2 stereo surround format

It has recently been proposed that a surround sound signal format should be standardised for use in applications such as high definition television (HDTV) sound. The main international standards organisations have consequently elected to

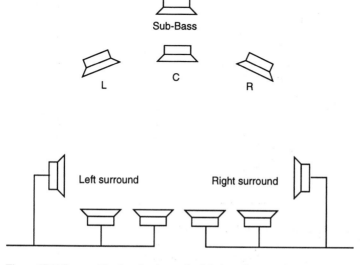

Figure 17.11 Proposed loudspeaker layout for 5.1 channel surround systems

compromise on a loudspeaker layout which incorporates three front channels and two rear channel, plus an optional sub-woofer channel for low frequency effects, amounting to what is often called 5.1 channel surround (see Figure 17.11). This allows for a stereo rear effects channel, which is judged by many listeners to be superior to the mono effects channel of conventional Dolby Stereo. This so-called 3–2 format now forms the basis for many of the proposed standards for surround sound broadcasting with television, and we may expect to see more of it in the future.

The proposed standard does not say anything about how source material should be made for the 3–2 format, it simply describes the loudspeaker layout. This makes is possible for it to be used for conventional cinema-style surround mixing where the rear channels are used principally for diffuse sound effects, and where there is not usually any attempt to reproduce an accurate all-round sound field. It could also be used for sound-only purposes with the rear channels either used to add diffuse reverberation to more precise frontal images, or to contribute to an ambisonic-style surround sound matrix of speakers.

17.5 Basic stereo signal formats

In the following section the nature of stereo audio signals will be discussed, together with definitions of the terms used to describe the various formats of stereo sound. Also included is a discussion of the effects of misalignment on stereo signals.

17.5.1 A, B, M and S signals

It is conventional in broadcasting terminology to refer to the left (L) channel of a stereo pair as the 'A' signal and the right (R) channel as the 'B' signal. In the case of some stereo microphones or systems these are called respectively the 'X' and the 'Y' signals. In a two channel stereo system the 'A' signal feeds the left loudspeaker and the 'B' signal feeds the right loudspeaker. In colour coding terms (for meters, cables, etc.), the 'A' signal is coloured red and the 'B' signal is coloured green. This may be confusing when compared with some domestic hi-fi wiring conventions which use red for the right channel, but it is the same as the convention used for 'port' and 'starboard' on ships. There is also a German convention which uses yellow for the left channel and red for the right.

It is sometimes convenient to work with stereo signals in the so-called 'sum and difference' format, since it allows for the control of image width and ambient signal balance. The sum or main signal is denoted 'M', and is based on the addition of L and R signals, whilst the difference or side signal is denoted 'S', and is based on the subtraction of R from L to obtain a signal which represents the difference between the two channels. The M signal is that which would be heard by someone listening to a stereo programme in mono, and thus it is important in situations where the mono listener must be considered, such as in broadcasting. Colour-coding convention holds that M is coloured white, whilst S is coloured yellow, but it is sometimes difficult to distinguish between these two colours on certain meter types.

17.5.2 Derivation of stereo signals

'LR', 'XY', or 'AB' stereo signals (they are all essentially the same thing) may be derived by many means. Most simply, they may be derived from a pair of coincident

---

FACT FILE

17.3

Stereo misalignment effects

Differences in level, frequency response and phase may arise between signals of a stereo pair, perhaps due to losses in cables, misalignment, and performance limitations of equipment. It is important that these are kept to a minimum for stereo work, as interchannel anomalies result in various audible side-effects. Differences will also result in poor mono compatibility. These differences and their effects are discussed below.

Frequency response and level
A difference in level or frequency response between A and B channels will result in a stereo image biased towards the channel with the higher overall level or that with the better HF response. Also, an A channel with excessive HF response compared with that of the B channel will result in the apparent movement of sibilant sounds towards the A loudspeaker. Level and response misalign-ment on MS signals results in increased crosstalk between the equivalent A and B channels, such that if

the S level is too low at any frequency the AB signal will become more monophonic (width narrower), and if it is too high the apparent stereo width will be increased.

Phase
Inter-channel phase anomalies will affect one's perception of the positioning of sound source, and it will also affect mono compatibility. Phase differences between A and B channels will result in 'comb-filtering' effects in the derived M signal due to cancellation and addition of the two signals at certain frequencies where the signals are either out-of- or in-phase.

Crosstalk
It was stated earlier that an inter-channel level difference of only 18 dB was required to give the impression of a signal being either fully left or fully right. Crosstalk between A and B signals is not therefore usually a major problem, since the performance of most audio equipment is far in excess of these requirements. Excessive crosstalk between A and B signals will result in a narrower stereo image, whilst excessive crosstalk between M and S signals will result in a stereo image increasingly biased towards one side.

---

directional microphones orientated at a fixed angle to each other. Alternatively they may be derived from a pair of spaced microphones, either directional or non-directional, with an optional third microphone bridged between the left and right channels. Finally they may be derived by the splitting of one or more mono signals into two, by means of a 'pan-pot', which is really a dual ganged variable resistor which controls the relative proportion of the mono signal being fed to the two legs of the stereo pair, such that as the level to the left side is increased that to the right side is decreased.

MS or 'sum and difference' format signals may be derived by conversion from the AB format using a suitable matrix (see Fact File 4.5), or by direct pickup in that format. A coincident pair of microphones may be used for direct pickup, such that the S signal is derived from a sideways-facing figure-eight, and the M signal is derived from one of a variety of polar patterns, depending on the desired final balance. For every AB stereo pair of signals it is possible to derive an MS equivalent, since M is the sum of A and B, whilst S is the difference between A and B. Likewise, signals may be converted from MS to AB formats using the reverse process.

The effects of signal misalignment in stereo systems are outlined in Fact File 17.3.

17.5.3 Conversion between signal formats

In order to convert an AB signal into MS format it is necessary to follow some simple rules. Firstly, the M signal is not usually a simple sum of A and B, as this will result in over-modulation of the M channel in the case where a maximum level signal exists on both A and B (representing a central image), therefore a correction factor is applied, ranging between –3 dB and –6 dB (equivalent to a division of the voltage by between √2 and 2 respectively).

e.g.: $M = (A + B) -3 \text{ dB}$ or $(A + B) -6 \text{ dB}$

The correction factor will depend on the nature of the two signals to be combined. If identical signals exist on the A and B channels (representing 'double mono' in effect), then the level of the uncorrected sum channel (M) will be two times (6 dB) higher than the levels of either of A or B, requiring a correction of –6 dB in the M channel in order for the maximum level of the M signal to be reduced to a comparable level. In a television production where most of the sound sources are panned centrally (e.g.: central dialogue) the 6 dB correction factor would be more appropriate, since there is a strong likelihood that a large number of operational situations will result in identical A and B signals close to maximum level. If the A and B signals are non-coherent (random phase relationship), then only a 3 dB rise in the level of M will result when A and B are summed, requiring the –3 dB correction factor to be applied. This is more likely with stereo music signals. As most stereo material has a degree of coherence between the channels, the actual rise in level of M compared with A and B is likely to be somewhere between the two limits for real programme material.

The S signal results from the subtraction of B from A, and is subject to the same correction factor:

e.g.: $S = (A - B) -3 \text{ dB}$ or $(A - B) -6 \text{ dB}$

S can be used to reconstruct A and B when matrixed in the correct way with the M signal (see below), since $(M + S) = 2A$ and $(M - S) = 2B$. It may therefore be appreciated that it is possible at any time to convert a stereo signal from one format to the other and back again. If either format is not provided at the output of microphones or mixing equipment it is a relatively simple matter to derive one from the other electrically, as was shown in Fact File 4.5.

17.6 Stereo microphone techniques

Stereo recordings can either be made using a single stereo microphone, such as those described in Chapter 4, or using two or more microphones configured in one of the arrangements described below. A stereo microphone typically comprises two or more directional capsules, and as such a stereo microphone is really at least two mono microphones in one housing. It is difficult to make a stereo microphone which operates as a spaced pair, unless the spacing is very small, for obvious physical reasons, and so stereo microphones tend to be based on the coincident pair principle. Spaced pair stereo is usually achieved using separate mono microphones, unless a 'dummy head' is used for binaural recording, in which case two small microphones may be buried in the ear canals of the head (see section 17.9).

17.6.1 Coincident pair principles

Coincident pairs are normally two directional microphones (or two capsules within one housing) which may be angled over a range of settings to allow for different configurations and operational requirements. The pair may be operated in either the AB or MS modes, and a matrixing unit is sometimes supplied with microphones which are intended to operate in the MS mode in order to convert the signal to AB format for recording. The directional patterns (polar diagrams) of the two microphones need not necessarily be figure-eight, although if the microphone is used in the MS mode the S capsule must be figure-eight (see below). Directional information is represented solely by the level differences between the capsule outputs, since the two capsules are mounted physically as close as possible, and therefore there are no phase differences between the outputs except at the highest frequencies where inter-capsule spacing may become appreciable in relation to the wavelength of sound.

17.6.2 AB pairs

AB pairs are normally mounted vertically in front of the 'sound stage', such that the two capsules are angled to point symmetrically left and right of the centre of the source stage (as shown in Figure 17.12). The choice of angle depends on the polar response of the capsules used. A coincident pair of figure-eight microphones at 90° gives a very accurate correlation between the actual angle of the source and the apparent position of the virtual image when reproduced on loudspeakers, but there

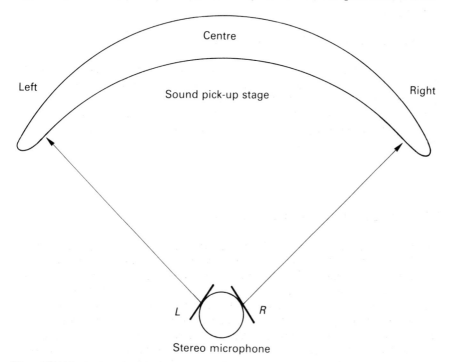

Figure 17.12 In the AB mode a coincident stereo microphone's capsules are oriented so as to point left and right of the centre of the sound stage

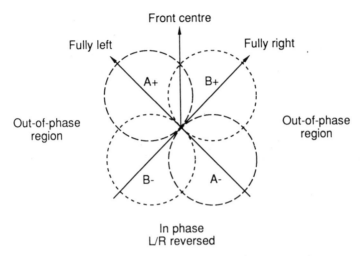

Figure 17.13 The polar diagram of a coincident pair of figure-eights shows that there is a large out-of-phase region in the side quadrants, and a left-right reversal at the rear

are also operational disadvantages to the figure-eight pattern in many cases. We shall proceed to investigate coincident pairs which use a number of different polar patterns, bearing in mind that the theoretical arguments about correct correlation between real and phantom images can only be taken so far when there are conflicting operational considerations.

Figure 17.13 shows the polar pattern of a coincident pair which is set to figure-eight pattern, and it serves as a useful example for the discussion of a number of phenomena which are not so easy to visualise with other pairs. Firstly, it may be seen that the fully left position corresponds to the null point of the right capsule's pickup. This is the point at which there will be maximum level difference between the two capsules. As a sound moves across the sound stage from left to right it will result in a gradually decreasing output from the left mic, and an increasing output from the right mic. Since the mics have cosine responses, the output at 45° off axis is 2 times the maximum output, or 3 dB down in level, thus the takeover between left and right microphones is smooth for music signals.

The second point to consider with this pair is that the rear quadrant of pickup suffers a left–right reversal, since the rear lobes of each capsule point in the opposite direction. This is important when considering the use of such a microphone in situations where confusion may arise between sounds picked up on the rear and in front of the mic, such as in television sound where the viewer can also see the positions of sources. The third point is that pickup in both side quadrants results in out-of-phase signals between the channels, since a source further round than 'fully left' results in pickup by both the negative lobe of the right capsule and the positive lobe of the left capsule. There is thus a large region around a crossed pair of figure-eights that results in out-of-phase information, this information often being ambient or reverberant sound. Any sound picked up in this region will suffer cancellation if the channels are summed to mono, with maximum cancellation occurring at 90° and 270°, assuming 0° as the centre-front.

The operational advantages of the figure-eight pair are the crisp and accurate positioning of sources, together with a natural blend of ambient sound from the rear.

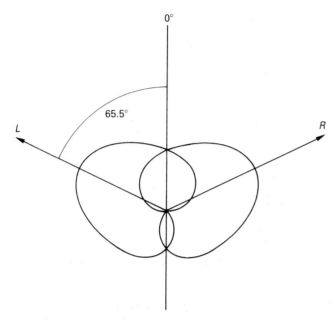

Figure 17.14 A coincident pair of cardioid microphones should theoretically be angled at 131°, but deviations either side of this may be acceptable in practice

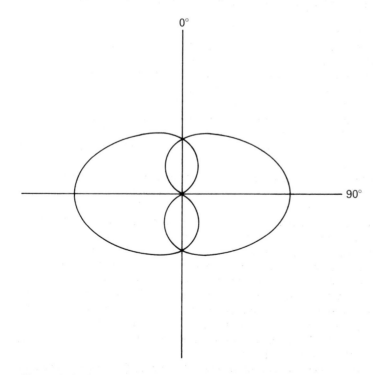

Figure 17.15 Back-to-back cardioids often work well in practice, and have no out-of-phase region

---

FACT FILE

17.4

Stereo
width
issues

With any AB crossed pair, fully left or fully right corresponds to the null point of pickup of the other channel's microphone, as already stated, and, as will be seen below, this corresponds also to the point where the amplitude of the M signal equals that of the S signal (where the sum of the channels is the same as the difference between them). It is very important to grasp the difference between the angle subtended by the two capsules and the angle between fully left and fully right. As the angle between the capsules is made larger, the angle between the null points will become smaller.

Operationally, if one wishes to 'widen' the stereo image one will widen the angle between the microphones which is intuitively the right thing to do. This results in a narrowing of the angle between fully left and fully right, which is also correct, since sources which had been, say, half left in the original image will now be further towards the left. A narrow angle between fully left and fully right results in a very wide sound stage, since sources have only to move a small distance to result in large changes in reproduced position. This corresponds to a wide angle between the capsules. Further coincident pairs are possible using any polar pattern between figure-eight and omni, although the closer that one gets to omni the greater the required angle to achieve adequate separation between the channels. The

hypercardioid pattern is often chosen for its smaller rear lobes than the figure-eight, allowing a more distant placement from the source for a given direct-to-reverberant ratio. Since the hypercardioid pattern lies between figure-eight and cardioid, the angle required between the capsules is correctly around 110°.

The theoretical foundation (as described earlier in this chapter) should not be forgotten in that it suggests the need for an electrical 'narrowing' of the image at high frequencies in order to preserve the correct angular relationships between low and high frequency signals, although this is rarely implemented in practice with coincident pair recording. A further consideration to do with the theoretical versus the practical is that although microphones tend to be referred to as having a particular polar pattern, this pattern is unlikely to be consistent across the frequency range and this will have an effect on the stereo image. Cardioid crossed pairs should theoretically exhibit no out-of-phase region since there should be no negative rear lobes, but in practice most cardioid capsules becomes more omni at LF and more hypercardioid at HF, and thus some out-of-phase components may be noticed in the HF range whilst the width may appear too narrow at LF. Attempts have been made to compensate for this in stereo microphone design, one example of which has been used by Sanken in an MS microphone (the CMS-2), by adding high and low frequency acoustic compensation mechanisms to maintain relatively constant directional response over the frequency range.

Some cancellation of ambience may occur, especially in mono, if there is a lot of reverberant sound picked up by the side quadrants. Disadvantages lie in the large out-of-phase region, and in the size of the rear pickup which is not desirable in all cases and is left-right reversed. Stereo pairs which are made up of capsules having less rear pickup may be preferred in cases where a 'drier' or less reverberant balance is required, and where frontal sources are to be favoured over rearward sources. In such cases the capsule responses may be changed to be nearer the cardioid pattern, and this requires an increased angle between the capsules to maintain good correlation between actual and perceived angle of sources.

The cardioid crossed pair is angled at approximately 131°, as shown in Figure 17.14, although angles of between 90° and 180° may be used to good effect depending on the 'width' of the sound stage to be covered. At an angle of 131° a centre source is 65.5° off-axis from each capsule, resulting in a 3 dB drop in level compared with the maximum on-axis output (the cardioid mic response is equivalent to $0.5(1 + \cos\vartheta)$, where ϑ is the angle off-axis of the source, and thus the output at 65.5° is √2 times that at 0°). A departure from the theoretically correct angle is often necessary in practical situations, and it must be remembered that the listener will not necessarily be aware of the 'correct' location of each source, neither may it matter that the true and perceived positions are different. A pair of 'back-to-back' cardioids (Figure 17.15) has often been used to good effect since it has a simple MS equivalent of an omni and a figure-eight, and has no out-of-phase region. Although theoretically 'fully left' is at 90° off-centre, it may be appreciated that there will in fact be a satisfactory level difference between the capsules to make fully left appear at a somewhat smaller angle.

AB pairs in general have the possible disadvantage that central sounds are off-axis to both mics, perhaps considerably so in the case of crossed cardioids. This may result in a central signal with a poor frequency response and possibly an unstable image if the mic's polar response is erratic. Whether or not this is important depends on the importance of the central image in relation to that of offset images, and will be most important in cases where the main source is central (such as in television, with dialogue). In such cases the MS technique described in the next section is likely to be more appropriate, since central sources will be on-axis to the M microphone. For music recording it would be hard to say whether central sounds are any more important than offset sources, and thus either technique may be acceptable.

In commercial 'side-fire' AB microphones, it is common for one of the capsules to be fixed and for the other to rotate in order to set the angle between the capsules. It is sometimes difficult to see where the centre-front position is with some mics, and this may have to be discovered through trial and error by getting an assistant to walk around the mic making a suitable noise until equal level is obtained from both capsules. Alternatively, some stereo mics have a small LED which moves with each capsule to identify the direction in which each is pointing.

17.6.3 MS pairs

Although some stereo microphones are built specifically to operate in the MS mode, it is possible to take any coincident pair capable of at least one capsule being switched to figure-eight, and orientate it so that it will produce suitable signals. The S component (being the difference between left and right signals) is always a sideways-facing figure-eight with its positive lobe facing left. The M or mono component may be any polar pattern facing to the centre-front, although the choice of M pattern depends on the desired equivalent AB pair (see below), and will be the signal which a mono listener would hear. True MS mics usually come equipped with a control box which matrixes the MS signals to AB format if required. A control for varying S gain is often provided as a means of varying the effective acceptance angle between the equivalent AB pair (see below).

It should be remembered that MS signals may not be monitored directly: they are sum and difference components and must be converted to AB format at a convenient point in the production chain. The advantages of keeping a signal in the MS format until it needs to be converted will be discussed below, but the major advantage of pickup in the MS format is that central signals will be on-axis to the M capsule, resulting in the best frequency response. Furthermore, it is possible to operate an MS mic in a similar way to a mono mic (so long as precautions are taken when moving the mic, as detailed in section 17.17.5) which may be useful in television operations where the MS mic is replacing a mono mic on a pole or in a boom.

To see how MS and AB pairs relate to each other, and to draw some useful conclusions about stereo width control, it is informative to consider again a coincident pair of figure-eight mics. For each MS pair there is an AB equivalent. This stands to reason, since M is simply the sum of A and B signals, whilst S is simply the difference between them. The polar pattern of the AB equivalent to any MS pair may be derived by plotting the level of $(M + S)/2$ and $(M - S)/2$ for every angle around the pair. Taking the MS pair of figure-eight mics shown in Figure 17.16, it

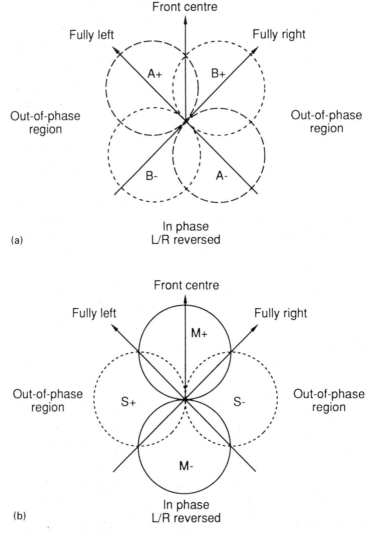

Figure 17.16 Every AB pair has an MS equivalent. An AB pair of figure-eights is shown in (a) and its MS equivalent in (b)

may be seen that the AB equivalent is simply another pair of figure-eights, but rotated through 45°. Thus the correct MS arrangement to give an equivalent AB signal where both 'capsules' are orientated at 45° to the centre-front (the normal arrangement) is for the M capsule to face forwards and the S capsule to face sideways.

A number of interesting points arise from a study of the AB/MS equivalence of these two pairs, and these points apply to all equivalent pairs. Firstly, fully left or right in the resulting stereo image occurs at the point where S=M (in this case at 45° off centre). This is easy to explain, since the fully left point is the point at which the output from the right capsule (B) is zero. Therefore $M = A + 0$, and $S = A - 0$, both of which equal A. Secondly, at angles of incidence greater than 45° off centre in either direction the two channels become out-of-phase, as was seen above, and this corresponds to the region in which S is greater than M. Thirdly, in the rear quadrant where the signals are in phase again, but left-right reversed, the M signal is greater than S again. The relationship between S and M levels, therefore, is an excellent guide to the phase relationship between the equivalent AB signals. If S is lower than M, then the AB signals will be in phase. If S equals M, then the source is either fully left or right, and if S is greater than M, then the AB signals will be out-of-phase.

To show that this applies in all cases, and not just that of the figure-eight pair, look at the MS pair in Figure 17.17 together with its AB equivalent. This MS pair is made up of a forward facing cardioid and a sideways facing figure-eight (a popular arrangement). Its AB equivalent is a crossed pair of hypercardioids, and again the extremes of the image (corresponding to the null points of the AB hypercardioids) are the points at which S equals M. Similarly, the signals go out of phase in the region where S is greater than M, and come back in phase again for a tiny angle round the back, due to the rear lobes of the resulting hypercardioids. Thus the angle of acceptance (between fully left and fully right) is really the frontal angle between the two points on the MS diagram where M equals S.

Now, consider what would happen if the gain of the S signal was raised (imagine expanding the lobes of the S figure-eight in Figure 17.17). The result of this would be that the points where S equalled M would move inwards, making the acceptance angle smaller. As explained earlier, this results in a wider stereo image, since off-centre sounds will become closer to the extremes of the image, and is equivalent to increasing the angle between the equivalent AB capsules. Conversely, if the S gain is reduced, the points at which S equals M will move further out from the centre, resulting in a narrower stereo image, equivalent to decreasing the angle between the equivalent AB capsules. This is neatly exemplified in a commercial example, the Neumann RSM 191i, pictured in Figure 17.18, which is an MS mic in which the M capsule is a forward-facing short shotgun mic with a polar pattern rather like a hypercardioid. The polar pattern of the equivalent AB pair is shown in Figure 17.19 for three possible gains of the S signal with relation to M (–6 dB, 0dB and +6 dB). It will be seen that the acceptance angle (φ) changes from being large (narrow image) at –6 dB, to small (wide image) at +6 dB.

The other thing to be noted about the effect of changing S gain is the size of the rear lobes of the AB equivalent. It may be seen that the higher the S gain the larger the rear lobes. Therefore, not only does S gain change stereo width, it also affects rear pickup, and thus the ratio of direct to reverberant sound.

MS microphones, therefore, are useful as a means of deriving signals in the sum and difference format, for subsequent matrixing back to A and B at a later stage. Whilst in the MS format, the proportions of M and S may be varied in order to affect

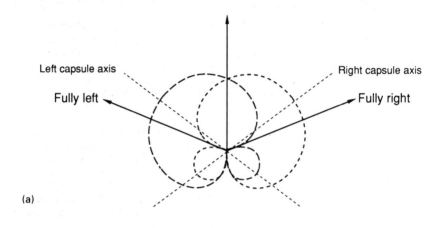

(a)

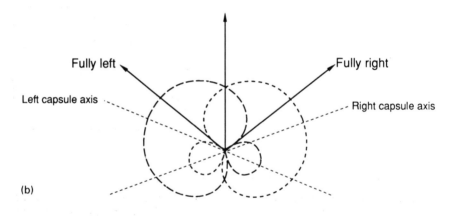

(b)

Figure 17.17 The MS equivalent of a forward-facing cardioid and sideways figure-eight, as shown in (a), is a pair of hypercardioids whose effective angle depends on S gain, as shown in (b)

the resulting stereo width and direct-to-reverberant balance. Any AB stereo pair may be operated in the MS configuration, simply by orientating the capsules in the appropriate directions and switching them to an appropriate polar pattern, but certain microphones are dedicated to MS operation simply by the physical layout of the capsules (see Fact File 17.5).

FACT FILE

17.5

End-fire and side-fire configurations

Physically, there are two principal ways of mounting the capsules in a coincident stereo microphone, be it MS or AB format: either in the 'end-fire' configuration where the capsules 'look out' of the end of the microphone such that the microphone may be pointed at the source (see the diagram), or in the 'side-fire' configuration where the capsules look out of the sides of the microphone housing. It is less easy to see the direction in which the capsules are pointing in a side-fire microphone, but such a microphone makes it possible to align the capsules vertically above each other so as to be time-coincident in the horizontal plane, as well as allowing for the rotation of one capsule with relation to the other. An end-fire configuration is more suitable for the MS capsule arrangement (see diagram below), since the S capsule may be mounted sideways behind the M capsule, and no rotation of the capsules is required, although there is a commercial example of an AB end-fire microphone for television ENG (electronic news gathering) use which houses two fixed cardioids side-by-side in an enlarged head.

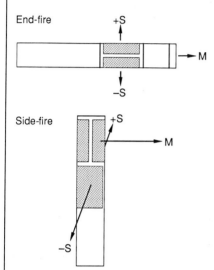

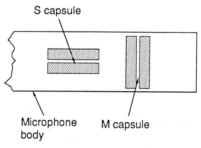

17.6.4 Operational considerations with stereo pairs

The control of S gain is an important tool in determining the degree of width of a stereo sound stage, and for this reason the MS output from a microphone might be brought (unmatrixed to AB) into a mixing console, so that the engineer may have control over the width. This in itself can be a good reason for keeping a signal in MS form during the recording process, although M and S can easily be derived at any stage using a conversion matrix. If the number of channels can be afforded on the mixing console, then it would be possible to have, say, one fader for the M signal, one for the S signal, and one for a phase-reversed S signal.

The diagram in Figure 17.20 shows how it is possible to derive an AB mix with variable width from an MS microphone using three channels on a mixer. It should be noted that no MS matrices are required. M and S outputs from the microphone are fed in-phase through two mixer channels and faders, and a post-fader feed of S is taken to a third channel line input, being phase-reversed on this channel. The M signal is routed to both left and right mix buses (panned centrally), whilst the S signal is routed to the left mix bus (M + S = 2A) and the –S signal (the phase-reversed

Figure 17.18 The Neumann RSM191i is an end-fire MS microphone whose control box allows control of S gain (courtesy of FWO Bauch)

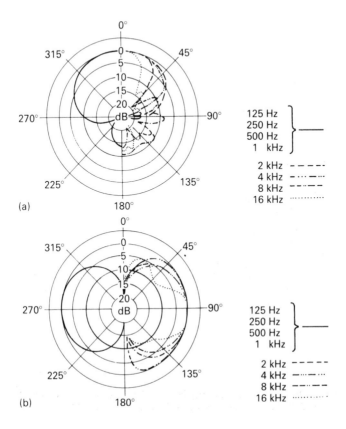

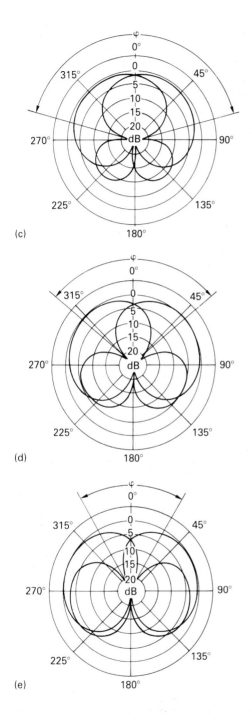

(c)

(d)

(e)

Figure 17.19 Polar patterns of Neumann RSM191i microphone. (a) M capsule, (b) S capsule, (c) AB equivalent with –6dB S gain, (d) 0dB S gain, (e) +6dB S gain (Courtesy FWO Bauch)

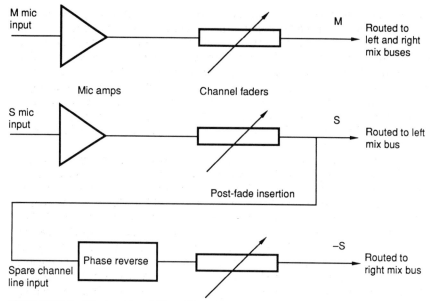

Figure 17.20 An AB mix with variable width can be derived from an MS microphone connected to three channels on a mixer (see text)

version) is routed to the right mix bus (M – S = 2B). It is important that the gain of the –S channel is matched very closely with that of the S channel. A means of deriving M and S from an AB format input is to mix A and B together with the B signal phase-reversed to get S, and without the phase reverse to get M. In order to convert back to A and B a similar method to the above could be employed, but this is wasteful of mixer channels and a matrix might be more convenient.

From a handling point of view a number of issued should be raised. Outdoors, stereo microphones will be more susceptible to wind noise and rumble than most mono microphones, as they incorporate velocity sensitive capsules which always give more problems in this respect than omnis. Most of the interference would reside in the S channel, since this has always a figure-eight pattern, and thus would not be a problem to the mono listener. Similarly, physical handling of the stereo microphone, or vibration picked up through a stand, will be much more noticeable audibly than with a mono pressure microphone. Stereo microphones should not be used close to people speaking, as small movements of their heads can cause large changes in the angle of incidence, and thus considerable movement in their apparent position in the sound stage. This is especially true in cases where the S gain is high (narrow angle of AB acceptance), as very small head or microphone movements result in large swings of the sound image.

17.7 Near-coincident microphone configurations

'Near coincident' pairs of directional microphones introduce small additional timing differences between the channels which may help in the localisation of transient sounds and increase the 'spaciousness' of a recording, whilst at the same time remaining nominally coincident at low frequencies. The spacing between the mics of such pairs is usually similar to the binaural spacing, and this may have something to do with the results from such pairs being a good compromise between loudspeaker and headphone listening. Subjective evaluations have shown that listeners are often considerably impressed by recordings made on near-coincident pairs. One comprehensive subjective assessment of stereo microphone arrangements, performed at the University of Iowa, consistently resulted in the near-coincident pairs scoring amongst the top few for their sense of 'space' and realism. These effects have been linked by one critic to 'phasiness' at high frequencies (which some people may like, nonetheless).

A number of examples of near-coincident pairs exist as 'named' arrangements, although there is a whole family of possible near-coincident arrangements using combinations of spacing and angle (resulting in various combinations of time and level difference). The so-called 'ORTF pair' (Figure 17.21) is an arrangement of two cardioid mics, deriving its name from the organisation which first adopted it (the Office de Radiodiffusion-Television Francaise). The two mics are spaced apart by 170 mm, and angled at 110°. The 'NOS' pair (Nederlande Omroep Stichting, the Dutch Broadcasting Company), uses cardioid mics spaced apart by 300 mm and angled at 90°. A pair of figure-eight microphones pointing forwards and spaced apart by 200 mm has been called a 'Faulkner' pair, after the British recording engineer who first adopted it. This latter pair can be found to offer good 'focusing' on a small-to-moderate-sized central ensemble with the mics placed further back than would normally be expected.

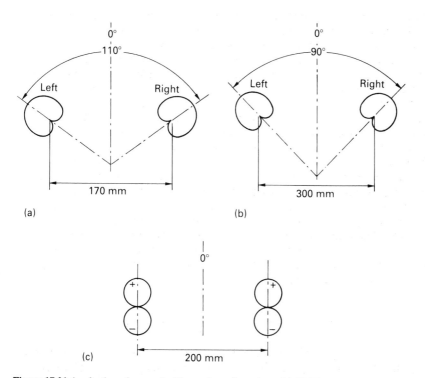

Figure 17.21 A selection of near-coincident pair configurations. (a) ORTF, (b) NOS, (c) Faulkner

17.8 Spaced arrays

17.8.1 Principles

Although spaced arrays have a historical precedent for their usage, since they were the first to be documented (in the work of Clement Ader at the Paris Exhibition in 1881) and have been widely used, they are perhaps less 'correct' theoretically. Spaced arrays rely principally on the precedence effect, in that the delays which result between the channels tend to be of the order of a number of milliseconds (as opposed to the binaural delay of up to 600 μs). With spaced arrays the level and time difference resulting from a source at a particular left-right position on the sound stage will depend on how far the source is from the microphones (see Figure 17.22), with a more distant source resulting in a much smaller delay and level difference. In order to calculate the time and level differences which will result from a particular spacing it is possible to use the following two formulae:

$$\Delta t = (d_1 - d_2)/c \quad ; \quad L = 20 \log_{10}(d_1/d_2)$$

where t is the time difference and L the pressure level difference which results from a source whose distance is d_1 and d_2 respectively from the two microphones, and c is the speed of sound (340 m/s).

When a source is very close to a spaced pair there may be a considerable level difference between the microphones, but this will become small once the source is more than a few metres distant. The positioning of spaced microphones in relation to a source is thus a matter of achieving a compromise between closeness (to achieve satisfactory level and time differences between channels), and distance (to achieve adequate reverberant information relative to direct sound). When the source is large

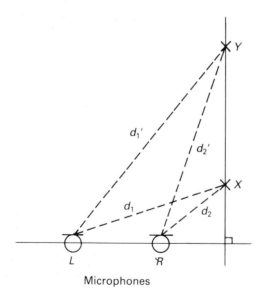

Microphones

Figure 17.22 With spaced omnis a source at position X results in path lengths d_1 and d_2 to each microphone respectively, whilst a source in the same LR position but at a greater distance (source Y) the path length difference is smaller, resulting in a smaller time difference than for X

and deep, such as a large orchestra, it will be difficult to place the microphones so as to suit all sources, and it may be found necessary to raise the microphones somewhat so as to reduce the differences in path length between sources at the front and rear of the orchestra.

Spaced microphone arrays do not stand up well to theoretical analysis when considering the imaging of continuous sounds, the precedence effect being related principally to impulsive or transient sounds. Because of the phase differences between signals at the two loudspeakers created by the microphone spacing, interference effects at the ears at low frequencies may in fact result in a contradiction between level and time cues at the ears. It is possible in fact that the ear on the side of the 'earlier' signal may not experience the higher level, thus producing a confusing difference between the cues provided by impulsive sounds and those provided by continuous sounds. The lack of phase coherence in spaced-array stereo is further exemplified by the test of phase inverting one of the channels on reproduction, an action which does not always appear to affect the image particularly, as it would with coincident stereo, showing just how uncorrelated the signals are.

It may thus be found that stereo perspective and accuracy of positioning are less realistic with spaced arrays, although many convincing recordings have resulted from their use. It has been suggested that the impression of 'space' that results from the use of spaced arrays is in fact simply the result of 'phasiness' and comb-filtering effects. Many recording engineers prefer spaced arrays because the omni microphones often used in such arrays tend to have a flatter and more extended frequency response than their directional counterparts, although it should be noted that spaced arrays do not *have* to be made up of omni mics (see below). Mono compatibility of spaced pairs is variable, although not always as poor in practice as might be expected. Coincident pairs, from a superficial point of view, appear to promise better mono compatibility due to the lack of phase differences between the capsules, but in reverberant surroundings there is often a lot of pickup in the out-of-phase region of such a pair, which suffers cancellation in mono.

17.8.2 'Decca Tree'

The so-called 'Decca Tree' is a popular arrangement of three spaced mono mics with an omnidirectional pickup pattern. The name derives from the usage of this

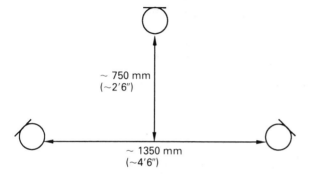

~ 750 mm
(~2'6")

~ 1350 mm
(~4'6")

Figure 17.23 The classic 'Decca Tree' involves three omnis, with the centre mic spaced slightly forward of the outriggers

technique by the Decca Record Company, although even that company would admit to not sticking rigidly to this arrangement in all circumstances.

A 'tree' of three omnis is configured according to the diagram in Figure 17.23, with the centre microphone spaced so as to be slightly forward of the two 'outriggers', although it is possible to vary the spacing to some extent depending on the size of the source stage to be covered. The reason for the centre microphone and its spacing is to stabilise the central image which tends otherwise to be rather imprecise, although the existence of the centre mic will also complicate the phase relationships between the channels, thus exacerbating the comb-filtering effects that may arise with spaced pairs. The advance in time experienced by the forward mic will tend to 'solidify' the central image, due to the precedence effect, avoiding the 'hole-in-the-middle' often resulting from spaced pairs. The 'outriggers' are angled outwards slightly, so that the axes of best HF response favour sources towards the edges of the stage whilst central sounds are on-axis to the central mic.

17.8.3 Other spaced pairs

Spaced microphones with either omnidirectional or cardioid patterns may be used in configurations other than the Decca Tree described above, although the 'tree' has certainly proved to be the more successful arrangement in practice. The precedence effect begins to break down for delays greater than around 40 ms (and often less, depending on the nature of the source), since the brain begins to perceive the two arrivals of sound as being discrete rather than integrated, thus beginning to sound like an echo or 'phasy' effect. It is therefore reasonable to assume that spacings between microphones which exceed this delay between channels should be avoided. This maximum delay, though, corresponds to a mic spacing of around 17 m, and such extremes have not proved to work well in practice owing to the great distance of central sources from either microphone compared with the closeness of sources at the extremes, resulting in a considerable level drop for central sounds and thus a 'hole in the middle'. Furthermore, the precedence effect for longer time delays than about 10 ms begins to result in the sound being firmly in the leading speaker.

Work by Dooley and Streicher has shown that good results may be achieved using spacings of between one-third and one-half of the width of the total sound stage to be covered, as shown in Figure 17.24, although closer spacings have also been used to good effect. Bruel and Kjaer manufacture matched stereo pairs of omni microphones together with a bar which allows variable spacing (Figure 17.25), and suggest that the spacing used is somewhat smaller than one-third of the stage width (they suggest between 5 cm and 60 cm, depending on stage width), their principle rule being that the distance between the microphones should be small compared with the distance from microphones to source. Informal experiments by the author suggest that interchannel delays of between 2 and 4 ms are adequate to give the impression that a transient sound is either fully left or right in the resulting image, depending on the nature of the signal. This agrees well with other sources who also point out that for a signal to appear half left or half right, only about one-quarter of the delay is needed compared with that needed for fully left or right, further reinforcing the likelihood of a hole-in-the-middle with spaced pairs.

Omni microphones are often used in addition to coincident pairs, as 'outriggers' to cover the extremes of an orchestra or choir, and to add some 'body' to the sound of the string section, as shown in Figure 17.26. This is hard to justify theoretically, and is beginning to move towards the realms of multi-microphone pickup, which is

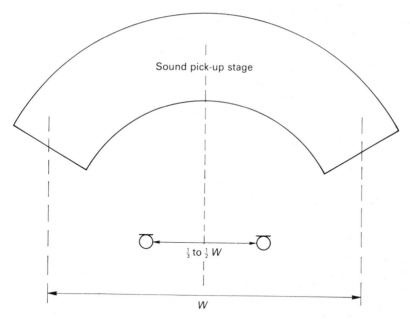

Figure 17.24 One source suggests that spaced omni microphones may be distanced at between one-third to one-half the width of the sound stage

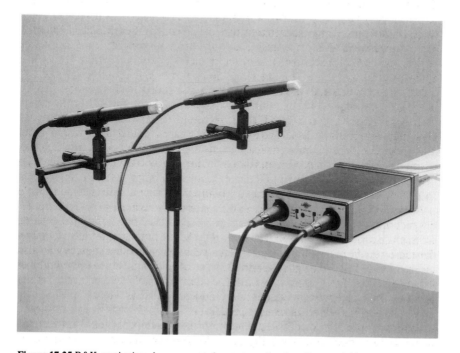

Figure 17.25 B&K omni microphones mounted on a stereo bar that allows variable spacing

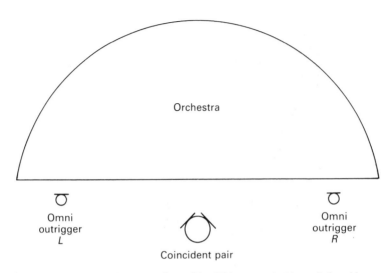

Figure 17.26 Omni outriggers may be used in addition to a coincident pair for wide sources

also hard to justify theoretically but which may be used for convenience and in order to produce a 'commercially acceptable' sound. When more than, say, three microphones are used to cover a sound stage one has to consider a combination of theories, possibly suggesting conflicting sonic information between the outputs of the different microphones. In such cases the sound balance will be optimised on a mixing console and theory can to a large extent be relegated to a back seat, since what is now being created is a synthetic balance subject to the creative control of the recording engineer.

17.9 Binaural pairs and 'dummy head' techniques

The binaural hearing and reproduction process has already been thoroughly explained, and it is now intended to look into the means of sound pickup for binaural reproduction on headphones. Binaural recordings can have an uncanny realism when monitored on headphones, although they do not tend to work well on loudspeakers unless equalised to remove the specific pinna transforms and other 'head' effects specific to dummy-head recordings, for reasons already stated.

Binaural mic techniques depend on the spacing of pressure microphones at the distance apart of a normal listener's ears, so as to produce two 'ear signals' which are as close as possible to a representation of the sound pressures at each ear. When these are reproduced at the same two ears on headphones, the result is a close approximation to the originally-recorded soundfield, and thus a very lifelike listening experience. Factors which must be considered are the effects of the pinna and ear canal on the sound to be picked up and reproduced, as well as the similarity (or lack of it) between different peoples' ears. The use of artificial 'heads' and baffles to simulate the shadowing effect of the head should also be considered.

Small pressure microphones (such as those used as 'tie-clip' mics, e.g.: Sony ECM-50) may be mounted either flush with the side of a real head on a headphone-

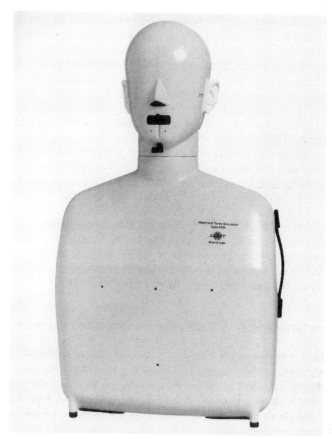

Figure 17.27 B&K Type 4128 head and torso simulator

like headband, this being useful if a person wishes to make a recording using his own ear spacing, perhaps whilst walking around with a portable tape recorder. The shadowing effects of his own head would then be introduced between the mics, although the effect of the pinna would be limited by the intrusion of the microphone and headband. Alternatively, a 'dummy head', such as the Neumann KU81i, may be used in which the microphones are mounted inside the ear canal, so that the response of the mic is modified by the resonant and reflective properties of the outer ear, as well as by the resonant cavities of the head in some cases, and also reflections off the torso in the case of a 'head and torso simulator' such as the B&K Type 5930. Dummy heads vary in their degree of 'reality', some having imitation skin as well as the correct resonant and absorptive properties as a real head, in order that the result is as 'life-like' as possible.

It is important to consider at which stage in the recording and reproduction process the pinna and ear canal effects will have been added - whether during recording or during reproduction. On replay, the existence of a headphone over the ear would largely suppress any pinna effects (although open-backed headphones may allow some to remain). Headphones particularly suppress the all-important conch response, the conch being the main resonant cavity of the pinna, disturbances of which have been shown to have a large effect on the perception of a sound as being 'at the ear'. To be accurate, the sound should only have been 'encoded' by the pinna and ear canal characteristics once, and thus it might be suggested that the most effective solution would be to mount microphones such that they are just at the entrance to the ear canal, in order that they are subject to pinna effects during recording, but not ear canal resonances (which are supplied on replay by the listener's own ears). For measurement purposes a slightly different approach applies, since it may be that the ear canal, pinna, and perhaps even inner ear responses are required to affect the microphone output, since the aim is often to measure noise levels in a situation as close to that of human perception as possible, without the intention to replay that noise again on headphones. For this reason B&K makes two 'head and torso simulators', one for binaural recording and one for measurement (Type 4128), as shown in Figure 17.27. As mentioned earlier, Theile has suggested a standard equalisation characteristic for headphones to be used in high-quality listening, in an attempt to solve the difficulties outlined above.

Although not simulating all the head effects, a convincing binaural recording may be made using two microphones spaced apart by the ear distance (15–20 cm) and separated by a wooden or perspex baffle with a diameter of around 25 cm. This approximates to the shadowing and delaying effects of a real head.

17.10 Binaural microphones suitable for loudspeaker listening

As briefly described earlier in this chapter, binaurally derived signals may be made more suitable for monitoring on loudspeakers by equalising them for a flat frontal response in the free field (some sources argue for a flat response averaged over the frontal hemisphere, whilst others argue for a flat diffuse-field response). This has the effect of removing the severe colouration otherwise perceived when monitoring dummy-head recordings on loudspeakers, and also results in improved imaging without a hole in the middle.

Schoeps has produced a microphone based on Theile's principles of 'head-related' stereo, the KFM 6U, which is essentially two pressure microphones mounted flush with the surface of a head-sized sphere in the ear positions. This microphone has a flat response to frontally-incident sounds, but introduces time delays and head-shadowing effects for off-centre sounds which, it is claimed, produce accurate images on loudspeakers. Preliminary subjective tests conducted by the author confirm that the microphone produces a pleasing image with depth and spaciousness, although low frequency images have been perceived to be less stable than those derived from coincident pairs. Crown has also produced a binaural microphone which produces signals suitable for loudspeaker reproduction, known as SASS – the Stereo Ambient Sampling System – again producing head-related effects without the specific dummy head equalisation which results from such features as pinnae and ear canals.

17.11 Multi-microphone arrangements

We have so far considered the use of a small number of microphones to cover the complete sound stage, but it is also possible to make use of a large number of microphones, each covering a small area of the sound stage and intended to be sonically 'independent' of the others. In the ideal world, each mic in such an arrangement would pick up sound only from the desired sources, but in reality there is considerable 'spill' from one to another. It is not the intention to provide a full resumé of studio microphone technique, and thus discussion will be limited to an overview of the principles of multi-mic pickup as distinct from the more 'theoretically-correct' techniques described above (sometimes known as 'purist' techniques).

In multi-mic recording each mic feeds a separate channel of a mixing console, where levels are individually controlled and the mic signal is 'panned' to a virtual position somewhere between left and right in the sound stage. The pan control takes the monophonic mic signal and splits it two ways, controlling the proportion of the signal fed to each of the left and right mix buses. Typical pan control laws follow a curve which gives a 3 dB drop in the level sent to each channel at the centre, resulting in no perceived change in level as a source is moved from left to right due to the summation of sound intensities from left and right loudspeakers, as shown in Fact File 6.2. The -3 dB panpot law is not correct if the stereo signal is combined electrically to mono, since the summation of two equal signal voltages would result in a 6 dB rise in level for signals panned centrally, and thus a -6 dB law is more appropriate for mixers whose outputs will be summed to mono (e.g.: radio and TV operations) as well as stereo, although this will then result in a drop in level in the centre for stereo signals. A compromise law of -4.5 dB is often adopted by manufacturers for this reason.

Multi-mic balances therefore rely on channel level differences, separately controlled for each microphone, to represent the positions of virtual sources in a synthesised sound stage, with relative level between channels used to adjust the 'strength' and thus perhaps the perceived distance of a source in a mix, although distance is more a function of the associated reflections and reverberant balance than of level. Artificial reverberation may be added to restore a sense of space to a multi-mic balance. It is common in 'classical' music recording to use close mics in addition to a coincident pair or spaced pair in order to reinforce sources which appear to be weak in the main pickup, these close mics being panned as closely as possible to match the true position of the source. The results of this are variable and can have the effect of 'flattening' the perspective, removing any depth which the image might have had, and thus the use of close mics must be handled with subtlety.

The recent development of cheaper digital signal processing (DSP) has made possible the use of delay lines, sometimes as an integral feature of digital mixer channels, to adjust the relative timing of spot mics in relation to the main pair so as to prevent this distortion of depth, and to equalise the arrival times of distant mics in order that they do not exert a precedence 'pull' over the output of the main pair. As described earlier in this chapter, it is possible to process the outputs of multiple mics to simulate binaural delays and head-related effects in order to create the effect of sounds at any position around the head, when the result is monitored on headphones.

Recommended further reading

Bartlett, B. (1991) *Stereo Microphone Techniques*. Focal Press
Eargle, J., ed. (1986) *Stereophonic Techniques – An Anthology*. Audio Engineering Society

Appendix 1

Understanding basic equipment specifications

The performance of an audio system may be measured to determine its effect on a sound signal passed through it, and it may also be assessed subjectively (in other words, by listening to it). Theoretically, if a system introduces an audible modification to the sound signal then one should also be able to measure it, provided that the right test can be devised and suitable equipment is available, but the difficulty of achieving this ideal is always increasing as the audible differences between systems become ever smaller, and the absolute fidelity of recording and reproduction improves. Digital recording and processing has brought with it the possibility for recording and transmission with zero degradation, and there is still considerable debate about just what can be heard and what cannot. In the following sections it is not the intention to get involved in the debates which always rage in hi-fi circles concerning minute differences in sound quality between systems (which do of course exist, but are often explained badly), but rather that the reader should gain an insight into the most commonly encountered system specifications and what they mean, as well as describing the audible effects of different distortions on sound signals.

A1.1 Frequency response – technical

The most commonly quoted specification for a piece of audio equipment is its *frequency response*. It is a parameter which describes the frequency range covered by the device – that is, the range of frequencies which it can record or reproduce. To take a simple view, for high-quality reproduction the device would normally be expected to cover the whole audio-frequency range, which was defined earlier as being from 20 Hz to 20 kHz, although some have argued that a response which extends above the human hearing range has audible benefits. It is not enough, though, simply to consider the range of frequencies reproduced, since this says nothing about the relative levels of different frequencies or the amplitude of signals at the extremes of the range. If further qualification was not given then a frequency response specification of 20 Hz – 20 kHz could mean virtually anything. It is important to compare devices' specifications on the same grounds, since otherwise little useful information can be gained.

The ideal frequency response is one which is 'flat' – that is, with all frequencies treated equally and none amplified more than others. Technically, this means that the gain of the system should be the same at all frequencies, and this could be verified by plotting the amplitude of the output signal on a graph, over the given frequency range, assuming a constant-level input signal. An example of this is shown in Figure A1.1(a), and it will be seen that the graph of output level versus frequency is a straight horizontal line between the limits of 20 Hz and 20 kHz – that is, a flat frequency response. Also shown in Figure A1.1 are examples of non-flat responses, and it will be seen that these boost some frequencies and cut others, affecting the balance between different parts of the sound spectrum (the audible effects of which are discussed in section A1.3). Typically, frequency response is quoted *with reference to the response at 1 kHz*. This means that the output level at 1 kHz is chosen as the level against which all other frequencies are compared, and would be given a relative level of 0 dB for this purpose. If the response at 5 kHz was said to be +3 dB ref. 1 kHz, this would mean that signals of 5 kHz would be amplified 3 dB more than signals at 1 kHz.

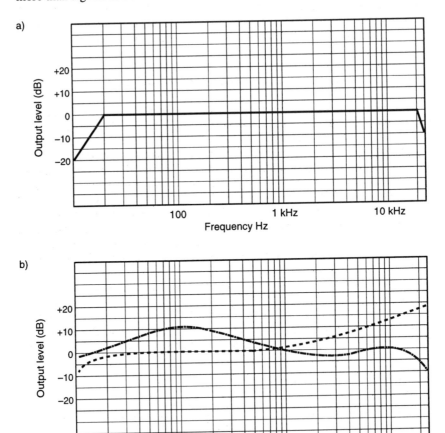

Figure A1.1 (a) Plot of a flat frequency response from 20 Hz to 20 kHz. (b) Examples of two non-flat frequency responses

Although a graph gives the most detail about frequency response, since it shows what happens at every point in the range, often only figures are given in specifications. It is common to quote frequency response as the upper and lower limits of the frequency range handled by the device, giving two frequencies at which the response is '3 dB down' – in other words, where the response is 3 dB lower than the response at 1 kHz. It is implied in such a case that the response between these points is moderately flat, although this cannot be taken for granted in practice. Thus a response of 45 Hz – 17 kHz (–3 dB) suggests that the device handles a frequency range between 45 and 17 000 Hz, at which extremes the response is 3 dB down. Below 45 Hz and above 17 000 Hz one would expect the response to fall off even further.

A more accurate way of specifying frequency response in figures, and one which leaves less room for misinterpretation, is to state a tolerance for allowed variations in level over the specified range. Thus a specification of 45 Hz – 17 kHz (±3 dB ref. 1 kHz) states that the response at any frequency between the limits will not deviate from the response at 1 kHz by more than 3 dB upwards or downwards.

A1.2 Frequency response – practical examples

Some practical examples may help to illustrate the above discussion of frequency response, and a table of typical specifications for a selection of devices is given in Table A1.1, so that they can be compared.

Firstly, purely electronic devices tend to have a flatter response than devices which involve a recording or reproduction process, since the latter usually incorporate mechanical, magnetic or optical processes which are more prone to distortions of all kinds. An amplifier is an example of the former case, and it is unusual to find a well-designed amplifier which does not have a flat frequency response these days – flat often to within a fraction of a decibel from 5 Hz up to perhaps 60 kHz. In the other category, though, there are LP turntables, tape recorders, loudspeakers and microphones, to name but a few, and these are all much more difficult to design for a flat response.

Devices which convert sound into electricity or vice versa (in other words, transducers) are most prone to frequency response errors, and some loudspeakers have been known to exhibit deviations of some 10 dB or more from 'flat'. Since such devices are also affected by the acoustics of their surroundings it is difficult to divorce a discussion of their own response from a discussion of the way in which

Table A1.1 Examples of typical frequency responses of audio systems

Device	Typical frequency response
Telephone system	300 Hz – 3 kHz
AM radio	50 Hz – 6 kHz
Consumer cassette machine	40 Hz – 15 kHz (±3 dB)
Professional analogue tape recorder	30 Hz – 25 kHz (±1 dB)
CD player	20 Hz – 20 kHz (±0.5 dB)
Good-quality small loudspeaker	60 Hz – 20 kHz (–6 dB)
Good-quality large loudspeaker	35 Hz – 20 kHz (–6 dB)
Good-quality power amplifier	6 Hz – 60 kHz (±3 dB)
Good-quality omni microphone	20 Hz – 20 kHz (±3 dB)

they interact with their surroundings. The room in which a loudspeaker is placed has a significant effect on the perceived response, since the room will resonate at certain frequencies, creating pressure peaks and troughs throughout the room. Depending on the location of the listener, some frequencies may be emphasised more than others, and this therefore makes it difficult to say what is the fault of the room and what is the fault of the speaker. A loudspeaker's response can be measured in so-called 'anechoic' conditions, where the room is totally absorbent and cannot produce significant effects of its own, although other methods now exist which do not require the use of such a room. A good loudspeaker will have a response which covers the majority of the audio-frequency range, with a tolerance of perhaps ±3 dB, but the LF end is less easy to extend than the HF end. Smaller loudspeakers will only go down to perhaps 50 or 60 Hz.

FACT FILE A1.1 Frequency response – subjective

Subjectively, deviations from a flat frequency response will affect sound quality. If the aim is to carry through the original signal without modifying it, then a flat response will ensure that the original amplitude relationships between different parts of the frequency spectrum are not changed. If, say, low frequencies are boosted with respect to high frequencies, then the original sound will be modified, making it sound more bass heavy. It is important not to be side-tracked by the fact that the human ear's frequency response is not flat (see Chapter 2), since this fact has no bearing on the need for a flat response in audio equipment. In audio equipment the important factor is that sounds come out of a system as they went in.

Some forms of modification to the ideal flat response are more acceptable than others. For example, a gentle roll-off at the high-frequency (HF) end of the range often goes unnoticed, since there is not much sound energy at this point. Domestic cassette machines and FM radio receivers, for example, tend to have upper limits of around 15 kHz, but are relatively flat below this, and thus do not sound unpleasant. Frequency responses which deviate wildly from 'flat' over the audio-frequency range, on the other hand, sound much worse, even if the overall range of reproduction is wider than that of FM radio.

If the frequency response of a system rises at high frequencies then the sibilant components of the sound will be empha-sised, music will sound very 'bright' and 'scratchy', and any background hiss will be emphasised. If the response is down at high frequencies then the sound will become dull and muffled, and any background hiss may appear to be reduced. If the frequency response rises at low frequencies then the sound will be more 'boomy', and bass notes will be emphasised. If low frequencies are missing, the sound will be very 'thin' and 'tinny'. A rise in the middle frequency range will result in a somewhat 'nasal' sound, perhaps having a rather harsh quality, depending on the exact frequency range concerned.

Concerning the effects of very low and very high frequencies, close to the limits of human hearing, it can be shown that the reproduction of sounds below 20 Hz does sometimes offer an improved listening experience, since it can cause realistic vibrations of the surroundings. Also, the ear's frequency response does not cut off suddenly at the extremes, but gradually decreases, and thus it is not true that one hears nothing below 20 Hz and above 20 kHz – one simply hears much less. Similarly, extended HF responses can sometimes help sound quality, and a gentle HF roll-off usually implies less steep filtering of the signal which in turn may result in improved quality.

Analogue magnetic tape recorders use a number of equalisation processes to ensure that the frequency response is as flat as possible, but this can usually only be achieved with one of a few specified tape types and formulations. An unsuitable tape may result in a non-flat response, unless the machine can be realigned for the new tape. Cassette machines have a number of different tape-type settings to ensure that the frequency response and other parameters are optimum for each tape formulation. The frequency response of an analogue tape machine is likely to vary with recording level, since at higher recording levels the high frequencies become 'compressed' because the tape is incapable of retaining them. For this reason the response of such a machine is usually quoted at a relatively low recording level, perhaps 20 dB below reference level (see section 8.5).

LP records are equalised before they are cut, intentionally to give them a non-flat response, but this is re-equalised before reproduction to restore the correct frequency balance. The reason for this is explained in section 11.2. If the RIAA equaliser in the amplifier which reproduces the recording is not properly designed then it will not restore the correct frequency balance, and this can occur in cheap hi-fi equipment.

Microphones vary enormously in their characteristics, and their frequency response depends a lot on their polar pattern and design (see Chapter 4). Cheap consumer microphones may have a response which only extends up to 10 or 12 kHz, whereas professional mics may cover a range at least up to 20 kHz. The LF end of the spectrum is equally variable, with omnidirectional microphones having a much more extended LF response than other pickup patterns. A microphone's response often varies with the angle of incidence of the source.

A1.3 Harmonic distortion – technical

Harmonic distortion is another common parameter used in the specification of audio systems. Such distortion is the result of so-called 'non-linearity' within a device – in other words, what comes out of the device is not exactly what went in. There are a number of types of non-linearity, but here it is the type which affects the shape of the waveform that is referred to. In Chapter 1 it was shown that only simple sinusoidal waveforms are completely 'pure', consisting only of one frequency without harmonics. More complex repetitive waveforms can be analysed into a set of harmonic components based on the fundamental frequency of the wave. Harmonic distortion in audio equipment arises when the shape of the sound waveform is changed slightly between input and output, such that harmonics are introduced into the signal which were not originally present, thus modifying the sound to some extent (see Figure A1.2). It is virtually impossible to avoid a small amount of harmonic distortion, since no device carries through a signal *entirely* unmodified, but it can be reduced to extremely low levels in amplifiers (see section A1.6).

Harmonic distortion is normally quoted as a percentage of the signal which caused it (e.g.: THD 0.1% @ 1 kHz), but, as with frequency response, it is important to be specific about what type of harmonic distortion is being quoted, and under what conditions. One should distinguish between *third*-harmonic distortion and *total* harmonic distortion, and unfortunately both can be abbreviated to 'THD' (although THD most often refers to total harmonic distortion). Total harmonic distortion is the sum of the contributions from all the harmonic components introduced by the device, assuming that the original wave has been filtered out, and

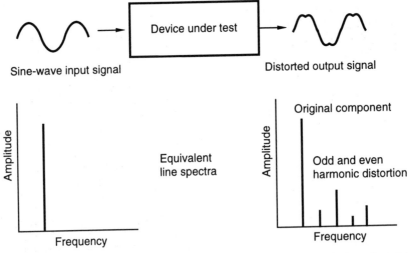

Figure A1.2 A sine wave input signal is subject to harmonic distortion in the device under test. The waveform at the output is a different shape to that at the input, and its equivalent line spectrum contains components at harmonics of the original sine-wave frequency

is normally measured by introducing a 1 kHz sine wave into the device and measuring the resulting distortion at a recognised input level. The level and frequency of the sine wave used depends very much on the type of device and the test standard used. Third-harmonic distortion is a measurement of the amplitude of the third harmonic of the input frequency only, and is commonly found in tape recorder tests since the third harmonic is the most prominent in magnetic recording systems.

It may be important to be specific about the level and frequency at which the distortion specification is made, since in many audio devices distortion varies enormously with these parameters (see section A1.6).

A1.4 Harmonic distortion – practical examples

In electrical devices such as amplifiers, distortion percentage does not usually change much with input level, but may vary slightly with frequency. With tape recorders distortion is very much a function of recording level and frequency, and can vary widely. In transducers, distortion usually remains at a moderately constant percentage with input level variation, but cheaper transducers can introduce fairly high levels of distortion.

Table A1.2 shows some typical quoted THD percentages for different audio devices, and it will be seen that they vary widely, with amplifiers and digital audio equipment having the lowest typical figures.

The distortion characteristics of digital audio systems are discussed in more detail in section 10.3, and will not be covered further here. Analogue tape machine distortion performance is often quoted in the form of a *maximum output level* or *MOL* figure, which is the recording level at which third-harmonic distortion reaches a certain percentage, considered to be the maximum sensible recording level. In professional recorders this is 3% at 1 kHz, and in consumer cassette machines it is

FACT FILE A1.2 Harmonic distortion – subjective

Harmonic distortion is not always unpleasant, indeed many people find it quite satisfying and link it with such subjective parameters as 'warmth' and 'fullness' in reproduced sound, calling sound which has less distortion 'clinical' and 'cold'. Since the distortion is harmonically related to the signal which caused it, the effect may not be unmusical and may serve to reinforce the pitch of the fundamental in the case of even-harmonic distortion.

The sound of third-harmonic distortion is easy to detect on pure tones, but less easy on music, and can be heard when recording a tone on to a tape recorder at high level whilst comparing the sound 'off tape' with the input signal. The tone no longer sounds 'pure', but has an edge to it. It contains a component one octave and a fifth above the fundamental tone.

Because distortion tends to increase gradually with increasing recording level in tape recorders, the onset of distortion is less noticeable than it is when an amplifier 'clips', for example, and many analogue tape recordings contain large percentages of harmonic distortion which have been deemed acceptable. Amplifier clipping, on the other hand, is very sudden and results in a 'squaring-off' of the audio waveform when it exceeds a certain level, at which point the distortion becomes severe. This effect can be heard when the batteries are going flat on a transistor radio, or when a hi-fi loudspeaker is driven exceedingly hard from a low-powered amplifier, and sounds like a serious breaking up of the sound on peaks of the signal. If tested with sine-wave sources, the result is as shown in Fact File 7.2.

Table A1.2 Typical THD percentages

Device	% THD
Good power amplifier @ rated power	< 0.05% (20 Hz – 20 kHz)
16 bit digital recorder (via own convertors)	< 0.05% (−15 dB input level)
Loudspeaker	< 1% (25 W, 200 Hz)
Professional analogue tape recorder	< 1% (ref. level, 1 kHz)
Professional capacitor microphone	< 0.5% (1 kHz, 94 dB SPL)

5% at 315 Hz. Typically, one might expect this distortion percentage to be reached at a recording level of around 10–12 dB above reference level in professional recorders with a good modern tape, and around 4–8 dB above reference level in cassette machines. Analogue tape machines and reference levels are discussed in more detail in section 8.5.

A1.5 Dynamic range and signal-to-noise ratio

Dynamic range and signal-to-noise (S/N) ratio are often considered to be interchangeable terms for the same thing. This may be true, but depends on how the figures are arrived at. S/N ratio is normally considered to be the number of decibels between the 'reference level' and the noise floor of the system (see Figure A1.3). The noise floor may be weighted according to one of the standard curves which attempts to account for the potential 'annoyance' of the noise by amplifying some

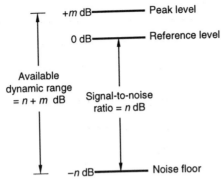

Figure A1.3 Signal-to-noise ratio is often quoted as the number of decibels between the reference level and the noise floor. Available dynamic range may be greater than this, and is often quoted as the difference between the peak level and the noise floor

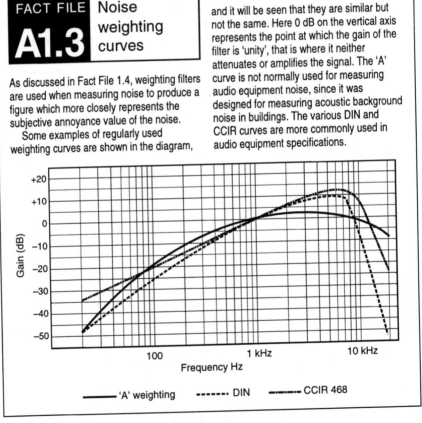

FACT FILE **Noise weighting curves**

A1.3

As discussed in Fact File 1.4, weighting filters are used when measuring noise to produce a figure which more closely represents the subjective annoyance value of the noise.

Some examples of regularly used weighting curves are shown in the diagram, and it will be seen that they are similar but not the same. Here 0 dB on the vertical axis represents the point at which the gain of the filter is 'unity', that is where it neither attenuates or amplifies the signal. The 'A' curve is not normally used for measuring audio equipment noise, since it was designed for measuring acoustic background noise in buildings. The various DIN and CCIR curves are more commonly used in audio equipment specifications.

parts of the frequency spectrum and attenuating others (see Fact Files 1.4 and A1.3). Dynamic range may be the same thing, or it may be the number of decibels between the *peak* level and the noise floor, indicating the 'maximum-to-minimum' range of

Table A1.3 Typical values for CCIR weighted S/N ratio

Device	S/N ratio
Consumer cassette machine (without noise reduction)	50 dB (ref. 315 Hz, 200 nWb m^{-1})
Professional analogue tape machine (without noise reduction) @ 15 ips (38 cm s^{-1})	65 dB (ref. 1 kHz, 320 nWb m^{-1})
16 bit digital audio recorder	94 dB (ref. peak level)
Professional power amplifier	108 dB (ref. max. output)

signal levels which may be handled by the system. Either parameter quoted without qualification is difficult to interpret.

For example, the specification 'Dynamic range = 68 dB' for a tape recorder means very little, since there is no indication of the reference points or weightings, whereas 'S/N ratio, CCIR 468-3 (ref. 1 kHz, 320 nWb m^{-1}) = 68 dB' tells the reader virtually all that is required. It says that the noise has been measured to the CCIR 468-3 weighting standard, and that it measures at 68 dB below the level of a 1 kHz tone recorded at a magnetic level of 320 nWb m^{-1}. This could at least be compared directly with another machine measured in the same way, even if the reference level was different, although the difference between the two reference levels would have to be taken into account. It is difficult, though, to compare S/N ratios between devices measured using different weighting curves.

In analogue tape recorders, dynamic range is sometimes quoted as the number of decibels between the 3% MOL (see section A1.6) and the weighted noise floor. This gives an idea of the available recording 'window', since the MOL is often well above the reference level. In digital recorders, the peak recording level is really also the reference level, since there is no point in recording above this point due to the sudden clipping of the signal (see section 10.3), and thus dynamic range and S/N ratio are often referred to this point, although some manufacturers have chosen to refer to a level 15 dB below it.

Table A1.3 gives some typical values for S/N ratio found in audio equipment.

A1.6 Wow and flutter

Wow and flutter are names used to describe speed (pitch) variations of a tape machine or turntable. Wow is applied to slow variations in speed and flutter is applied to faster variations in speed. The figures depend very much on the mechanical quality of the device and its state of wear and cleanliness. Again a weighting filter (usually to the DIN standard) is used when measuring to produce a figure which closely correlates with one's perception of the annoyance of speed variations. Specifications are usually quoted as WRMS (Weighted Root-Mean Square, a form of average), but occasionally peak figures may be used which will be worse than the RMS figures. Long-term speed accuracy is also quoted in many cases – this being the anticipated overall drift in speed of the machine over a reel of tape. These days speed drift is less of a problem than it used to be, with machines remaining stable to within hundredths of a per cent over the length of a reel. Drift is only really a problem if two machines are to be synchronised.

A typical figure for a good analogue tape machine would be better than 0.02% WRMS, and good cassette machines can approach this figure also. Cheap cassette machines coupled with poor tapes can cause the figure to rise considerably, with some examples approaching 0.5% or more. A good LP turntable may achieve 0.02% results, but again cheaper models will be worse. Digital audio recorders and CD players do not suffer from wow and flutter in the same way as analogue transports, since the audio data from the tape or disk is first passed through a so-called *timebase corrector*, prior to conversion, which removes any speed variations resulting from mechanical instability in the transport (see Fact File 10.4).

A machine with poor W&F results will sound most unpleasant, with either uncomfortable 'wowing' in the pitch of notes, or fast flutter which gives rise to a 'roughness' in the sound aptly described by the word 'flutter', and possibly some intermodulation distortion (see section A1.9).

A1.7 Intermodulation (IM) distortion

IM distortion results when two or more signals are passed through a non-linear device (see section A1.4). Since all audio equipment has some non-linearity there will always be small amounts of IM distortion, but these can be very low.

Unlike harmonic distortion, IM distortion may not be harmonically related to the frequency of the signals causing the distortion, and thus it is audibly more unpleasant. If two sine-wave tones are passed through a non-linear device, sum and difference tones may arise between them (see Figure A1.4). For example, a tone at

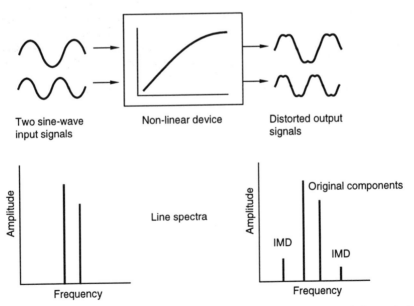

Figure A1.4 Intermodulation distortion between two input signals in a non-linear device results in low level sum-and-difference components in the output signal

$f_1 = 1000$ Hz and a tone at $f_2 = 1100$ Hz might give rise to IM products at $f_1 - f_2 = 100$ Hz, and also at $f_1 + f_2 = 2100$ Hz, as well as subsidiary products at $2f_1 - f_2$ and so on. The dominant components will depend on the nature of the non-linearity.

IM distortion can also arise when speed variations of a tape transport or LP turntable modulate the signals reproduced from them. For example, a tape transport with speed variations at 25 Hz modulating a reproduced signal at 1000 Hz could give rise to IM products at 975 Hz and 1025 Hz.

Low IM distortion figures are an important mark of a high-quality system, since such distortion is a major contributor to poor sound quality, but it is less often quoted than THD (see section A1.4).

A1.8 Crosstalk

Crosstalk figures describe the expected amount of break-through from one channel of a device to another. For example, in a stereo tape recorder crosstalk may arise between the left and right channels, or in a multitrack recorder between adjacent tracks. In general, crosstalk is undesirable. It is usually quoted either as negative decibels relative to the causatory signal (e.g.: –53 dB), or as decibels channel separation (e.g.: 53 dB).

Crosstalk may arise within the electronics of a device (such as by electromagnetic induction between tracks on a printed-circuit board), magnetically (by induction within the heads of a tape machine), or externally (say between cables in a duct). For good performance in a multitrack tape machine and mixer it is essential that crosstalk is very low, since the operator will not want components of one channel being audible on another. For stereo recorders and reproducers the requirement is not so stringent.

A typical figure for a good multitrack analogue recorder in reproduce mode is between 40 and 50 dB, but this is usually much worse between tracks in record and adjacent tracks in sync reproduce (see Fact File 8.3). In digital equipment the crosstalk between channels is exceptionally low (around –90 dB), since crosstalk is rejected as part of the replay decoding process. In analogue LP cartridges separation is quite poor, as it is also in analogue stereo FM radio and TV broadcasting (around 25–30 dB), but is normally adequate for maintaining stereo separation.

For further details of Focal Press books contact:

Focal Press
Linacre House
Jordan Hill
Oxford OX2 8DP
UK

Tel: +44 (0)1865 310366
Fax: +44 (0)1865 310898
Email: Margaret.Riley@bhein.rel.co.uk

Focal Press titles are available from all good booksellers. In case of difficulty please phone our UK direct order line on 01865 314080 with your credit card details ready. In the USA please phone 800-366-2665.

General further reading

Alkin, G. (1989) *Sound Techniques for Video and TV.* Focal Press

Alkin, G. (1996) *Sound Recording and Reproduction*, 2nd edition. Focal Press

Ballou, G. (1991) ed. *Handbook for Sound Engineers – The New Audio Cyclopedia.* Focal Press

Borwick, J. (1995) ed. *Sound Recording Practice.* Oxford University Press

Capel, V. (1994) *Newnes Audio and Hi-Fi Engineers' Pocket Book*, 3rd edition. Butterworth-Heinemann

Eargle, J. (1992) *Handbook of Recording Engineering.* Van Nostrand Rheinhold

Eargle, J. (1990) *Music, Sound, Technology.* Van Nostrand Rheinhold

Huber, D. and Runstein (1995) *Modern Recording Techniques*, 4th edition. Focal Press

Nisbett, A. (1995) *The Sound Studio*, 6th edition. Focal Press

Roberts, R. S. (1981) *Dictionary of Audio, Radio and Video.* Butterworths

Talbot-Smith, M. (1995) *Broadcast Sound Technology*, 2nd edition. Focal Press

Talbot-Smith, M., ed. (1994) *Audio Engineers Reference Book.* Focal Press

Woram, J. (1989) *Sound Recording Handbook.* Howard W. Sams and Co.

Index